B. P. Pratten

The family physician

Every man his own doctor

B. P. Pratten

The family physician
Every man his own doctor

ISBN/EAN: 9783337225001

Printed in Europe, USA, Canada, Australia, Japan

Cover: Foto ©berggeist007 / pixelio.de

More available books at **www.hansebooks.com**

THE HUMAN VISCERA.

THE

'AMILY PHYSICIAN;

OR,

EVERY MAN HIS OWN DOCTOR,

An Encyclopedia of Medicine,

CONTAINING KNOWLEDGE THAT WILL

:OMOTE HEALTH, CURE DISEASE AND PROLONG LIFE

DESCRIBING ALL DISEASES,

\D TEACHING HOW TO CURE THEM BY THE SIMPLEST MEDICINES.

ALSO,

'I ANALYSIS OF EVERYTHING RELATING TO COURTSHIP, MARRIAGE, AND THE PRODUCTION, MANAGEMENT, AND REARING OF HEALTHY FAMILIES,

TOGETHER WITH

APTER ON THE PREPARATION OF MEDICINES, GIVING PRESCRIPTIONS AND VALUABLE RECEIPTS, FULL AND ACCURATE DIRECTIONS FOR TREATING WOUNDS, INJURIES, POISONS, ETC.

COMPILED BY

LEADING CANADIAN MEDICAL MEN.

TH NUMEROUS EXPLANATORY ILLUSTRATIONS, AND A COMPLETE INDEX.

PRICE $1.25,

Toronto:

ROSE PUBLISHING COMPANY.
1889.

THE PUBLISHERS' PREFACE.

Believing in the sound principle that "an ounce of prevention is worth a pound of cure," the publishers confidently present this volume to the public, in the hope it will fill a long felt want in the community.

The purpose of the work is to fully inform every man and woman what are the real causes of sickness and what are the most approved methods of treatment. And moreover it is especially intended as an aid to home or domestic treatment in the thousand cases where a physician cannot readily be reached, and when, if ordinary common sense is used, his services are not required.

The compilers, all leading medical men in Canada, wish this work to make an honored mark in the Dominion from its merits, and its method and arrangement, and, above all, from the circumstances that it comprises all that is valuable in every other popular medical guide, added to many new principles and new methods—modes of cure that have been tried in every day practice, and have seldom been known to fail.

TORONTO, June, 1889.

TABLE OF CONTENTS.

——:o:——

CHAPTER X.

THE
FAMILY PHYSICIAN.

PART I.

CHAPTER I.

THE SICK ROOM AND THE NURSE.

Choice of the Sick Room—Its furniture—Air and ventilation—Hints in regard to its warmth—How it should be lighted—The importance of cleanliness—Duties and deportment of those nursing the sick--Personal attentions required by the sick—Management of those recovering from sickness—Household remedies—Family thermometry.

Choice of the Sick Room. The room in which a person is confined to bed with a serious and protracted ailment should be large, lofty, and well ventilated ; the window sashes ought to lower from the top as well as raise from below, and work with as little noise as possible. Walls painted in oil are better than those in plaster or paper. The door must, be noiseless, with a ventilator over it or made in one of the panels. A painted or polished floor is better than a carpeted one. It should be swept, or at the most lightly wiped, but not frequently washed, as the slow evaporation from the wet boards renders the room injuriously damp.

The room should be a light one, exposed to the direct rays of the morning and mid-day sun. Blinds and shades should be provided, for modifying the light when desirable.

Warmth of the Sick Room. As we have said, a thermometer should always be in the invalid's room, and by it the temperature must be regulated. The best temperature is that of 60°. If, however, the patient feels chilly at this temperature, it may be raised three or four degrees. In fevers and diseases of the nervous system a lower temperature is preferable, about 50° ; while in consumption and other affections of the chest an atmosphere of 65° is the most agreeable to the inflamed air-passages.

In the early morning hours a sick person is apt to be chilly, which is just the time when the room is usually the coolest—a fact that must be borne in mind, and precautions taken, lest the patient suffer serious injury from this cause.

A low-down grate is much the best way of warming the sick room.

Furniture of the Sick Room. All unnecessary furniture must be banished from the room. There is no need of any wardrobe, bureau, trunks, or bandboxes. Two *tables*, at least, are wanted; one of them should be small, and on casters, so as to roll easily to the side of the bed, for the immediate use of the patient; the other, a larger one, for the reception of medicine bottles, spare glasses, cups, spoons, and other articles in constant use. In the drawer of the latter table there should be constantly kept a sponge, a bundle of soft old linen, pair of large, and another of small scissors, a full pin-cushion, needles and thread, a piece of adhesive plaster, and oiled silk for covering poultices. A third table, if there be room for it, is useful as a dressing table, on which to place the brush, comb, and other toilet articles. Over it a looking-glass may be hung, but never in such a position as to permit of the invalid seeing himself in it as he lies in bed. A movable *washstand*, on casters, so that it can be readily shoved to the side of the bed, is very useful. An ample supply of clean towels should be constantly on hand. There ought not to be more than three *chairs*, of which one should, if possible, be an easy or reclining chair. A *lounge* or *sofa* is of great utility, particularly during recovery, when the sitting posture cannot be long maintained; it is of service, also, for the patient to recline upon, when his strength permits, while his bed is being made. The *bedstead* should not be too wide. A greater width than three and a half feet renders it often difficult to reach and move the patient. Two bedsteads are much better than one. Each should be provided with its own sheets and cover. The patient passing half the time in one, permits the entire bedding of the other to be thoroughly aired out of the room, a very important measure in prolonged illness. The bedstead must be low, so as to permit of the patient getting in and out easily, and of his being lifted and moved with facility. Iron bedsteads are much to be preferred to wooden ones. The bed should be without curtains, and placed with its head to the wall, so as to admit of access on each side. It is best placed between two windows, or at the side of a window. The *bed clothing* must be light. Heavy cotton counterpanes and Marseilles coverlets must be discarded from the sick bed, and only good light blankets employed. The pillow-cases and sheets are better of cotton than of linen, and should be frequently changed; daily, in the case of infectious fevers. The pillows must be firm and elastic, and arranged so as to support the back, and not piled up in such a manner as to thrust the head forward upon the chest, and so increase the difficulty of breathing. The bed should be level and not too hard. *Curtains* and heavy drapery of all descriptions are objectionable in a sick room, as they harbor dust and contagious matter. Besides these main pieces of furniture, there are a number of articles of use in every sick room. Among these we may mention the following, viz.: a *thermometer* (one so constructed that it can be put in water to get the heat of a bath) should hang on the wall, at the height of the bed; a *foot warmer*, which may be made of a common bottle filled with hot water, or of a rubber bottle or cushion, or an earthenware bottle sold for the purpose, should be at hand when wanted; a *stomach warmer*, made of an India-rubber bottle or bag filled with hot water, is often of use; *air and water cushions*, or a cushion in the shape of a ring, filled with bran, are serviceable in warding off pressure from prominent parts, giving support, etc.; a *pillow rest*, made of a bag of the same width, but twice the length, of a pillow-case, stuffed with hay, straw, or oat-chaff, in such a manner as to make it taper down like a writing desk, from back to front, is an excellent thing for propping up pillows; a *cradle*, to support the bed-clothes, and prevent them pressing upon a painful part, is useful in many ailments; *folding rests for the legs*, made in the shape of an open book, and covered by a blanket, give comfort to the weary limbs; a *medicine glass*, marked so as to measure exactly a teaspoonful, a desertspoonful, and a tablespoonful, enables medicine to be given with much greater exactitude than can be done with spoons, which vary in size; a *medicine spoon*, which can be obtained at most druggists', permits of medicines being administered to very weak and partially unconscious patients, as it is covered, and has an opening near the end, so as to avoid all spilling: a *feeder*, consisting of a half-covered vessel furnished with a spout, for giving soups, teas, and other fluids, is convenient for administering drinks to the patient when lying in bed; a *bed-side pocket*, to be pinned at the side of the bed, makes a convenient place for keeping the pocket-handkerchief, a bottle of scent, etc., within the reach of the invalid; a *sick tray*, consisting of a board hollowed out in front to fit the body, and supported on four short legs, to be placed on the bed before the patient, is often much more convenient than an ordinary waiter on the knees or a table at the side of the bed; a *bed-chair* is often of service; a *bed-pan*, or *slipper*, is necessary for those too weak to rise from the bed; two *baskets*, with divisions or compartments, are useful for medicine bottles, one containing medicines for internal use, the other, of another color and form, containing external applications; and two baskets for cups and dishes, one for those needing washing, which are to be quickly sent out of the room and exchanged for the other basket with its clean ware. The following articles must *not* be kept in the sick room, viz.: dirty linen, implements of cooking, prepared food, medicines discontinued by the physician, soiled cups or glasses, slop basins or pails.

Air of the Sick Room. Fortunately the prejudice against admitting fresh, pure air into the chambers of the sick, particularly of fever patients, is to a great extent a thing of the past. The custom of almost hermetically sealing the rooms of patients in fever was very prevalent among nearly all classes, some forty or fifty years ago. Dr. James Gregory, of Edinburgh, the celebrated professor of medicine, used to mention in his lectures, that as no argument was of avail in procuring the admission of fresh air into the sick rooms of the poor, he generally pushed his cane through the panes of the windows. This, however, was not always adequate to insure the intended effect, as he often found the broken panes pasted over with paper on his next visit. No such forcible measures are now, usually, required. People generally err from carelessness rather than ignorance or prejudice, as few can be unaware of the fact, the subject having been so widely discussed, that foul air is poisonous for sick and well. In many cases, it is useful, once or twice a day, to cover the patient well in bed and protect him from drafts, and then open wide all the doors and windows, so as to wash out the room with fresh air. An excellent test for foul air is the nose ; pure air is tasteless and free from all smell.

The air of the room should never be permitted to become too dry, as is the tendency in heated rooms in winter. A basin of boiling water occasionally brought into the room, and replaced by another when it becomes cold, will usually give sufficient moisture.

A constant supply of pure air is important for others in the sick chamber as well as for the patient, as it lessens their c'.... of contracting the disease. The confinement of a contagious p   ti.l in a close room increases its force ; the more poisonous emanation: .e diluted by admixture with fresh air the less dangerous they bec   2. Hence certain diseases are more frequently propagated in winter,   en doors and windows are kept shut, than in summer, when the outside air is freely admitted.

While keeping the air that the sick breathe fresh and pure, by ventilation, care must be taken not to chill the patient. There is little danger, however, with ordinary precautions, of taking cold when lying in bed. Sick people usually catch their colds by rising for a moment from a warm bed without throwing any wrap around them, or by sitting up in bed without a covering over the shoulders.

Of course, no air is pure which contains smoke from a badly made fire, or from the chimney of an oil lamp. Not only smoke, but gas, frequently escapes from a stove, and proves a source of trouble to the sick, particularly those affected with chest ailments.

Of all the sources of contamination of the air of the sick chamber, one of the worst is the chamber utensil. On this subject, FLORENCE NIGHTINGALE, with her usual good sense and thoroughness writes : " The use of any chamber utensil *without a lid* should be utterly abolished, whether among sick or well. You can easily convince yourself of the necessity of this absolute rule, by taking one with a lid, and examining the under side of that lid. It will be found always covered, whenever the utensil is not empty, by condensed offensive moisture. Where does that go when there is no lid ? But never, never should the possession of this indispensable lid confirm you in the abominable practice of letting the chamber utensil remain in a patient's room *unemptied*, except once in twenty-four hours. Yes, impossible as it may appear, I have known the best and most attentive nurses guilty of this. Earthenware, or if there is any wood, highly polished and varnished wood, are the only materials fit for patient's utensils. A slop pail should never be brought into a sick room. It should be a rule invariable, that the utensil should be carried directly to the water-closet, emptied there, rinsed there, and brought back. There should always be water and a cock in every water-closet for rinsing. But even if there is not, you must carry water there to rinse with. I have actually seen, in the private sick room, the utensils emptied into the foot pan and put back under the bed. I can hardly say which is most abominable, whether to do this or rinse the utensil *in* the sick room. In the best hospitals, it is now a rule that no slop-pail shall ever be brought into the wards, but that the utensils shall be carried direct, to be emptied and rinsed at the proper place. I would it were so in the private house."

In typhoid fever, dysentery, and other similar affections in which the infectious principle of the disease resides in the discharges, the chamber utensil, so soon as it is cleaned, which should be at once after use, should have poured into it about half a tumblerful of a strong solution of sulphate of iron, kept on hand for the purpose in a large bottle or jug, and made by dissolving a pound of copperas in a gallon of water. Directly after the vessel is used, and before its contents are emptied, about a tumblerful of the same solution should be poured in, to destroy the fetor and lessen the liability of infection.

Light of the Sick Room. The sick require plenty of light as well as an abundance of air. It is only in the beginning and violent stages of certain diseases, and in some nervous affections, that it is advisable to partially darken the room.

It is a fact in regard to which there can be no doubt, that the sick do better and recover more quickly in a sunlit room, than in one into which the rays cannot enter, or from which they are always excluded. An agreeable view from the window is also desirable, especially during convalescence, to relieve the tired eye, and gratify the craving, nearly always experienced, for looking out of the window.

The night light must be so placed as not to throw a shadow on the ceiling over the patient, or on the wall in front of him. Shadows of things or persons in the room often assume, to the fearful gaze of the weakened patient, forms of terror, and may excite delirium or even convulsions. If oil be burned, care must be taken that the lamp does not smoke nor smell ; if gas, that the gas does not escape.

Cleanliness of the Sick Room. Cleanliness and order are important in the chamber in which the sick lie, but the dust and noise made in sweeping and arranging the room are frequently a source of great annoyance to the silent sufferer. As after a night's rest the patient is best able to bear a little bustle, the morning should be selected for putting the room to rights. If there be a carpet on the floor, it is to be sprinkled with moist tea-leaves before being lightly swept with a hand-brush and dust-pan.

The furniture, the bed and bedding, the floor and walls, and every article in the sick room, should be always clean. No vessel or implement used by the patient should be suffered to remain in the apartment, but be at once taken out, to be returned as soon as cleaned. Every glass, cup or plate in which food is administered, must be taken immediately out of the room and washed. Neither ought any glass or spoon in which medicine is given be suffered to remain with the small portion which is always left in it ; it must be instantly and well cleansed.

Duties and Deportment of those Nursing the Sick. Those women make the best nurses for the sick who are between the ages of twenty-five and fifty-five, active and vigorous, in good health, of happy cheerful disposition, with kind feelings, and a temper not easily ruffled and of orderly, clean and neat habits.

The nurse should cook nothing in the sick room ; move about without noise, fidget, hurry or bustle ; keep the room in order, sweet and clean ; take out of the room instantly all evacuations of the patient slops, soiled linen and wet towels ; avoid eating anything that gives a bad smell to the breath ; order food so as to have it ready promptly at the time for giving it ; observe minutely the orders received for administering food and medicine ; see that the patient's mouth and nostril remain uncovered during sleep : never express a doubt in regard to the propriety or efficacy of the treatment employed ; always look confidently for recovery; make no comparisons with other cases ; say nothing to dis courage or alarm the invalid ; never whisper ; give no prohibited food and permit of no forbidden indulgencies ; see that there is a supply of fresh air, and that the patient is kept out of all drafts ; preserve a proper uniform temperature by means of the thermometer.

She should also keep in writing, jotting down at the moment, the exact time, by the watch, of all the important events in the day, such as the taking of food and medicine, the coming on of sleep, or delirium, or restlessness, the dejections of the patient, any remarkable change in the symptoms, and other matters of interest bearing upon the progress of the ailment and the condition of the patient. Fortunate is he who has wife, daughter, mother or sister who can fulfil these requirements ; and sensible is the woman who tries to fit herself, by thought, reading and observation, to take care in a proper manner of those dear to her, in the hours of pain and weakness, under the shadow of death.

The dress of the attendant on the sick is deserving of attention. It should be neat and clean, of a soft, warm color, and of a material which will wash. Black is always nasty to the delicate sense of smell of the sick. A dress that rustles must never be worn in the sick room. Creaking shoes or slippers are of course improper. In many cases of illness, particularly if the room be not carpeted, it is well to have a pair of large loose slippers outside the door, to be slipped on over their shoes by those entering, not excepting the doctor.

Personal Attention Required by the Sick. When there is a tendency, which exists in many diseases, to coolness of the surface of the body, heat should be kept up by means of hot water bottles, warm bricks, or tin cans filled with hot water, etc., applied externally.

The *skin* of the patient must be kept clean by sponging with tepid water, to which a little whiskey or vinegar may be added. Care must be taken while thus cleansing the skin not to expose the person to a draft, and not to uncover the body more than is necessary at one time. This sponging of the entire person may be repeated, in most cases with advantage, every day. In any event, the feet at least twice a week. Recovery is retarded in cases in which the disease is assisted by the presence of dirt.

The *bed* should be tidied and put straight every day, and the sheet upon which the patient lies should be kept smooth and free from crumbs.

The *body linen* must be changed at least once a week, in most cases oftener. The following rules in regard to changing are useful :—

1. Do not begin to change until all that is likely to be needed is ready.

2. See that there is no draft from an open window or door.

3. Have the fresh linen well aired and warmed beforehand.

4. Avoid moving or uncovering the patient more than is absolutely necessary.

5. Do not call upon the patient for too much help.

Delirious patients must, of course, never be roughly dealt with, neither must they be argued with or contradicted in their assertions. It is best to appear interested in their conversation, while watching over and controlling with gentleness but firmness all their actions likely to do them injury. Such patients should never be left alone, nor should one person be in attendance without being able to call for immediate assistance at a moment's notice.

For rules in regard to cooking for the sick, and for special receipts for the invalid's table, see Chapter IX, commencing on page 320.

For directions in regard to giving baths, and applying blisters, cups, leeches, poultices, etc., see Chapter X, page 332.

Management of those Recovering from sickness. Quiet and rest are essential in every recovery from sickness. The mind as well as the body demands repose after suffering. Any prolonged or violent exertion or mental excitement is, for a long time after recovery has set in, injurious, and may occasion a relapse.

After every illness, as soon as the invalid becomes strong enough, a change of air and scene are of the utmost service. To those living in the city, a visit to the country is specially useful. The sick must not, however, be deprived of home care and comforts too soon. A journey, however short, is always fatiguing to the feeble, and a removal to a new abode, always a source of excitement and some discomfort to those not long from a bed of sickness. When the proper time arrives, the return of health is greatly facilitated by a judicious change of residence for a while.

Household Remedies. The articles which are used to restore the sick to health are often and conveniently divided into "household remedies" and "drugs." By the former, those articles are understood which are found in every grocery store and household, which are bought and kept for use in health, but possess medical properties, sometimes very valuable ones. Such, for example, are salt, mustard, vinegar, oil, spices, ginger, etc. An acquaintance with their virtues as medicines, and their proper use in curing disease, is an acquisition which every one ought to make. Often such simple articles answer the purpose much better, with more promptness, and at less cost, than the more unknown products of the apothecary shop.

In the latter, the numerous bottles and drawers, with their labels in large letters, presenting strange words and mysterious abbreviations, bewilder the ordinary visitor. He is deeply impressed with the difficulties in the study of physic ; and the danger of meddling with what he knows is often poisonous is present to his mind. These sentiments are just, for many of these drugs are dangerous, others are rare and costly, others difficult to prepare and administer. It is not wise for any one who has not seriously studied medicine to tamper with them, and hence, in such a manual of domestic practice as this, it would be inexcusable to recommend them.

Fortunately, it is not necessary. Although there are so many hundred drugs in every pharmacy, their multiplication is a matter of trade rather than necessity. The most learned and successful physicians do not use very many remedies. Probably the essential parts of a hundred prescriptions from their hands would, in ninety cases, be found in less than ten different drugs.

We propose, in this work, to take advantage of the same principle of selection, and shall make the reader familiar with a limited number of drugs of wide application, and which have the further advantages of being not costly, not dangerous, and of decided virtues. They can be used in many different complaints, and in the forms which we shall recommend are peculiarly suitable for those to use who have not studied the details of the druggist's trade and physician's calling. They will be often referred to, and when they are once provided in the family, we shall en-deavour to render them sufficient for the treatment of nearly all ordinary diseases, supplemented, as we shall presume them to be, with those " household remedies " to which we have already referred.

It would be advisable, therefore, for every one who intends to made practical use of this book to obtain these remedies, a list of which we shall presently give. They should be carefully put up and kept in a locked chest or box, where children cannot get at them. All the articles can be had of any druggist, though, as some of them are often adulter-ated, and others spoil if kept too long, care should be exercised in their purchase and preservation.

Sometimes, when several drugs are mingled, they act differently and more successfully than either of them (or any other substance) when taken alone. This is why, in most prescriptions or recipes, one sees three or four or more articles mentioned. Another advantage, and a very great one in children and delicate persons, is that it gives the chance to conceal the taste of a disagreeable drug, and often to render it quite palatable.

Many of these firms keep on hand medicine chests containing a select assortment of such drugs as are most commonly used by emigrants, mis-sionaries, ship-captains, and others remote from apothecary stores.

The number of these offers a wide choice, as not all are selected with reference to the great principles which should govern the selection of medical agents for popular use, which we take to be (1), that they are efficient ; (2) that they are not dangerous ; (3) that they do not readily spoil by time and change of temperature ; (4) that they are not distress-ingly unpalatable ; (5) that they are not too bulky ; (6) that they are not very costly. Of the " family medical chests " which have come under notice, we like, as well as any, and in some respects better than any, that which has been sold by the " Medical Commission Agency," in Phila-delphia (115 South Seventh Street). It is called the " Traveller's Medi-cal Kit," and having been carefully selected with a view to the above re-quirements, may serve as a sample of them. Its contents have been more or less changed, from time to time, as improvements in the manu-facture of drugs seemed to suggest. Lists can be obtained from the Agency, giving their prices and contents.

At the end of this work a number of selected receipts will be given which will answer the above requirements, and which will be freely re-ferred to under the treatment of diseases. Any of these can be obtained from the apothecary store, and if used according to directions, will answer a good purpose. A number of them might be prepared and kept on hand ready for use. Others, which will be spoken of, are the following :—

Ammonia, or Hartshorn. The preparations of ammonia which will be referred to are, first, the "aromatic spirits of ammonia," a valuable stimulant, serving in place of alcoholic fluids. The dose of it, for an adult, is twenty drops, in a tablespoonful of water, repeated when necessary. Second, the solution of the acetate of ammonia. A tablespoonful of this, in a wineglass of water, is the quantity for an adult. It acts on the skin and kidney, and is very cooling in fevers, and similar conditions.

Bromides. These are the bromide of potassium or of sodium. They are given to allay pain and nervousness, in doses of 20 or 30 grains.

Iodine. Tincture of Iodine, which is iodine dissolved in alcohol, is frequently used as an external application, to cause gentle counter-irritation, and remove swollen glands, chilblains, stiffness of joints, slow rheumatism, etc. The *iodide of potassium*, dose ten grains, and *syrup of iodide of iron*, dose half a teaspoonful, are very valuable "alteratives," that is, they "alter" the general conditions of the body.

Opium, one grain of which is an adult's dose, is generally given as *laudanum*, which is opium dissolved in alcohol, one grain in twenty drops ; or *morphia*, which is a chemical extract of opium, one quarter of a grain of which equals one grain of opium ; or *paregoric*, which contains one grain of opium to an ounce ; or as *Dover's powder*, which has one grain of opium in each ten. All these must be used with caution.

A Thermometer is useful, not only to maintain the air of a sick room at a proper temperature, and to test the heat of warm baths, foot baths, and the like, but in the hands of a judicious person, is a great aid in deciding the severity of a case of disease, by ascertaining the temperature of the patient. For this purpose, the instrument is so made that its bulb can be placed under the tongue or in the arm-pit. Hold it there three or four minutes, and you have the temperature of the skin and blood of the patient. This should not be above 99°, nor below 97°. Any material variation from these limits indicates danger, although the symptoms in other respects do not appear threatening. "Family Thermometers" have been devised, with a zero, showing the animal heat in health, so that any deviation from it is perceptible by the most inexperienced hand.

CHAPTER II.

DISEASES OF THE AIR PASSAGES.

Cold in the Head—Catarrh of the Throat—Bronchitis—Diphtheria—Asthma—Pleurisy—Inflammation of the Lungs—Consumption—Influenza—Quinsy—Ulcerated Sore Throat.

Cold in the Head. *Symptoms.*—The patient's eyes and nose run freely; at first, however, the eyes are suffused and watery, but the nose dry, swollen and irritable; the air will not readily pass through it, the smell is lost, and constant efforts are involuntarily made to clear the passage. Sneezing occurs on every access of cold air. The distress extends up to the forehead—there are pain, headache, chillness and slight fever. As the cold seems to break, a free, watery discharge commences to flow from the nose, which may be enormous in quantity.

How Treated.—The attack is generally readily broken up by a mustard foot bath, warm drinks and ten grains of Dover's powder at bed time. When the fever is high, the solution of the acetate of ammonia, two tablespoonfuls in a tumbler of water at bed-time will usually relieve this symptom. Many persons abort such an attack by one large dose of quinine, or by a hot lemonade at bed-time and sweating it off.

Catarrh of the Throat. By catarrh we understand an *inflammation;* the blood vessels of the part increase in size, there is an increased flow of liquid and sooner or later there is a casting off of some part of the membrane itself, causing a thick ropy discharge. This is the history of catarrh, no matter where it appears—that is, no matter what mucous membrane be affected. The liability to catarrh varies extremely among different persons; a slight exposure to cold, or rapid transition from one temperature to another, is quite sufficient to excite a catarrhal inflammation somewhere in many individuals, while other persons exposed to the same influences suffer from no such inconvenience. Then, too, among those who are liable to catarrh there is considerable diversity as to the part most easily affected; in some exposure to cold is followed by discharge from the nose; in others bronchitis and cough occur; while still others are apt to suffer from inflammation of the lungs or of the kidneys. All these things may result from "taking cold," though the catarrh inflammation more frequently affects the mucous membranes situated nearest the surface, among them that of the larynx. We may say in general that poorly nourished people are more prone to a catarrh as a result of exposure to cold than healthy and robust individuals; and those who are constantly exposed to the weather in the performance of their usual avocations are less easily affected than those whose pursuits are seden-

11

tary. Another factor is important in determining the liability to catarrh; namely, that after repeated attacks a mucous membrane becomes more readily affected ; it is a weak spot, and yields most readily to any influences tending to derange the individual's health.

Among the causes of catarrh are the breathing of very cold air or dust ; excessive efforts at singing, shouting, screaming and violent coughing. But, in addition, agents which do not affect the throat directly, may also cause a catarrh of this organ ; thus, chilling of the skin, especially that of the feet, is frequently the direct cause of the catarrh.

Beside these various causes which may induce a catarrh in any individual, there are certain other factors which frequently induce the disease. Some very obstinate cases of catarrh are due to constitutional taints, such as syphilis ; and finally there are numerous instances of the disease in which the cause is to be found in the growth of a tumor in the throat.

Symptoms.—Acute catarrh does not usually cause serious constitutional symptoms ; though in some instances slight shiverings mark the outset, and some fever accompanies the disease The patient complains of a sensation in the throat, which he describes usually as tickling, or in more severe cases, as burning or soreness ; this sensation is aggravated by coughing or speaking.   At the same time the voice is changed in character, grows deeper, hoarse, perhaps cracked, and finally may be lost altogether. This change in the voice indicates that the mucous membrane covering the vocal chords is in a state of inflammation or *catarrh*.   Another constant symptom is a violent cough, due to the irritation in the inflamed mucous membrane.

Treatment.—Since the acute attack usually subsides spontaneously within a week, treatment with medicines is usually unnecessary.   It will suffice to direct the patient to remain so far as possible in a uniform temperature, and not expose himself to sudden changes of atmosphere : he should avoid the effort to talk, so far as possible, and should especially resist the inclination to cough. He will of cou se say that he can't help it, but he *must* help it. The irritation of the throat, and hence the tendency to cough, can be much diminished by one of the following prescriptions :

Hydrocyanic acid (dilute),   -   -   -   Half a drachm.
Sulphate of morphia,   -   -   -   -   Half a grain.
Syrup of tolu,   -   -   -   -   -   One ounce.
Water,   -   -   -   -   -   One ounce.
Mix and take a teaspoonful every two hours.

Or relief may be obtained from the following :

Syrup of wild cherry, -   -   -   -   - One ounce.
Syrup of squills,   -   -   -   -   - One ounce.
Camphor water,   -   -   -   -   - One ounce.
Mix and take a teaspoonful every two hours.

Should the disease become chronic the throat should be gargled thoroughly twice daily with a solution of salt and water (2 or 3 teaspoonfuls to the pint) and the patient should take besides the following mixture highly recommended by the famous Dr. Da Costa :

Acetate of potash - - - -	One ounce.
Muriated tincture of iron - - -	One ounce.
Dilute acetic acid - - - -	One-half ounce.
Simple elixir - - - - -	One ounce.
Water to make six ounces.	

Take a teaspoonful four times a day.

Bronchitis. This is an inflammation affecting the lining membrane of the bronchial tubes—that is the tubes which form the continuation of the windpipe.

Symptoms.—Acute inflammation of the bronchial tube usually begins with a cold in the head ; that is, a catarrh of the nose. Thence it extends down the throat and larynx to the bronchial tubes. Thus a variable period, from a few hours to two or three days, may elapse from the beginning of the catarrh in the nasal passages to the establishment of the bronchitis. The inflammation in the bronchial tubes begins with a sense of tightness across the chest, accompanied with a feeling of rawness or soreness. These sensations are aggravated by every act of coughing ; there is at the same time a certain amount of fever, usually slight. The constitutional disturbance may be so considerable as to prevent the patient from attending to his usual vocation, or may be almost imperceptible. The cough is at first dry, but very little mucus being expectorated. During, and especially at the end of, each act of coughing a painful sensation is felt under the breast bone ; this is also the case upon breathing cold air, or upon drawing a long breath.

After some days the cough becomes " looser ; " the expectoration becomes easier, more profuse and less painful ; the matter expectorated being frothy, viscid, and often streaked with blood. After three or four days it becomes thick and yellow, or green. By this time most of the disagreeable symptoms have subsided ; the patient is troubled by no other symptoms than the necessity for frequent, sometimes violent, coughing. The usual duration of the attack is twelve to fourteen days, though this period is often prolonged by carelessness or neglect on the part of the patient.

Treatment.—It is often possible to cut short a " cold on the chest " within twenty-four hours, by taking, at bed-time, a hot foot-bath, a glass of hot toddy or lemonade, and ten grains of Dover's powder. This attempt will, however, be unsuccessful unless made after the first indication that the individual has taken cold—that is, before the sense of tightness in the chest occurs. If this measure be omitted, a brisk saline cathartic, such as the citrate of magnesia, should be given. The pain and soreness in the chest will be relieved by a light mustard-plaster over the breast-bone ; the cough may be "loosened" by taking a half-spoon-ful of the compound syrup of squills every two hours. If this remedy provoke nausea, it may be replaced by the following :

Tartar emetic,	- - -	Two grains.
Syrup of wild cherry,	- -	
Water,	- - - -	Each three ounces.

Mix ; take a teaspoonful every two hours.

So soon as expectoration becomes easier, the removal of the mucus from the bronchial tubes may be promoted by giving a quarter of a teaspoonful of the syrup of ipecac every hour ; or, if the cough be some-what violent, the following prescription :

Nitrate of potash,	- -	Two drachms.
Syrup of squills,	- -	Two ounces.
Tincture of digitalis,	- -	Half a drachm.
Sugar,	• - • -	
Gum arabic,	• - -	Each two drachms.

Water enough to make six ounces. A teaspoonful of this may be put in a wineglassful of water and sipped every ten or fifteen minutes.

Should the disease become chronic, croton oil, tincture of iodine, plasters of hemlock or pitch may be applied to the chest, while tonics, stimulating expectorants and alteratives are used. In these cases, pure fresh air is valuable, and can do no harm, though this is generally pro hibited, lest they should take fresh cold.

How Prevented.— Cold in the chest may be prevented by keeping the chest and arms well protected in cold weather, and the feet from damp-ness. Cod-liver oil and syrup of the iodide of iron continued for a long time will strengthen the system of one predisposed to this affection, and secure immunity from repeated attacks.

Diphtheria. *Symptoms.*—Prior to the appearance of the peculiar symptoms of the disease, there are languor, uneasiness, sore throat, and swelling of the glands of the throat. Fever sets in, headache, etc., and difficulty of swallowing. Examining the throat shows the parts swollen, red and purple, followed, in a day or two, by a coating of dirty or yellow white, like we buckskin. At the end of eight or nine days, this begins to loosen, cleans off, and recovery commences. In the severe cases, all these symptoms are aggravated ; in children, the membrane causes a croupy cough, and the most energetic efforts to obtain breath, the patient becoming livid, and evidently strangling. In what is known as the ma-lignant form, the first onset is with intense pain in the head, high fever, nausea and vomiting, and even bleeding from the mucous surfaces. The coating is very dark, ash-colored, leathery, and exhales an offensive odor. Prostration rapidly ensues, followed by stupor and death.

Treatment.—Dr. A. Brondel writes, in the *Bulletin Général de Thérapeutique* of November 15th, 1886, concerning the treatment of diphtheria by benzoate of sodium, and asserts that of two hundred consecutive cases he has not lost a single one. He admits the possibility of a mistaken diagnosis in some instances, but even excluding fifty per cent. on this account he has still one hundred cases without a death. His method is as follows : Every hour the patient takes a teaspoonful of a solution of benzoate of sodium, fifteen grains to the ounce, and at the same time one-sixth of a grain of sulphide of calcium in syrup or granule. In addition to this the throat is thoroughly sprayed every half-hour with a ten per cent. solution of benzoate of sodium. This is done religiously at the regular intervals, day and night, but no other local treatment is employed : no attempt is made to dislodge the false membrane, and no pencilling nor painting of the fauces is resorted to Nourishment consists of beef-juice, tender rare meat, milk, etc., but bread and all other articles which may cause irritation of the throat are forbidden The sick room is kept filled with steam from a vessel containing carbolic acid, turpentine and oil of eucalyptus in water.

Asthma. Asthma occurs in paroxysms, at irregular intervals. The paroxysm seems to be brought on both by certain conditions of the patient himself and by certain conditions of the atmosphere. The attack may begin without warning, though those individuals who have long suffered from the disease can usually predict some hours in advance the approach of the paroxysm. The attack may come on slowly, requiring, that is, two or three hours for its development ; or it may attain its full intensity in a few minutes. The paroxysm usually occurs at night, or early in the morning.

Treatment.—The treatment of asthma consists of two parts : The management of the patient during the paroxysms, and the treatment during the intervals After he has been placed in a large airy room, one of the quickest remedies to afford him relief is stramonium taken in the shape of cigarettes, or the leaves may be burned and the smoke inhaled. If no stramonium leaves can be obtained, the same effect may be produced by tobacco leaves or cigars, though this means is far less reliable. Next to stramonium in efficacy is the nitrate of potassium or *nitre*. Pieces of paper which have been soaked in a solution of nitre may be burned, and the smoke inhaled ; or, if preferred, they may be made into cigarettes and smoked.

A remedy that may be easily procured and always used without danger is the following :—

Tinct. of Lobelia,	1 oz.
Iodide of Ammonia,	2 dr.
Bromide of Ammonia,	3 dr.
Syrup of Tolu,	3 oz.

A teaspoonful every one, two, three, or four hours.

Dr. Roberts Bartholow.

Of this prescription Dr. Bartholow says, "It gives relief in a few minutes."

In the intervals, keep the system in good order, build up, if necessary ; give bromide of potassium, in quite full doses, say fifteen to thirty grains, three times a day, for many weeks. Where practicable, change of climate is, perhaps, the surest remedy, and often relieves permanently.

Pleurisy. By this we mean an inflammation of the membrane, lining the inner chest wall and covering the lungs.

Symptoms.—First a chill some time after exposure, then a sharp pain in the chest, with fever and a short cough. When fluid begins to collect in the chest the pain diminishes, but the difficulty of breathing increases.

Treatment.—Early in the disease, restrain the inflammation and lessen the pain by 10 grains of Dover's powder, repeated if necessary in 4 or 5 hours. When the fluid has collected the following is an excellent remedy :

Powdered Squills,   -   -
Powdered Digitalis,   -   -   Each 30 grains.
Blue Mass,   -   -   -   -   20 grains.

Mix and divide into 30 pills ; take one three times a day.

The side may be painted with iodine and a linseed poultice applied or a Spanish fly blister may be used if necessary when the case is obstinate.

Acute Pleurisy. At the onset of the attack :—

Tincture of Aconite Root,   -   -   -   - 20 drops.
Camphorated Tinct. of Opium,   -   -   - ½ ounce.
Nitrous Spirits of Ether,   -   -   -   - ½ ounce.
Solution of Acetate of Ammonia,   -   -   - 5 ounces.

Mix.—A tablespoonful every three hours.

J. H. Ripley.

Inflammation of the Lungs. *Symptoms.*—It usually begins with a chill and fever, followed by a dull pain in the chest, and more or less cough. The spittle is mixed with blood and hence looks rusty. This rusty spittle is a sure sign of the disease. The height of the disease is reached about the ninth day, when, in favorable cases the patient begins to recover, but if unfavorable, the change for the worse now takes place.

Treatment :—

Muriate of Ammonia,   -   -   -   -   - 3 drachms.
Tartar Emetic,   -   -   -   -   -   - 2 grains.
Sulphate of Morphia,   -   -   -   -   - 3 grains.
Syrup of Licorice,   -   -   -   -   - 4 ounces.

Mix.—A teaspoonful every two hours.

Dr. N. S. Davis.

Calomel, - - - - - - - 6 grains.
Powdered Ipecac., - - - - - - 6 grains.
Powdered Opium, - - - - - - 3 grains.
White Sugar, - - . - - - - 30 grains.
Divide into six powders.

Take one powder every four hours alternately with the preceding prescription. At the same time cover the chest with emollient poultices.
Dr. N. S. Davis.

At the end of twenty-four hours he omits the powders, and if the bowels have not been moved he gives a mild laxative. If the symptoms are not favorably modified in 3 or 4 days, a blister is placed on the side of the chest most affected.

Should the pulse become soft and frequent, the breathing abdominal and the lips of a leaden hue.

Quinine, - - - - - - - - 2 grains.
Carbonate of Ammonia, - - - - - 4 grains.

Mix.—Take at a dose. If delirium becomes troublesome add 10 minims of chloroform to the ammonia mixture. If there is indication of malarial influence, quinine may be given during the remissions.

Consumption. In this disease every effort should be made to build up the system generally by the use of tonics, and a liberal nourishing diet Stimulants, such as whiskey (in the form of eggnog) may be used, but only to a limited extent. Plenty of fresh air is of first importance. Let the patient be warmly dressed and then remain out of doors the most of the time. The fear of "catching cold" in this disease often keeps the patient in over-heated rooms, and thus in reality hastens the end. Cod Liver Oil or its preparations should be taken regularly. Beyond this, the disagreeable symptoms of the disease should be treated as they arise. These symptoms are night-sweats, sleeplessness, diarrhoea, bleeding from the lungs, and cough and pain in the chest.

The Night Sweats of consumption may be quite profuse and exhausting, even before the later stages of the disease are reached. They are best and soonest relieved by improving the patient's strength and vigor, since the night sweats are merely indications of the general debility and exhaustion induced by the disease. It may be possible, also, to materially reduce this perspiration and its ill effects, either by using some astringent wash externally, or a preparation of belladonna as a medicine. For the former purpose, alum may be dissolved in alcohol, and the patient lightly sponged with this, before retiring ; at the same time minute quantities of atropia—one one-hundredth of a grain—may be given as a pill, at night. Or the following mixture may be given :

Aromatic sulphuric acid, - - - Three drchms.
Sulphate of quinine, - - - - Fifteen grains.

B

Water to make two ounces. Mix, and give a teaspoonful in water at night.

The Pains in the Chest are often so troublesome as to disturb the patient's rest extremely. These can usually be relieved by the use of belladonna plaster applied to the painful spot, or friction with chloroform liniment or a light mustard plaster may also prove beneficial.

For the Sleeplessness :—

Bromide of Potash,	- - -	6 drachms.
Water to	- - - - -	6 ounces.

Take 3 teaspoonsful at bed time with a little wine and water.

Dr. Brown-Séguard.

For the Diarrhœa :—

Resin turpentine,	- - - -	3 grains.
Nitrate of silver,	- - - -	¼ grain.
Opium,	- - - - -	¼ grain.

Make one pill. Take when needed.

This formula is used in these cases at Roosevelt Hospital with uniform good results, the diet being milk boiled with mutton suet until it is as thick as cream.

For the bleeding from the Lungs :—

Acetate of lead,	- - - -	40 grains.
Powdered Digitalis,	- - -	20 grains.
Powdered Opium, -	- - -	10 grains.

Mix and make into 20 pills. Take one every 4 hours.

Dr. Roberts Bartholow.

For the Cough :—

Dilute hydrocyanic acid,	- - -	Half a drachm.
Sulphate of morphia,	- - -	Half a grain.
Syrup of tolu, water,	- - -	Each one ounce.

Mix, and take half a teaspoonful every hour.

Influenza. It has lately been very much the fashion to call any kind of cold which is accompanied by catarrhal symptoms, Influenza ; but this, in nine cases out of ten, is a misnomer. The true disease seldom occurs, except as an epidemic, attacking many persons at once. It comes on quite suddenly.

Symptoms.—Its symptoms are those of a general fever. There is great prostration of strength, generally showing loss of appetite, heat and thirst, cough and difficulty of breathing, owing to the air valves and bronchial passages being clogged with mucus ; there is also running at the nose and eyes, weight across the brow with throbbing pain and great depression of spirits. The febrile symptoms do not commonly last more than four or five days, sometimes but one or two, but the cough generally remains for a considerable time, varying according to circumstances, such as exposure to cold or wet, predisposition to cough, &c.

Treatment.—With the strong and healthy this is not a dangerous disease, but aged or weakly persons are frequently carried off by it. In the former case but little medical treatment is required. Keep the patient in bed, and let the temperature of the room be warm and equable; open the bowels with a gentle aperient, such as rhubarb and magnesia, or senna mixture, and follow this up with weak-wine whey, or some warm diluent drink, in a pint of which a grain of tarter emetic and a drachm of nitrate of potash has been dissolved; give a wineglassful of this about every four hours. It is not generally safe to practice much depletion; but where there is great difficulty of breathing, and irritation of the throat, a few leeches may be applied above the breast bone, in the hollow of the neck. Stimulating liniments may also be applied to the chest, and mustard poultices, but blisters are scarcely to be recommended. Hot fomentations may also be useful, and medicated inhalations, such as a scruple of powdered hemlock or henbane, sprinkled in the boiling water, from which the stream ascends into the throat. The fresh leaves of the above plants may be used, or a drachm of the tincture, if these cannot be procured. When the fever is subdued, if there is still cough and restlessness, a five-grain Dover's powder may be given at bedtime, or one-eighth of a grain of acetate of morphine, with a five-grain squill pill, for the cough if required. If there is great feebleness, tonics must be administered; infusion of calumbia, cascarilla, or gentian, with carbonate of ammonia; one ounce of the former with five grains of the latter, three times a day, with a mildly nutritious diet,—broths, arrowroot, sago, and a small quantity of wine. Such is an outline of the course to be taken in most cases of influenza which really require medical treatment at all; generally warmth, rest, and good nursing, will do all the business. Should the cough be very obstinate, and resist all efforts to remove it, change of air will generally prove effectual, and this is beneficial in most cases.

Quinsy. An inflammation of the throat, principally occupying the glands. This kind of inflammatory sore throat generally commences with cold chills, and other febrile symptoms. There is fulness, heat, and dryness of the throat, with a hoarse voice, difficulty of swallowing, and shooting pains towards the ear. When examined, the throat is full of florid red color, deeper over the tonsils, which are swollen and covered with mucus. As the disease progresses, the tonsils become more and more swollen, the swallowing becomes more painful and difficult, until liquids return through the nose, and the viscid saliva is discharged from the mouth. Very commonly the fever increases also, and there is acute pain of the back and limbs.

Causes.—Exposure to cold, wearing damp clothes, sitting in wet rooms, getting wet feet, coming out suddenly of a crowded and heated room into the open and cold air. It may also be brought on by violent exertion of the voice, and by suppressed evacuations.

Treatment.—When the case is not severe, it may be treated, in the early stages, like catarrh ; but, when it is, more active measures will be required. An emetic, followed by a strong purgative ; a blister outside the throat, and warm bran or linseed poultices ; a cooling regimen, with acid drinks, or pieces of rough ice put into the mouth and allowed to dissolve ; leeches at the side of the throat if it swells much ; inhaling the steam of hot water through a teapot or an inverted funnel ; and the continuation, every four hours or so, of a saline aperient. These will be the proper measures to adopt. When the abscess has burst, and the inflammatory symptoms have subsided, a generous diet will be necessary, with tonic medicines. If the tonsils continue swollen, they should be rubbed outside twice a day with stimulating liniments. Turpentine and opodeldoc, equal quantities, will be as good as any ; and the throat gargled with salt and water, a teaspoonful of the former put into a tumbler full of the latter.

When there is chronic soreness of the throat, with hoarseness and cough, there is commonly also a relaxed and elongated uvula, which closes the passage when the patient lies down, and causes a sensation of choking. In this case, a gargle made with salt and Cayenne pepper (about a tablespoonful of the former and a teaspoonful of the latter, in a pint of boiling water) should be tried ; the throat should be kept uncovered, and sponged with vinegar twice a day. If these means are unsuccessful, it may be necessary to have part of the uvula cut off. This must be done by a surgeon. Also, the application of caustic must sometimes be made when the throat has a granulated appearance.

Ulcerated, or Putrid Sore Throat.

This sort of sore throat shows itself by white specks, covering ulcers, appearing in the throat, together with great debility, and an eruption on the skin.

Causes.—Contagion (infection) ; from a humid state of the atmosphere, it becomes epidemical, and will sometimes rage through families, villages, or towns ; and is also produced by similar causes to typhus, or malignant fevers, to which it seems akin in its nature.

Symptoms.—It commences with cold shiverings, anxiety, nausea, and vomiting, succeeded by heat, thirst, restlessness, and debility ; also, much oppression at the chest ; the face looks flushed, the eyes are red, a stiffness is perceived in the neck, with a humid breathing, hoarseness of the voice, and soreness in the throat. After a short time the breath becomes offensive, the tongue is covered with a thick brown fur, and the inside of the lips is beset with vesicles, containing an acrid matter ; upon inspection of the throat, a number of sloughs, between a light ash and a dark brown color, are to be seen. From the first attack of the complaint, there is a considerable degree of fever, with a small irregular pulse, and the fever increases towards the evening. About the second or third day, large patches make their appearance about the face and neck, which by degrees become dispersed over every part of the body. As the sloughs in the throat spread, they generally become of a darker color, and the whole throat is soon covered with thick sloughs, which, when they fall off, discover deep-seated ulcers.

Treatment.—The bowels should be opened with a dose of Rochelle salts or sulphate of magnesia. To cleanse the throat, use the following gargle :—

Honey of Roses,   -   -   -   - 1 ounce.
Tincture of Myrrh, -   -   -   - ½ ounce.
Vinegar,   -   -   -   -   - 1 ounce.
Decoction of Barley,   -   -   - 10 ounces.

Mix and use frequently. Or the following :—

Muriatic Acid,   -   -   -   - 1 drachm.
Compound Tincture of Cinnamon, - ½ ounce.
Tincture of Myrrh, -   -   -   - 1 ounce.
Decoction of Peruvian Bark,   -   - 6 ounces.

Mix and use frequently. Breathe the steam of hot vinegar and water into the throat.

The following is a good astringent draught :—

Aromatic Confection,   -   -   - 1 drachm.
Tincture of Catechu,   -   -   - 1 drachm.
Laudanum,   -   -   -   -   - 30 drops.
Chalk Mixture,   -   -   -   - 2 ounces.
Cinnamon Water   -   -   -   - 2½ ounces.

Mix. Take two tablespoonsful every four hours. Shake well always before taking. Sometimes bleeding from the mouth, nose or ears, takes place in the latter stages of the disease : and, becoming alarming, use the following as a wash :—

Sulphate of Copper,   -   -   - 1½ drachms.
Alum,   -   -   -   -   -   - ½ drachm.
Rectified Spirit of Wine, -   -   - 1 ounce.
Pure Water,   -   -   -   -   - 7 ounces.

Mix, and apply internally with a tent, or on linen cloths. The diet must be light and nourishing—tapioca, sago, rice, and the like : the drinks must be acidulated ; free air, but not cold ; the room sprinkled with vinegar, and generally as is laid in acute or typhus fever, use the following in the room, as a purifying anti-infectious gas :—

Take a pound of common salt, put it into an earthen dish, occasionally pour a tablespoonful of sulphuric acid (oil of vitriol) ; stir up with a stick, avoid breathing over it when the fumes are rising. Do this four or six times a day, whenever infectious diseases are raging. It is a great preventive.

CHAPTER III.

DISEASES OF THE FOOD AND BLOOD PASSAGE.

Mumps—Ulcerated sore throat—Dyspepsia or indigestion—Biliousness—Heartburn—List of foods—Diarrhœa—Dysentery—Constipation—Colic—Inflammation of the Bowels—Tapeworm—Dropsy—Jaundice—Piles—The heart—Neuralgia of the Heart.

Mumps. This is an inflammation of one of the glands which secrete saliva—the *parotid gland*, situated at the angle of the jaw, just below and in front of the ear.

Treatment.—The only treatment required is the palliation of the pain, which is often very severe. The application of cloths wrung out in hot water, or of linseed poultices, frequently changed, will usually relieve the severity of the pain. If the painful feeling be felt especially in the ear, it may be well to syringe the ear gently with warm water three or four times a day, yet there is no danger that the ear itself is involved in the inflammation. The use of mild liquid diet—milk and soups—completes the care necessary during the disease.

Ulcerated Sore Throat. *Treatment.*—When ulcers form in the throat, a solution of Lunar Caustic 40 grains, water 1 ounce, should be made, and with a camel's hair brush dipped in this solution the ulcers should be touched once a day. At the same time tonics should be given.

Dyspepsia or Indigestion, with its accompanying train of discomforts, is the result of haste or carelessness in eating, not chewing the food, excess or deficiency of food, fatigue, excitement, study, the use of tobacco, or of ardent spirits, and last, but not least, the ignorant use of medicine.

Symptoms.—The most prominent symptom is a feeling of uneasiness, not exactly pain, at the stomach. Pain would be indicative of ulceration. Nausea, and even vomiting, occasionally occur. There is a clammy feeling in the mouth, with a bitterness or sourness. The skin is generally sallow. The bowels are generally costive, but this may be alternated by diarrhœa. When biliousness is present, the stools are apt to be scanty, and of light clay color. The following symptoms appear from time to time : water-brash, a running of tasteless fluid from the mouth, heartburn, caused by the presence of acid in the stomach, and extending up the œsophagus ; palpitation of the heart ; headache, depression of spirits, melancholy, or the occurrence of foolish ideas ; disorders of taste, vision, etc. ; more or less vertigo in every case. Dyspepsia is always an obstinate affection, but can scarcely be regarded as dangerous.

22

Treatment.—Medicine is less important than the regulation of the quantity and quality of food, and the hours for its ingestion. Plain, easily digested food, is required. The stomach should never be loaded ; the meals should not be at too long or too short intervals. Beef, mutton, fowl, oysters, roasted, stewed, or panned, never fried, with bread rather stale ; crackers, kiln dried ; rice ; stewed or fresh fruit ; and milk, as a beverage, will be the proper diet for the dyspeptic, and will give a sufficient variety. Pastry must be scrupulously avoided. Some prefer, and may have, milk tea, or milk and warm water, as a drink with the meals. Some seem to require a slight stimulant with dinner, and here, ginger and water, or cider, or sherry wine and water may be allowed in small quantities. Where the digestion is greatly impaired, the food may be given in very small quantities, at short intervals. To use the remark of a distinguished authority, "sixty meals a day." Of course, this is only to be enforced until the stomach will bear a larger quantity at a time, when the plan should be abandoned. Daily exercise, in the open air, and bathing, are very important, as well as rest at meal time. Never eat under an excitement. At a meal, the family chat aids the digestion ; hence the better appetite while travelling ; the mind is relieved of cares, etc.

As regards medicine in this disease, the bowels should be kept free by magnesia, rhubarb, senna, or some such mild laxative ; to aid diges- tion the following is excellent :

Dilute Hydrocianic acid, - - - 25 drops.
Subnitrate of Bismuth, - - - ½ drachm.
Syrup of Orange, - - - - 1 ounce.
Infusion of Gentian to - - - 8 ounces.

Take a teaspoonful 3 times a day before meals.

Dr. Farquharson.

Biliousness. *Symptoms.*—Variable appetite, dizziness, bad taste in the mouth, a feeling of weariness, especially in the morning, and some- times a pain between the shoulders. The patient's tongue is coated and he feels "sick."

Treatment.—At bedtime take 5 grains of blue mass, and the following morning, enough magnesia to move the bowels. After the bowels have moved begin to take the following :

Dilute Nitro-Muriatic Acid, - - Six drachms.
Liquor Strychnia, - - - 1½ drachms.
Compound Tincture of Gentian, - One ounce.
Syrup of Orange, - - - - One ounce.
Water to eight ounces, - - -

Take a tablespoonful 3 times a day before meals.

Heartburn. This is best relieved by taking ten grains of bicarbonate of soda in water before meals. Remember this symptom often depends upon the presence of dyspepsia, which must be cured before the heartburn will disappear.

In order to afford some assistance in the selection of diet for dyspeptic patients, the following table, adapted from Hartshorne, is added:

Easy of Digestion.	Moderately Digestible.	Hard to Digest.
Mutton,	Beef,	Pork,
Venison,	Lamb,	Veal,
Chicken,	Rabbit,	Goose,
Turkey,	Duck,	Salt meats,
Hare,	Pigeon,	Sausages,
Beef tea,	Snipe,	Salt fish,
Mutton broth,	Soups,	Lobster,
Milk,	Eggs,	Herring,
Most fresh fish,	Raw oysters,	Salmon,
Turbot,	Stewed oysters,	Shrimps,
Sole,	Potatoes,	Oils,
Haddock,	Beets,	Cheese,
Roasted oysters,	Turnips,	Fresh bread,
Rice,	Cabbage,	Toast,
Tapioca,	Lettuce,	Pastry,
Sago,	Celery,	Cakes,
Arrowroot,	Apples,	Nuts,
Asparagus,	Raspberries,	Pears,
Cauliflower,	Bread,	Plums,
Baked apples,	Puddings,	Cherries,
Oranges,	Rhubarb,	Cucumbers,
Grapes,	Chocolate,	Onions,
Strawberries,	Coffee,	Carrots,
Peaches,	Porter.	Parsnips,
Ale.		Pickles.

Diarrhœa. The acute form of this disease is frequently due to undigested or improper food, or to exposure to cold, and usually subsides of itself after a dose of castor oil has been given to carry away the irritating substance. After two days, should the discharge still continue, the following will be found useful:

Prepared Chalk,	- - -	1½ drachm.
Powdered Acacia,	- - -	1 drachm.
White Sugar,	- - ..	1 drachm. ♦
Tincture of Opium,	- -	10 drops.
Water,	- - - -	3 ounces.

Take a teaspoonful every hour.

Dr. Dewees.

Or

Tincture of Opium,   -   -   - ⎫
Tincture of Capsicum, -   -   - ⎪
Aromatic Tinct. of Rhubarb, - - ⎬ An equal quantity
Spirits of Peppermint,   -   - ⎪   of each.
Spirits of Camphor,   -   -   - ⎭
Take from 20 to 40 drops as required.

Dr. Ruschenberger.

Dysentery, or bloody flux, or bloody stools, is an inflammation of the large bowel, the colon, hence it is also called colitis.

Cause.—This disease appears to occur most frequently about the last of August and September, and is due to the use of unripe fruit, improper food, exposure to cold aud damp after the heat of the day, impure water, etc. It often prevails in a locality as an epidemic, attacking the majority of those exposed to it.

Symptoms.—It is characterized by pain in the lower part of the abdomen ; tenderness on motion and on pressure ; frequent desire to evacuate the bowels ; small, bloody, mucous passages ; great tendency to strain or bear down, which is called " tenesmus ; " griping, called " tormina," and moderate fever. There may be debility, generally not at first ; ulceration, and purulent, shreddy discharges may occur later.

Treatment.—Where the strength will bear it, and the inflammation is great, leeches may be usefully applied, at first, to the abdomen, at the seat of the greatest soreness, followed by warm poultices of flaxseed, hops, Indian meal. A tablespoonful of castor oil, with fifteen to twenty drops of laudanum, is the usual dose before anything else, and this is generally very useful. A large injection of warm water is often of great service.

When the disease has become fully established the following prescriptions have been found to be eminently successful.

Powdered Catechu,   -   -   - 2 drachms.
Powdered Acacia,   -   -   - ½ ounce.
Water,   -   -   -   -   - 6 ounces.
Take a tablespoonful every two hours.

Dr. Niemeyer.

Also

Sulphate of Copper,   -   -   - ½ grain.
Sulphate of Magnesia, -   -   - 1 ounce.
Dilute Sulphuric Acid, -   -   - 1 drachm.
Water,   -   -   -   -   - 4 ounces.
Take a tablespoonful every four hours.

Dr. Bartholow.

Of course, great care should be paid to the diet, which should be mainly of rice, arrowroot, chicken or beef tea ; and for thirst, iced drinks, as slippery elm water, gum arabic water, etc.

With great debility, quinine may be given, or other tonics and even stimulants, if there is great weakness.

Constipation or Costiveness. *Treatment.*—The first thing to be done is to establish the habit of trying to evacuate the bowels every day at a certain hour ; the best time for most people is just after breakfast. It matters not if the bowels do not act ; the practice of attempting should be persisted in, and in time it will break up the confined state of the bowels. Adopt a diet free from all astringents, taking care especially that there is no alum in the bread, and using a coarser kind. Let the food consist of a due admixture of meat and vegetables for dinner ; the beverage water. For breakfast, stale bread or dry toast, with a moderate quantity of butter, honey, fish, or bacon ; cocoa, perhaps, is preferable to tea or coffee ; and porridge made with Scotch oatmeal, probably better still. Regular exercise, either by walking or on horseback, should be taken. Roasted or boiled apples, pears, stewed prunes, raisins, gruel with currants, broths with spinach, leeks and other soft pot-herbs, are excellent laxatives. If the above mode of living fail to relax the bowels, inject warm water by means of enema

It is a great mistake in this disease to use powerful medicine. The milder the better.

One of the best remedies for habitual constipation is the following :

Senna leaves,	- - -	Three ounces.
Liquorice root,	- - -	" "
Sulphur,	- - - -	Two "
Fennel seed,	- - -	One ounce and a half.
White sugar,	- - -	Six ounces.

Pulverize thoroughly and mix.   Take from a teaspoonful to a table-spoon, either dry or in water.

This powder has the advantage that it can be used for a considerable time without weakening the bowels, and thus creating a demand for more powerful laxatives.

Another prescription which will be found beneficial in constipation resulting from indigestion, is the following :

Powdered rhubarb,	- - - . -	12 grains.
Podophylline,	- - - - -	4 "
Extract of nux vomica,	- - -	8 "

Mix and make 24 pills.   Take one at night.

Another formula which has been much used is the so-called " safety pill : "

Extract of hyoscyamus,	- - -	10 grains.
Extract of nux vomica,	- - -	6 "
Extract of aloes,	- - - -	30 "
Powdered ipecac,	· - - - -	2 "

Mix and make 20 pills   Take one at night.

The last prescription we will mention in this connection is also perhaps the best :

Flued extract of cascara sagrada,   -   -   4 ounces.

Take a teaspoonful at bedtime, and repeat if necessary next morning.

Colic.—Wind Colic.   Wind colic is a severe and distressing pain in the bowels, sometimes a stoppage, and swelling about the pit of the stomach and the navel.   The complaint may be caused by weakness in the digestive organs, by indigestible food, unripe fruit, or costiveness.

Treatment.—If the pain is caused by having eaten anything indigestible, an emetic should be immediately taken.   If this does not bring relief a dose of salts, or sweet tincture of rhubarb, may.   If there is no sickness at the stomach, a little essence of peppermint in water, or brandy, or gin, in hot water, may be sufficient to expel the wind and give relief.   If there be costiveness, and continued pain, a stimulating injection should be given.

Bilious Colic.   Bilious colic is a dangerous disease.   There is griping, twisting, tearing pain about the navel, or sometimes over the whole belly.

Causes.—It is caused by irritating articles taken into the stomach, vititated bile, long exposure to cold, torpid liver and skin, great unnatural heat, &c.

Symptoms.—It comes and goes by paroxysms.   Sometimes the abdomen is drawn in, at other times swelled out, and stretched like a drumhead.   At first the pain is relieved by pressure, but after a time the belly grows tender to the touch.   There is thirst and heat, and a discharge of bilious matter from the stomach.   In the worst cases, the pulse is small, the face pale, the features shrunk, and the whole body covered with a cold sweat.

Treatment.—Administer an active purgative injection immediately.   Give a mixture of pulverized camphor, four grains ; cayenne, twelve grains ; white sugar, one scruple.   This, divided into four powders, and given once in fifteen minutes, will relieve the pain,—at the same time a mustard-poultice should be laid upon the belly.   The sickness of stomach may be allayed by hot fomentations over the stomach, in which are a few drops of laudanum ; also on the feet.

Inflammation of the Bowels.   *Symptoms.*—There is usually great pain and tenderness of the abdomen, accompanied with fever, and a frequent, small, hard pulse.   Sometimes at first, the pain is confined to one spot ; but it generally soon extends over the whole of the abdomen.   It is very severe, and much increased by any motion, even coughing, sneezing, or drawing a long breath.   Even the weight of the bed-clothes is sometimes unbearable.   It is acute and cutting, and sometimes occurs in paroxysms ; and the patient usually lies on his back with his knees drawn up. The bowels are usually constipated, but sometimes the reverse ; and commonly there are present nausea, vomiting and hiccough.

Treatment.—The most important agent in the treatment of this disease is *opium*. As the patient usually vomits, the drug must be given either by injection into the rectum or by insertion under the skin. The patient can usually endure, without danger, an amount of opium which could not safely be administered to a healthly person ; thus half a grain to a grain of opium can be given every three or four hours, according to the severity of the pain ; the general plan is to administer the drug until the pain is subdued. Hot applications should also be made to the abdomen ; these may consist of light mustard poultices, or of cloths wrung out in hot water ; the latter may be sprinkled with turpentine. It should be remembered that the bowels must not be disturbed during peritonitis; even though the patient have no evacuation for several days or a week, it is advisable to avoid the use of carthartics. Whenever it becomes necessary to secure a passage of the bowels, this may be done by a rectal injection of hot water.

Tape Worm. *Symptoms.*—There are no signs by which the exist￼ence of a tape worm can be positively asserted. Numerous symptoms are supposed to indicate the presence of the animal—dizziness, ringing in the ears, impairment of vision, flow of saliva, itching about the nose, impairment of appetite and digestion, colicky pains in the abdomen and emaciation. These, however, may all exist from other causes in cases where no tape worm is present ; while on the other hand the worm may be discovered in individuals who consider themselves perfectly well.

The only positive proof of the existence of a tape worm is the passage of some of its joints from the intestine. If the worm have attained considerable size, such fragments are passed daily, or at least at short intervals.

Treatment.—One of the commonest, and certainly a very efficient, means for expelling the worm is *turpentine.* This is given in quantities varying from one to two tablespoonfuls for adults, usually mixed with the same quantity of castor oil, and taken floating on milk. This dose may be repeated every second or third day, until the fragments of the worm cease to appear. The objection to the use of turpentine is that it sometimes causes difficulty in passing water, and it may induce a state of intoxication in some cases.

Another popular remedy is the oil of *male fern.* This may be given in doses of one or two teaspoonfuls, either in mucilage or acacia. Two hours after this dose the patient may take a teaspoonful of Turpentine in tablespoonful of castor oil. A still more familiar remedy is made of pumpkin seeds. Two ounces of the seeds are pounded in a mortar with six ounces of water, and the mixture strained. Half of this may be taken in the morning and half in the evening. It will probably be necessary to repeat this treatment for several days.

Whatever remedy may be selected, it is important that certain preparatory treatment should be adopted before the worm remedy itself is given. This treatment consists in abstinence from food for several hours, or a day ; or the patient may employ light diet, such as broth and milk, for two or three days previous to the use of the remedy.

Dropsy. Dropsy consists of an unnatural accumulation of serous or watery fluid, in various parts of the body. Persons of all ages are liable to it. It is divided into five kinds, according to the part affected : first, dropsy of the skin, generally called *anasarca ;* second, dropsy of the belly, called *ascites ;* third, dropsy of the chest, called *hydrothorax ;* fourth, dropsy of the head, or water in the brain, called *hydrocephalus ;* fifth, scortal bag, called *hydrocele.*

Causes.—Excessive and long-continued evacuations, weakening the system ; a free use of fermented spirituous liquors ; confirmed and incurable indigestion ; diseases of the liver, spleen, pancreas, mesentery, or others of the viscera ; preceding diseases, as asthma, scarlet fever, &c. ; anything debilitating the digestive organs ; sometimes from family predisposition.

Symptoms.—This disease generally commences with swelling of the feet and ankles toward night, which for a time disappears in the morning. The swelling, when pressed will pit ; it gradually ascends till the whole body is swelled, in the first sort, and the belly in the second sort ; the urine scanty, thick and high-colored ; thirst is great, breathing difficult, especially in the third sort, and a troublesome cough ; the flesh wastes, and the patient weakens ; in the fourth sort, pains on the top of the head, and often convulsion or apolexy ; in the fifth sort, the scrotal bag is much enlarged, and much pain in consequence.

Treatment.—The diet must be of a dry heating nature, using pungent vegetables, as garlic, mustard, onions, cresses, horseradish, shalots, &c., and the flesh of wild animals. Avoid drinks as much as possible ; quench the thirst with acid liquors, mustard whey, and the like ; and take some of the following :—

Cream of Tartar, - - - - -	1 drachm.
Sulphate of Potass, - - - - -	10 grains.
Rhubard in Powder, - - - - -	5 grains.

Take in pumpkin-seed two or three times a day Or use the following :—

Powder of dried Squill-Root, - - -	2 grains.
Blue Pill, - - - - - -	5 grains.
Opium, - - - - - -	½ grain.

Make into a pill and take one at bedtime for 4 or 5 nights followed by the above powder the morning after.

THE FAMILY PHYSICIAN.

Another excellent remedy in general Dropsy, is the following:—

Infusion of Digitalis,	-	-	-	-	3½ ounces.
Vinegar of Squills,	-	-	-	-	½ ounce.

Take a tablespoonful 2 or 3 times a day.

Dr. Barthrow.

Jaundice.   A disease arising from obstruction to the passage of the bile into the intestines, from disorders of the liver.

Treatment.—The diet should be cool, light and diluting,—consisting chiefly of ripe fruit and mild vegetables; the drink, barley water or linseed tea, sweetened with liquorice; the bowels must be kept gently open.   When the disease has abated, constant doses of Peruvian bark should be given, with good port wine: plenty of exercise taken, and a mustard poultice occasionally placed over the liver.   The following has been of great benefit : Remain in a warm bath, of one hundred degrees, for twenty minutes.   Take, every other night, five grains of blue pill, and five grains of compound aloe pill on those nights when the blue pill is not ordered.

Piles.   *Treatment.*—Keep the bowels well opened by a mild aperient such as castor oil, or senna tea, and apply to the part one of the following ointments :

Hemlock bark finely powdered,	-	-	-	1 ounce
Fresh Lard,	-	-	-	-	6 ounces.

Mix and apply.

Powdered nut gall,	-	-	-	-	2 drachms.
Camphor,	-	-	-	-	1 drachm.
Melted wax,	-	-	-	-	1 ounce.
Tincture of Opium,	-	-	-	-	2 drachms.

Mix and apply.

Should the piles be internal, the bowels should be kept lax, and after each motion one ounce Pond's Extract should be gently injected into the bowel, the patient lying down for half an-hour-afterwards.

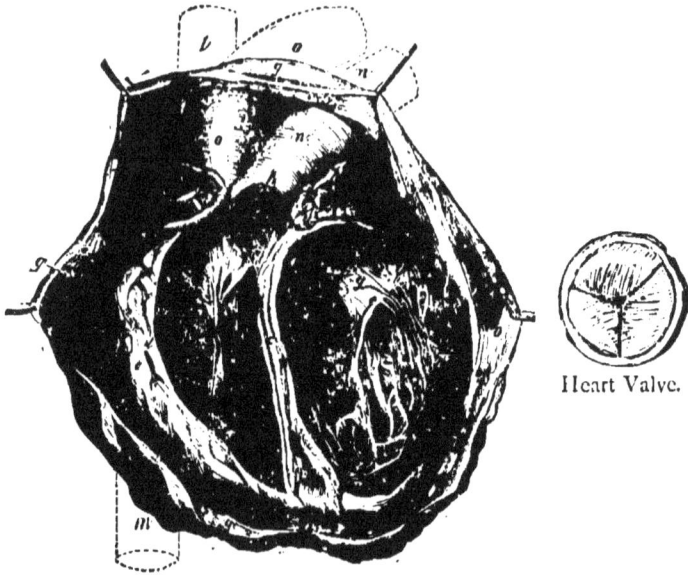

Heart Valve.

Front View of the Heart, with the Walls of the Right and Left Ventricles removed.

a, b, Right and left walls of the ventricles.   *c,* Septum ventriculorum.   *d,* Cavity of the right ventricle.   *e,* Cavity of the left ventricle.   *f,* Valves of the right ventricle.   *g,* Valves of the left ventricle.   *h,* Entrance of the pulmonary artery.   *i,* Entrance of the aorta. *l, m,* Upper and lower vena cave.   *n,* pulmonary artery.   *o,* Aorta.   *q,* Heart case or pericardium.

The Heart.   Palpitation of the heart is an increase in the force or frequency of the heart's action.   It is frequently produced by physical action or mental emotion, and is sometimes the result of disease.   Sometimes the palpitations are loud and clear and regular ; at others they are faint and intermittent ; now, a distinct throb or several and then, a tremulous flutter.   Usually these symptoms merely indicate functional derangement of the heart and not organic change.

Causes.—A disordered stomach may be the cause, although there may be no other symptoms of this.   We have known cases in which a very slight irregularity in the mode of living has produced palpitation of the heart, and that, too, in an otherwise healthy person.   In some, almost any strong nervous stimulant will produce it, and we recollect one instance in which it always came on after a cup of tea, and was never troublesome when this beverage was not taken.

We mention this to show that palpitation is not always, nor indeed commonly, symptomatic of heart disease ; and need therefore cause no unnecessary alarm, although its frequent recurrence should set the patient inquiring as to what is the real cause. Young women with whom there is derangement of the menstrual functions, in whom the blood is watery and 1..or, wanting the red corpuscules ; the listless, the pallid, the hysterical, in these we meet with palpitation in its most aggravated forms ; as also in the indolent, the susceptible, and the delicate; those who dwell on morbid fancies, and excite the imagination with sensual thoughts, or horrible pictures. To such every beat of the pulse seems like a call from the world of spirits, every flutter and palpitation like a brush from the wings of the angel of death, or the whispering voice of an accusing conscience.

Treatment.— In these cases the only treatment likely to be of service must be directed towards removing the predisposing and exciting causes, and establishing a more healthy nervous condition : gentle exercises, tonics, change of air and scene, an endeavour to occupy the mind in some useful and moral pursuit ; a well regulated and generally frugal, although sufficiently nourishing diet ; and a strict avoidance of all that can excite or stimulate either mind or body.   By this means palpitations not connected with organic disease, may generally be got rid of. If the patient is of a full habit, and has a tolerably strong pulse, a course of general purgatives may be necessary. They should not be salines, but of a cordial nature, something like this :—

Pill of Aloes and Myrrh,	-	-	-	½ drachm.
Compound Galbanum Pill,	-	-	-	½ drachm.

Divide into twelve pills, and take one at bedtime.

Compound Infusion of Senna,	-	-	3 ounces.
Decoction of Aloes,	-	-	-	3 ounces.
Spirits of Sal Volatile,	-	-	-	1 drachm.
Compound Tincture of Cardamums,		-	2 drachms.
Tartrate of Potash,	-	-	-	½ ounce.

Mix and take two tablespoonsful occasionally.

Neuralgia of the Heart (*Angina Pectoris*). A disease which is commonly connected with ossification, or other morbid affections of the heart.

Symptoms.—It is characterized by a sudden and most violent pain across the chest, which extends down the arms, and seems to threaten immediate dissolution. It sometimes comes on during rest, but most usually after violent exertion. The paroxysm does not commonly last long, but it has been known to continue for an hour or more.

Treatment.—An anodyne combined with ammonia has sometimes been found very effectual in relieving the spasm. The following is a good formula :—

Fetid Spirits of Ammonia,	- -	½ ounce.
Solution of Morphine,	- - -	3 drachms.
Camphor Mixture,	- - -	6 ounces.

Take a tablespoonful every half hour until relieved. If the paroxysm is very violent, a little hot brandy and water may also be taken ; or a teaspoonful of sal volatile or ether in water, and repeated at intervals. If the pain continue, frictions and mustard plasters applied to the chest, soles of the feet, and calves of the legs. Where there is extreme faintness, the horizontal posture should be adopted. Persons subject to these attacks would do well to provide themselves with the following, as a medicine in case of need :—Half an ounce each of sulphuric ether, spirits of ammonia, and sal volatile ; two drachms of tincture of opium. Mix, and take a teaspoonful in water ; and repeat at the end of an hour if relief be not afforded.

CHAPTER IV.

DISEASES OF THE EYE AND EAR.

Squinting—Inflammation of the Eye—Short Sight—Long Sight—Old Sight—Weak Eyes—Ingrowing Eyelashes—Blindness—Cataract—Inflammation of the Eyelids—Styes—Earache—Wax in the Ear—Runnings from the Ear—General Cautions regarding the Ear.

THE EYE.

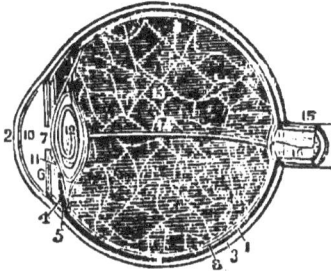

Longitudinal Section of the Globe of the Eye.

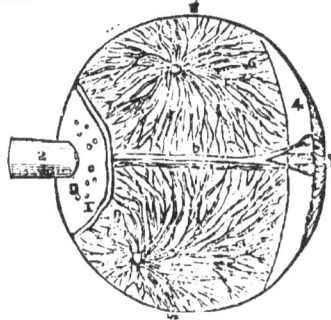

The Eyeball divested of its first tunic.

1 marks the course of the outer tunic, called the sclerotic, which invests four-fifths of the globe, and gives it its peculiar form. It is a dense fibrous membrane, thicker behind than in front, where it presents a bevelled edge, into which fits like a watch-glass the *cornea* (2), which invests the projecting portion of the globe, and is composed of four layers, viz., the *conjunctiva* or *cornea propria*, consisting of thin lamellæ, or scales, connected by extremely fine areolar tissue; the *cornea elastica*—an elastic and excessively transparent membrane, which lines the inner surface of the last; and the *lining membrane* of this front vestibule of the Eyeball, whose second tunic is formed by the *choroid* (3), represented by the dark line; the *ciliary ligament* (4), which develops from its inner surface the *ciliary processes*, and the *iris* (6), of which the opening at 7 represents the *pupil*. We shall go more into details presently as to the nature of these several constituents of the tunics and other parts of the Eye; at present we will keep to general outlines, as represented in the diagram. The third tunic then, is the *retina* (8), which is carried forward to the *lens* (12), by the *zonula ciliaris*, a prolongation of its vascular layers passing along the front of the *Canal Petit* (9), which entirely surrounds the lens. In the space marked 10, is contained the *aqueous humor*; 11 is the *posterior chamber*; 12 the lens more convex behind than before and enclosed in its proper capsule; 13 marks the inner

1, part of the outer tunic, the sclerotic; 2, the optic nerve, communicating with the ball at the back; 3 3, distinguish the outline of the choroid coat; 4, the ciliary ligament, a dense white structure which surrounds, like a broad ring, the circumference of the iris (5). This ligament serves as a bond of union between the external and middle tunics of the Eyeball, and serves to connect the cornea and sclerotica at their lines of junction with the iris and external layer of the choroid; 6 6, mark the *venæ vorticosæ*: and 7 7, the trunks of these veins at the point where they have pierced the sclerotica; 8 8, the posterior ciliary veins, which enter the Eyeball in company with the posterior ciliary arteries, by piercing the sclerotica at 9. The course of one of the long ciliary nerves, accompanied by a vein, is marked by 10.

area of the globe filled with a thin membrane called the *hyaloid* and containing the vitreous humor ; 14 is the tubular sheath of the membrane, through which passes an artery connected with the capsule of the lens, and, at the back of the eye, with the optic nerve, as represented at 16. Of this nerve, 1; marks the *neurileuma*, or sheath.

Squinting. Squinting is, in the vast majority of cases, the result of defective formation of the eyes. It may result either from an undue depth or an undue flatness of the eyes—that is, from myopia or from hypermetropia. If the child be short-sighted, he is apt to have that form of squint in which one eye turns outward ; if, on the other hand, he be far-sighted in a high degree he is prone to the common form of squint in which one eye turns inward.

Treatment.—The remedy for squinting lies, during its early stages, in the use of glasses. It is often possible to correct a squint entirely, before it has existed more than two cr three years, by the judicious employment of spectacles. The glasses act, of course, by correcting the near-sightedness or far-sightedness upon which the squint depends.

After the individual has been in the habit of squinting for several years, it is rarely possible to correct the deformity entirely by the use of glasses. Yet there should be no hesitation whatsoever in submitting the eye to an operation whereby it can be straightened ; since, even though the sight of the eye be somewhat impaired, the vision can often be restored, at least so as to make the eye practically valuable.

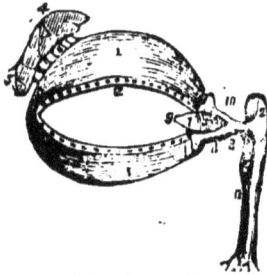

Eye-Appendages.   Meibomian Glands.

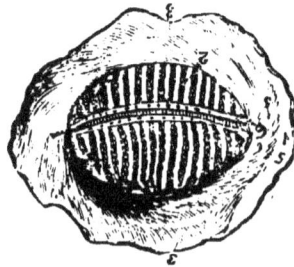

1, the *superior* or upper *tarsal cartilage*, along the lower border of which (2) are seen the openings of the *Meibomian glands* ; 2, the *inferior* or lower *tarsal cartilage*, along the upper edges of which are also openings of the above-named glands ; 4, the superior or orbital portion of the *Lachrymal gland*, from which come tears ; 5, its inferior or palpebral portion ; 6, the *Lachrymal ducts*, or channel through which the tears pass to the outer surface of the eye ; 7, the *Plica semilunaris*, containing a small plate of cartilage, which appears to be the rudiment of a third lid, such as developed in some animals ; 8, the *Caruncula lachrymalia*, the source of the whitish secretion which so constantly collects in the corner of the eye ; it is covered with minute hairs, which can sometimes be seen without the aid of a microscope ; 9, the *Puncta*

1 2, the inner side of the eyelids ; 3 3, the *Conjunctiva* ; 4, the apertures of the glands, along each corner of the lids ; 5 5, 6 6, the *Papillæ lachrymales* and the *Puncta lachrymalia* ; 7, the apertures of the ducts of the *Lachrymal gland*.

lachrymalis, the point or external commencement of the ducts, which terminate at the *lachrymal sac*, the position of which is marked by 12 ; as are the *superior* and *inferior lachrymal canals* by 10 and 11. The *nasal duct*, marked by 15, and 14 is its dilation with the lower meatus of the nose.

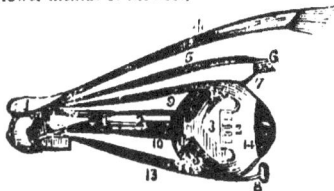

Muscles of the Eye.

1, Sphenoid Bone ; 2, the Optic Nerve ; 3, the Globe of the Eye ; 2 the Upper Muscle, called the *Levator palp....e*, the Lifter of the Eyelids : 5, the *Superior Oblique*, so called from the direction in which it draws the Eyeball ; we see its cartilaginous pully (6), and the reflected portion passing downward to its point of connection with the ball, beyond which the *Inferior Oblique* has its bony origin—the point of which is marked by the little square knob. The other four muscles are called *Recti*, straight ; the *Superior Rectus*, sometimes called the *Levator Oculi*, erector of the eyes, and sometimes *Superbus*, because its action gives an expression of pride ; its opposite, 13, the *Inferior Rectus*, sometimes called *Deprimu oculi*, depressor of the eye, and *Humulus*, as giving an expression of humility ; 10, the *Rectus Internus*, sometimes called *Adductor Oculi*, from its drawing the Eyeball toward the nose, and *Bibitorius*, a sort of punning name, in allusion to the cup, or orbit, towards which it directs the glance ; 11 and 12, *Rectus Externus*, the one showing its two heads of origin, and the other its termination ; the intervening portion of muscle [having been removed] has the name of *Abductor Oculi*, because it turns the ball outwards; *Indignabundus* is another name for it, as giving an expression of scorn. In our diagram, the internal rectus passes behind the optic nerve, which partly conceals it ; 14, the *tunica albigania*, or white tunic, formed by the expansion of the tendons of the four *Recti* muscles.

Inflammation of the Eye.

Symptoms.—It usually begins with a sensation of grating in the Eye as though particles of sand were under the lids. The Eye reddens and the discharge from it increases. On waking in the morning the lids are glued together, and the pain gradually becomes worse.

Treatment.—As a rule the patient requires no constitutional treatment since the local inflammation is not commonly associated with symptoms of general disturbance. If the patient be hot and thirsty he should have a saline laxative, such as a teaspoonful of Rochelle salts or of the citrate of potash.

The eyes may be bathed every hour or two with one of the following lotions, care being taken to allow a little of the solution to flow into the eyes at every application :

Alum, - - - Ten grains.
Water, - - - Two ounces.

If the case be severe the following lotion may be employed :

Alum, - - - Eight grains.
Sulphate of zinc, Two grains.
Water, - - - Two ounces.

In the intervals between the application of the lotion the eye may be washed with water to secure the escape of the discharge ; the patient will derive much comfort from the application of a cloth, such as a soft handkerchief, soaked in cold water and allowed to rest upon the eye. In making such applications to the eyes, only thin cloth should be used, and not more than three thicknesses should be applied to the eye, since otherwise the heat produced may aggravate the pain.

If there be much swelling and puffiness of the conjutiva the following solution may be prepared :

| Nitrate of silver, | - | - | - | - Two grains. |
| Water, | - | - | - | - Two ounces. |

Two or three drops of this should be dropped into the eye morning and night. If this solution be used the alum mixture above given is not required, but the eye should be cleansed and kept cool as directed above.

In order to prevent the lids from sticking together the edges may be smeared with vaseline when the patient retires at night.

Short Sight and Long Sight. That structures so delicate as those which make up the eye should be sometimes defective in their action, will not excite surprise. Two defects, in particular, are quite common, namely, *short sight* and *long sight*. Most persons see objects, as, for instance, the type of this page, in the most perfect degree at a distance of about sixteen inches from the eye. The short-sighted hold their books nearer than this, sometimes very close to the eye, while the long-sighted hold them two feet or more away. These defects are occasioned by the quantity of water in the antechamber of the eye ; thus, in the short-sighted, the cornea is observed to be too round or bulging, while, in the long-sighted, it is too flat. Naturally, in youth, the eye is fuller than in middle life and in old age, so that the tendency of the young is to short sight that of the old to long sight. It is a fact deserving attention that the number of short-sighted children and of far-sighted adults is increasing. Parents should know the causes of short sight, and take care that their children are not exposed to them It is very frequently developed by the habit of holding books and other objects unnecessarily close to the eyes, or bending down the head very near to them. Therefore, small print, bad ink and paper, delicate drawing, fine needlework, imperfectly lighted rooms, high seats and low desks or tables, are to be avoided, as leading to the habit in question.

Those who are short-sighted would do well to observe certain simple rules for the preservation of their sight. They should naturally choose books with large type, and hold the book up to the eyes, not the eyes down to the book. Bending over a flat table in writing must be avoided by the use of a high desk. No close nor fatiguing work should be undertaken, and the eyes must never be overtasked by too continuous labor, but relieved by frequent rests. In regard to the use of spectacles, those who are only slightly affected may usually dispense with them, but those who are affected to a considerable degree will find safety as well as advantage in their use. In no case, however, must too strong glasses be employed, as they are injurious. The short-sighted need feel no alarm, unless the trouble is evidently increasing from year to year, when rest, and, perhaps, treatment, become imperative.

Old sight is different from long sight. Old people do not see better in the distance than young ; they are merely unable to see so well objects close at hand. This impairment of vision sometimes begins at the early age of thirty-five, without attracting notice ; ten years after, the book is held, instinctively, several inches further from the eyes than in youth. The impairment gradually increases with advancing years, until at last the book has to be held so far that the letters can no longer be distinguished, and all reading without glasses becomes impossible. This defect is remedied by spectacles, which have to be made stronger and stronger with invading years. When the glasses habitually employed have to be put at a distance, and kept on the tip of the nose in order to render them a service, stronger ones should at once be obtained, and the eyes not fatigued by attempting to make the old ones do. The glasses should be sufficiently strong to permit of the book or newspaper being held within a foot of the eye.

Weak eyes frequently result from reading, drawing or sewing, in a light either too bright or too dim. The full glare of the sun and twilight are equally injurious. The practice of reading in bed, or when lying on a lounge, is also often a cause of weak eyes. It is better, therefore, always to sit up to read.

We give below some excellent washes for ordinary weakness of the eyes :—

Sulphate of Copper, - - -	15 grains.
Camphor, - - - - - -	4 grains.
Boiling Water, - - . - -	4 ounces.

Mix ; strain, and when cold, make up to four pints with water. Bathe the eye night and morning with a portion of the mixture. Or the following :—

Spirit of Mindererus, - - - -	1 ounce.
Rose Water, - - - - - -	7 ounces.

Mix, and use occasionally ; or this :—

White Vitriol, - - - - -	10 grains.
Elder-Flower Water, - - - -	8 ounces.

Mix, and apply as occasion may demand.

Ingrowing Eyelashes. Children who have suffered from inflammation of the eyes, and others who have not, are occasionally much annoyed by the tendency of some of the hairs which form the eyelashes to turn inwardly and thus irritate the ball of the eye. It leads to a sense of pricking and an irritable and watery state of the eye. To remedy this, the offending hairs must be carefully plucked out from time to time. This seems a simple and trifling matter, but, in fact, few manipulations require more care. The hair must not he broken, as its stiff stump will cause far more distress than its natural fine point. A pair of delicate forceps should be used, the hair grasped firmly between their points, and never be sharply jerked out, but removed with a slow, steady pull.

Blindness. Deprivation of sight may proceed from various causes, such as one of the diseases which affect the eyeball, or deficiency of power in the optic nerve, local or general paralysis, or any disease whose seat is in the brain or the nervous system ; the formation of a speck on the eye, or of a film over the lens. Sometimes the affection of the brain or nerves, from which loss of sight proceeds, is sympathetic, arising from a disordered stomach. In this case, as in many others, it is but transient ; and matters may be set right by a blue pill and senna draught, with low diet, and avoidance of the exciting causes of the disorder. If these do not have the desired effect, a surgeon should be consulted, as there is reason to suspect some organic mischief. Leeches on the temples, blisters behind the ears, cupping in the neck,—either or all of these may be tried, should there be a sense of fulness, headache, or giddiness, accompanying indistinctness of vision. In this case, too, more powerful medicines, such as colocynth and calomel pills, should be taken, and a course of depletion vigorously carried out.

Proceeding, as blindness does, from such a variety of causes, few general directions can be given for its treatment. When it is owing to a change in the structure of the eye itself, its approaches will be very gradual, unless this change is the result of active inflammation. Temporary loss of sight is a frequent symptom of apoplexy. It also results from diseases of exhaustion, and sometimes occurs after copious bleeding ; its total loss may be effected by a blow on or about the region of the eye. For the blind from birth there is no hope of recovery.

Cataract. A disease of the eye causing opacity of the crystalline lens, which prevents the passage of the rays of light, and so produces blindness.

Causes.—The real cause of this disease does not appear to be well understood. It may proceed from external violence, but more commonly it has some internal and occult origin. It is of slow growth, and can only be operated on at a certain stage, when the opaque body in the pupil has assumed a sufficient density.

Symptoms.—A dimness and mistiness of vision, which may generally be noticed when any opacity can be perceived on the lens itself. Then there are optical illusions, like specks or motes floating before the eye. This is succeeded by the gradual falling as it were, of a curtain upon the outward view, which is finally obscured altogether. Sometimes the progress of the disease is slow and gradual, but frequently it is rapid, especially in the latter stages. Persons who have passed the middle age are most likely to be affected by it, and sometimes it has made considerable progress in one eye before the patient is made aware of it by some accidental circumstance which for a time prevents the use of the other.

Treatment.—There is no medicinal remedy that is known to have any effect upon this disease ; nor is it at all likely, from the structure of the

parts, that any such remedy exists.   All palliative measures,' therefore,
are confined to attention to the general health of the patient, and the
removal of any inflammatory symptoms that may exist along with it.
The only mode of cure is actual removal by an operation ; but so long
as one eye remains unaffected, the operation may be delayed,

Inflammation of the Edge of the Eyelids. The edges of the
eyelids are sometimes very red and stiff, in consequence of the inflammation of the small follicles or ducts which open there.

Treatment.—The best remedy is a little red precipitate ointment
rubbed into the roots of the lashes, when the lids are closed on retiring
to rest.   This may be repeated every night until no longer required. A
little grey powder, combined with rhubarb, should be given, and the
patient kept quiet and somewhat low.   When inflammation has been
going on in the eyelids for a time, their insides, when inverted, will often
present a rough granular appearance.   In this case, they should be gently rubbed over with a smooth piece of dry sulphate of copper.   The
lid should be kept open after the application until the eyeball is syringed
with warm water, to remove from it any of the solution caused by the
flow of tears acting on the sulphate.   There will probably be great
smarting of the eye, and increased redness of the white portion, which
must be suffered to subside before the application is repeated, which it
will, most likely have to be many times.

Styes. Styes are little inflammatory tumors which freqently make
their appearance on the edges of the eyelids of children.   They rarely
affect grown persons ; and, although troublesome, are not at all dangerous locally, nor prejudicious to the general health.   They run the same
course as boils, which in reality they are.

Treatment.—Generally they require no medical treatment, but when
very large and painful, a hot water fomenting will prove beneficial. When
once the matter has escaped, they heal very quickly.   A simple dressing
of spermaceti ointment is sometimes required, but not often.

Dr. Abadil gives the following :

Boracis acid	-	-	-	-	¼ drachm.
Distilled Water	-	-	-	-	2 ounces.

With a wetted piece of wadding drop some of this solution on the stye
several times a day.   It is said not only to effect a cure, but to prevent a
return of the annoyance.

THE EAR. Ear-Ache (*Otalgia*). Ear-ache may proceed from
abscess in one or more of the passages, or may be altogether neuralgic.
In children it is not uncommon during the period of dentition, and is
especially severe in cutting the permanent teeth.   Grown persons sometimes suffer from it when producing their wisdom-teeth.   It is often
brought on by exposure to cold or draughts.   There is not often much
constitutional derangement, although the pain is sometimes excruciating, unless it is long continued.

THE EAR

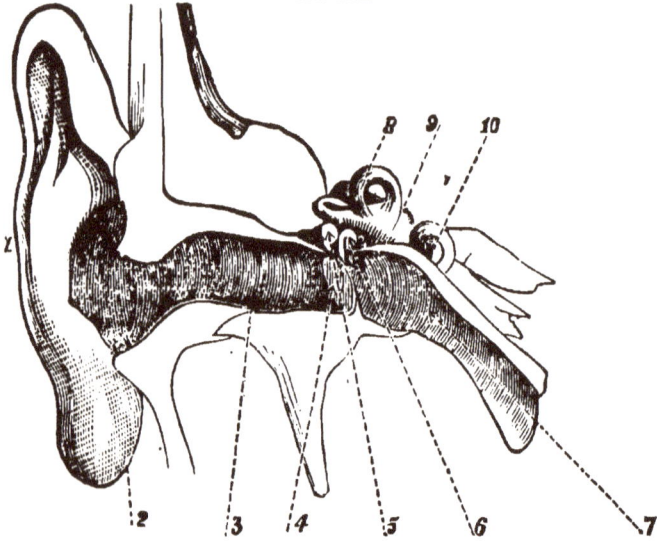

1, pinna; 2, lobule; 3, tube; 4, tympanic membrane; 5, incus, or anvil; 6, malleus, or hammer; 7, Eustacian tube; 8, semicircular canals; 9, vistibule; 10. cochlea.

Section showing the hollow of the Cochlea.

Treatment.—In children, during dentition, lancing the swollen gums will often afford relief, especially if an aperient is given, such as rhubarb or magnesia combined with a little ginger. Elder children may have three or four drops of olive or almond oil, with one or two drops of laudnum dropped into the ear, and take compound senna mixture, repeated until the bowels are freely opened. Should these remedies not prove effectual, a fomentation of camomiles and poppies should be applied, and a warm poultice afterwards. The heart of a roasted onion applied warm to the external orifice will sometimes afford relief. If the case is very obstinate, two or three leeches behind the ear, followed by a blister, may be tried, with an anodyne salient aperient something like this :—

Acetate of Morphine - - -	$\frac{1}{2}$ grain.
Solution of Acetate of Ammonia - -	3 ounces.
Sulphate of Magnesia - - •	1 ounce.
Water or Camphor Mixture - -	5 ounces.

Mix, and take two tablespoonsful every four hours.

When ear-ache is caused by an abcess, and is attended with much swelling and severe pain, hot fomentations and poultices will be the treatment, syringing the external passages with warm water; and, after the abcess has discharged, with a solution of sulphate of zinc, in the proportion of eight grains to the ounce of plain, or rose water, attention being paid to the bowels. With some persons any derangement of the general health will cause the formation of these abscesses, and in each case the treatment must be rather general than local. Ear-ache, no doubt, often proceeds from derangement of the digestive organs, and may be relieved by active purgatives and emetics.

Where a tonic is required, the following will be found very good :—

Citrate of Iron, · - - -	1 drachm.
Strychnine, · - · -	1 drachm.
Syrup of Orange-peel, - • -	2 ounces.
Soft Water, - - - -	$\frac{1}{2}$ pint.

Mix. Dose, one teaspoonful three times a day.

Wax in the Ear. When this substance becomes too hard, or accumulates too much, there will be a sense of contraction, with cracking or hissing noises, and generally deafness to a considerable extent. In this case the ear should · be syringed with warm soap-suds, the instrument used being a proper one for the purpose, holding about four ounces, and having but a small tube or pipe which does not fill the whole passage, but allows the escape of the back-water, for catching which a handbasin should be held close against the neck. As many as a dozen syringe fulls may be injected at one time. A strong lotion should be put into the ear-passage over-night, and kept there by means of cotton wool or wadding. Almond oil and laudanum, in the proportion of two ounces of the former to one of the latter, is a good application in this case, as in many other kinds of ear disease.

Runnings of the Ear.—Delicate and scrofulous children are liable to a yellow discharge, which suddenly comes on, and is at first often stained with blood, and accompanied by feverishness and great pain in the parts. There is generally redness and swelling of the passages of the meatus, and inflammation of the surrounding skin. This may arise from an inflamed state of the membrane which lines the passages, or from an abscess formed beneath it, or between the cells of the bones of the mastoid process. The discharge may be caused by some foreign substance thrust in the ear.

Treatment.—For the purulent discharge from the ear, which is induced by this or any other cause, a lotion made with two drachms of solution of chlorinated soda to six ounces of rose, or elder-flower water, should be injected, but not with any force. The best method is to let it flow into the ear, held so as to receive it fairly, from a small sponge saturated with the lotion.

Counter-irritation will sometimes have a good effect on purulent discharges from scrofula or other causes. A small blister behind the ear is the best application, but it should not be kept open for any length of time, or it will weaken the system too much. When the discharge is the result of active inflammation, and is attended by febrile symptoms, a spare diet and aperients must be the treatment ; but weakly, scrofulous systems require a generous diet and tonic medicines.

General Cautions.—From the outer ear there is a tube or canal, the *auditory canal,* leading to the middle ear. At the bottom is a tightly-stretched membrane, which makes the outer wall of the middle ear, or the head of the drum. The auditory canal is about an inch and a quarter long, and slightly curved upon itself, so as to be higher in the centre than at either end. It carries in, just as an ear-trumpet does, the waves or undulations of the air, collected by the outer ear, to the middle ear. The hairs at the entrance of this canal, and the wax which keeps it soft, are for the purpose of excluding dust, insects and other substances. Children frequently fall into the habit of picking the ear with the finger, thus irritating it, and exciting inflammation and running. Many adults, also, cannot let their ears alone, but are constantly thrusting toothpicks, pins and needles into them, to clean them out. This habit often leads to permanent injury, and not unfrequently, by an accident, punctures the drum of the ear, and destroys the hearing. The practice which some people have of stopping up this tube with pieces of wool, " to keep out the cold," is absurd and hurtful ; absurd, because if nature intended the ear to be thus shut up, she would have done it herself; hurtful, as it heats the ear, makes it very sensitive and liable to take cold and become inflamed.

CHAPTER V.

DISEASES OF THE SKIN.

Pimples or Acne—Dandruff—Itch—Barber's Itch—Salt Rheum or Eczema—Chilblains—
Baldness—Erysipelas—Nettle-Rash or Hives—Prickly Heat or Lichen—Irritation or
Itching—Freckles—Lice—Mother's Marks—Other External Diseases—Hang Nails—
Ingrown Nails—Blood Blisters—Boils or Feruncles—Bunions—Carbuncles—Corns—
Felon or Whitlow—Goitre—Sores and Ulcers—Fetid or Foul Feet—Warts—Ring-
worm—Diet in Skin Diseases.

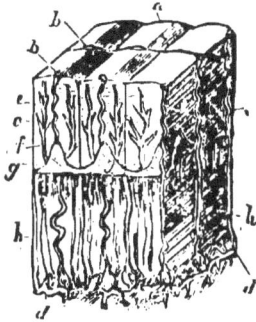

Diagram of the Structure of the Skin.

a, Epidermis ; *b b*, Pores ; *c c*,
Layers of epidermis and rete mu-
cosum ; *f*, Inhalent vessels ; *g g*,
Papillæ of the skin ; *h h*, Corium
or true skin ; *d d d*, Bulbs of se-
doriferous glands opening i·
glands *b b*.

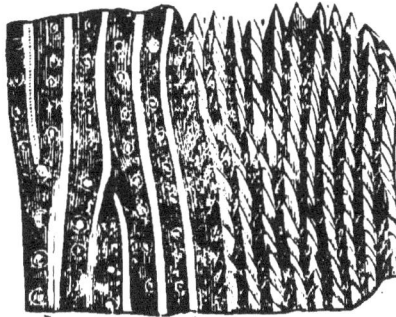

Pores and Papillæ of the Skin.

On the left is a magnificent View of the Ridges of
the Cuticle, as seen in the Palm of the Hand, with
the openings of the Pores in their Furrows. On the
right, the Cuticle has been removed, leaving corresponding rows of Papillæ.

Dandruff. *Tre. .t.*—For the removal of dandruff it is necessary
not only to keep the scales brushed out of the hair, but also to correct, if
possible, the unhealthy action of the sebaceous glands. The hair may
be gently brushed with a soft brush, and then washed with a little soap
and water. After this, the yolks of two eggs may be applied to and
thoroughly rubbed into the scalp. The repetition of this process daily
is often of itself sufficient to remove the difficulty. If the scales still
form, there may be substituted for the eggs the following prescription :

Tannic acid, - - - - - 1 drachm.
Simple ointment, - - - - 1 ounce.

Mix and rub thoroughly into the scalp.

44

Another valuable remedy is the following :

Hydrate of chloral,	-	-	-	-	2 drachms.
Water,	-	-	-	-	-	4 ounces.

If there be much oily matter in the hair, the following prescription may be used :

Aromatic spirits of ammonia,	-	-	4 drachms.
Glycerine,	-	-	-	-	-	½ ounce.
Rosemary water,	-	-	-	-	4 ounces.

Pimples or Acne. This is the most common of skin diseases, and is usually due to some disorder of digestion.

Treatment.— Guard against constipation by the use of citrate of magnesia and occasionally 4 grains of Blue pill.

Avoid all articles of diet which are found to be followed by pimples, such as alcohol, pastry, buckwheat cakes, etc.

Numerous lotions and ointments are recommended and sold for the cure of acne ; yet none of these can be relied upon, unless proper measures are taken to remove the condition of the stomach, bowels, liver, etc., upon which the difficulty depends. Among the best of these are the following :

Flowers of sulphur, -	-	2 drachms.
Tincture of camphor,	-	3 drachms.
Glycerine, -	-	-	-	-	1 ounce.
Rose water,	-	-	-	-	4 ounces.

This may be applied over the affected spots, and may be rubbed gently on those parts of the skin affected with the disease two or three times daily.

Vertical Section of the Skin Largely Magnified.

h, Sweat Gland ; *i,* a Hair enclosed in its Follicles, and showing its pair of Sebaceous Glands ; *p,* a Sebaceous Gland.

Borax,	-	-	-	-	-	1 drachm.
Carbonate of soda,	-	1 drachm.
Glycerine,	-	-	-	-	4 drachms.
Tincture of camphor,	1 ounce.
Distilled water,	-	-	To make 6 ozs.

This may be applied in the same way.

Precipitated sulphur,	-	-	-	-	Half an ounce.
Carbonate of potash,	-	-	-	-	Four drachms.
Glycerine,	-	-	-	-	-	Two ounces.
Sulphuric ether,	-	-	-	-	One ounce.
Alcohol,	-	-	-	-	One ounce.

This mixture should be carefully applied to the pimples and to those parts of the skin that seem liable to exhibit an eruption. In all these cases no more friction should be employed than is necessary to apply the lotion.

In all cases success can be hoped for only after persevering use of the remedies, both those for application to the skin and those which are designed to improve the condition of the digestive organs. Cases are found which seem to resist almost all measures of treatment ; such individuals have at least the consolation that the affection will yield to time, even if all medicines prove unavailing.

Itch. *Symptoms.*—The most prominent symptom of this disease is a constant and intolerable itching. It never comes on of itself ; but is always the result of contact with an affected person. It first shows itself in an eruption of small bladders or vesicles filled with a clear watery fluid, occurring principally on the hand and wrist, and in those parts most exposed to friction, such as the spaces between the fingers, and the flexures of the joints, etc. ; after a time it extends to the legs, arms, and trunk, but it rarely appears on the face.

Treatment.—The itch is never got rid of without medical treatment ; but to that it will always yield, provided proper cleanliness be observed. Sulphur is the grand specific for it ; it may be applied in the form of ointment, prepared as follows :

Flowers of Sulphur,	- - -	2 ounces.
Carbonate of Potash,	- - -	2 drachms
Lard,	- - - - -	4 ounces.

To be rubbed well in, whenever the eruption appears, every night and morning—washing it off with soap and flannel before each fresh application. The most effectual plan is to anoint the whole body, from the nape of the neck to the soles of the feet, and out to the ends of the fingers ; put on socks, drawers, flannel wrapper, and gloves, and so remain in bed for thirty-six hours, repeating anointing operation twice during that time ; then take a warm bath, and wash the whole person with soap and flannel.

In mild cases, a sulphureous vapor bath taken twice in twenty-four hours, with warm soap and water washing, will generally be sufficient.

Barbers' Itch is the development of pimples and pustules in the beard and whiskers. Each pimple, if examined, will be found to have a hair passing through it. There is but one way to set about curing this troublesome and disfiguring disease, and that is as follows : Remove the crusts by oil and poultices. Cut the beard short with scissors. Then pull out every hair that is seen to come from a yellow point or pimple. A small forceps should be used for this. When this is completed, rub thoroughly with the following ointment :

Flowers of Sulphur,	-	-	-	-	1 scruple.
White Precipitate of Mercury,		-	-	1 scruple.
Carbolic acid, pure,	-	-	-	-	10 drops.
Fresh Lard,	-	-	-	-	1 ounce.

Mix well.

Repeat this treatment, plucking out the hairs, as directed, and continue as long as there are any yellow points. Do not shave for at least six months after the disease has left.

Salt Rheum, (*Eczema*). This is one of the commonest, most troublesome and hence most important of all the diseases of the skin. In adults, eczema occurs in almost all parts of the body; the forehead, the cheeks, the eyelids, the nose, the lips are very often affected. When the disease occurs in the face, especially around the mouth, it is apt to prove obstinate, in consequence of the constant movement of these parts. In men, too, the presence of the beard is an additional obstacle to the cure of the affection, since, if the beard be allowed to grow, it is impossible to reach the disease satisfactorily, while if the face be closely shaven, the eczema is often thereby aggravated.

In women eczema is very common around the nipples and on the genitals. Many of the affections of the nipples during nursing are simply cases of eczema. Very many women who are afflicted with diseases of the womb, or of the vagina, suffer constantly from eczema on the neighboring skin, as the result of irritating discharges. In some cases, too, eczema frequently appears periodically during the menstrual flow.

Eczema is also frequently found in adults on the leg, especially just above the ancle. This is especially often the case with those who are troubled with enlarged or "varicose" veins. In this location the eczema is very obstinate and is apt to proceed in the course of time to the formation of ulcers.

Treatment.—In those cases in which the disease is traceable to external irritation, the source of this irritation should be of course removed. In some instances this is, under the circumstances, impossible, since the patient is unable to abandon his employment; thus the most obstinate cases occur in women who are compelled to have their hands constantly in water. In such instances much good can be derived from the use of rubber gloves. So, too, if the eczema depends upon an irritating discharge from the vagina, it will be necessary to adopt such treatment as will stop this discharge before the eczema can be cured. In general in the treatment of eczema in which there is considerable watery discharge and the formation of crusts, the surface should be softened by oiling the skin thoroughly, or by the use of a light flax-seed poultice. After this is accomplished and the crusts removed, one of the following ointments may be applied :—

| Oxide of Zinc, | - | - | - | - | 2 drachms. |
| Lead Water, | | | | - | 2 drachms. |

Glycerine,	-	-	-	- 4 drachms.
Lime Water,	-	-	-	- To make 8 oz.

This may be applied to the surface by means of a soft cloth, such as an old handkerchief, which should be saturated with the lotion and laid upon the raw surface. If there should be much burning and itching, one of the following remedies may be used :—

Carbolic Acid,	-	-	-	- ½ ounce.
Water,	-	-	-	- 1 pint.

This may be applied frequently on soft cloths.

The following quotation shows how this disease is managed in Vienna, the headquarters of the world for the treatment of skin diseases :—

" In eczema of the scalp or of the ears, the crusts are to be removed by inunctions with oil, which are to be made twice a day, about three ounces of oil being used each time. The crusts are thus removed, and raw places are covered with flannel. If the skin be not much thickened, the salves which have been mentioned above can be applied at once. If the skin be found red and ' weeping,' it will be necessary to apply recti-fied spirits repeatedly. When the swelling subsides, if the skin continues to secrete freely, tar may be applied in the case of adults ; this agent should not, however, be used for children, because their skin is much more sensitive, and swelling, perhaps even suppuration, of the glands in the neck can be induced very easily. If the eczema extends into the ear, one may use injections or stringents ; but in order to bring the salve into contact with the entire surface of the ear, a piece of sponge, properly shaped and covered with lint, should be smeared over with the salve and inserted into the ear. If the eczema extends high into the nostril, we may use suppositories. Each of these may contain :

Cocoa Butter,	-	-	-	- 8 grains.
Oxide of Zinc,	-	-	-	- 5 grains.

This should be inserted into the nostril.

" A child suffering from eczema of the face should have a piece of linen smeared with diachylon ointment laid upon the raw place; this may be fastened with a flannel bandage and renewed constantly until the crusts have been removed."—*Neumann.*

Chilblains. To prevent this annoyance, avoid cold or wet feet. Sleeping with the stockings on is also liable to produce them. A nightly foot bath of cold or tepid salt and water with plenty of rubbing with a rough towel, and exercise during the day will be most likely to prevent chilblains.

Treatment.—Should they come—as sometimes they will—in spite of all precautions, let them be gently rubbed every night with some stimu-lant application ; alcohol, spirits of turpentine, or camphorated spirits of wine, are all good for this purpose : but the application which we have found most efficacious is a lotion made of alum and sulphate of zinc—2 drachms of each to half a pint of water, rubbed in warm ; it may be made more stimulating by the addition of one ounce of camphorated spirits.

Baldness. *Treatment.*—Take of

Pure Glycerine,	-	-	-	- 3 drachms.
Lime Water,	-	-	-	- 4 ounces.

To be applied to the scalp, night and morning, with a soft tooth brush, after the head has been cleaned by gently washing with Castile or Sulphur soap and warm water.

This is an excellent treatment to commence with for slight scurfiness of the head, and falling of the hair, and baldness.

After several weeks' use the preparation may be changed to the following :—

Take of

Tincture of Cantharides,	-	-	- $\frac{1}{2}$ ounce.
Pure Glycerine,	-	-	- 3 drachms.
Lime Water,	-	-	- 4 ounces.

To be rubbed into the skin, briskly, twice a day.

The following hair tonic is also often of service for scurf and commencing baldness :—

Take of

Rock Salt,	-	-	- As much as will dissolve.
Pure Glycerine,	-	-	A tablespoonful.
Flour of Sulphur.	-	-	A tablespoonful.
Old Whiskey,	-	-	A tumblerful.

Mix.

Erysipelas. This disease has been popularly known as the Rose, from its red color ; and as St. Anthony's Fire, partly from its burning heat, and partly because the saint whose name it bore was supposed to have the power of curing it with a touch. There are several species of this disease ; but without going into the particular characteristics of each, it will be sufficient for us to state what are the general symptoms of erysipelatous inflammation, and best remedial measures.

Symptoms.—Chill, headache, furred tongues, quick pulse and often disordered stomach for a day or two previously, then a tingling burning sensation with stiffness and pain at some particular part followed by a discoloration of the skin ; the red or purplish tint is at first confined to one part, but soon extends itself sometimes over the whole part affected, frequently this is the head which, with the face becomes much swollen and tender ; the eyelids puff out and close the eyes, and the lips swell.

Treatment.—The treatment must always be adapted to the patient ; and the constitutional treatment is, therefore, of far more consequence than the application of remedies to the inflamed skin.

If the individual be at the beginning of the attack somewhat debilitated, or if he be evidently much exhausted by the onset of the disease,
it is extremely important that his strength should be sustained in every
possible way. For this purpose we rely upon iron and quinine. The
following prescription may be given :

Tincture of the chloride of iron	- -	One ounce
Sulphate of quinine	- - -	One drachm
Tincture of nux vomica,	- - -	Half an ounce.
Syrup of orange peel,	- - -	Two ounces.
Water,	- - - - -	To make four ounces.

Mix, and take a teaspoonful in water every four hours.

Sometimes the pain is so intense that it becomes necessary to administer opium. Twenty drops of laudanum may be given every three or
four hours until the pain is somewhat allayed.

For application to the skin itself one of the best remedies is the following.

Tincture of opium,	- - -	- One ounce.
Liquor plumb sub acetatis (lead waterr)		- Five ounces.

Mix and apply by saturating soft clothes with the lotion and laying
them upon the inflamed skin.

In the early stage of the inflammation, the application of cloths wrung
out in ice water, or of the ice itself, will often be grateful to the patient,
though it has probably no influence in arresting the disease ; after the
first day or two, cloths wrung out in hot water will usually be found more
agreeable to the patient than the ice.

Nettle-Rash or Hives. Nettle-rash appears in the shape of
elevated patches, or "wheals," which are of irregular shape, flat upon
the top, hard and usually of a pale red color ; in some cases, however,
the elevated portion of the skin is whiter than that which surrounds it.

It occurs most frequently in women and children. It is often the result of some indiscretion in diet. Oysters, fish, pickles, honey and
strawberries are among the articles which seem especially apt to induce
an attact of urticaria. Certain medicines also occasion nettle-rash in
some individuals. Thus it has been known to follow the use of turpentine, copaiba, chloral and morphine. Some individuals learn by experience to avoid certain articles of food and certain drugs, knowing the indulgence in them is followed by an outbreak of nettle-rash.

Treatment.—An emetic should be first administered, if the eruption
is caused by anything recently taken into the stomach ; it should be
followed by a saline aperient—senna mixture, with salts, is perhaps best,
and this repeated until the bowels are freely moved ; if the febrile symptoms do not subside, a mixture composed of sweet spirits of nitre, 2
drachms ; liquor of acetate of ammonia, 1 ounce ; and camphor mixture,

5 ounces should be given, two tablespoonsful every four hours. In the chronic form, a simple diet, active exercise, an avoidance of any articles of diet likely to excite the eruption ; keeping the bowles regular by gentle aperients, combined with anti-acids ; a five grain rhubarb pill an hour before dinner, or a small piece of the root chewed, are good remedial means ; the tepid bath should be occasionally used, or sponging, to keep the skin in a healthy state ; to allay the irritations, dust starch powder over the irruptions, or use a lotion made of rose or elder-flower water in half a pint of which has been dissolved 1 drachm of carbonate of ammonia, and ½ a drachm of sugar of lead.

Prickly Heat or Lichen, is a disease caused by intense and long-continued heat ; but it may be excited by the same causes which produce the nettle-rash, when the system is prepared for it. It is one of the most annoying plagues of a tropical climate.

Treatment.—For the relief of the itching and burning sensation attendant on prickly heat, which in tropical countries are often absolutely unbearable, the best remedy is cold water—ing caution when the patient is perspiring. Live sparingly and take a few doses of

Powdered Aloes	-	-	-	- 2 drachms
Powdered rhubarb	-	-	-	- 1 drachm
Powdered Jalap	-	-	-	- 2 drachms.
Powdered Cream of Tartar,	-	-	-	4 drachms.
Magnesia,	-	-	-	- 1 drachm.
Best Honey,	-	-	-	- 1 ounce.

Mix and divide into 120 pills, take 2, 3 or 4 on going to bed.

Irritation, Itching. A papulous affection of the skin, attended with troublesome itching. Sometimes it is attended with a sensation as of ants or other insects creeping over and stinging the skin, or of hot needles piercing it. This disease, although not dangerous, is a cause of great discomfort, and sometimes even misery ; it attacks persons of all ages, and is not easily got rid of, sometimes lasting for months, or even years.

Treatment.—Wash well, every evening before going to bed, with Castile soap, and allow it to dry in. Brandy or alcohol may be used in the same manner. An ounce of lemon juice in a pint of water, or vinegar used in the same proportion, will be found useful ; also, water and spirits of camphor. The diet should be carefully regulated, and all stimulants avoided.

This disease often affects the private parts, especially in pregnant females. In this case one of the following lotions will be useful :—

Hyposulphite of Soda,	-	-	- 4 drachms.
Glycerine,	-	-	- 2 drachms.
Distilled Water,	-	-	- 6 ounces.

Apply as a lotion. *Dr. Fox.*

Or,

Carbolic Acid,	-	-	.	- 6 grains.
Water,	-	-	-	- 1 ounce.

Apply as a lotion.

Freckles. To remove freckles use one of the following :—

Fresh Lemon Juice,	-	-	A wineglassfull.
Rain Water,	-	-	One pint.
Attar of Roses,	-	-	A few drops.

Mix, and put in a well corked bottle.

Wash the face and head with this several times daily, letting it dry into the skin.

Dr. Wilson, London.

Or,

Juice of cucumber pressed from the fruit.

Boil over a quick fire, cool rapidly and bottle.

Apply a tablespoonful diluted with two tablespoonsful of water, night and morning.

Dr. Cheery.

Or,

Juice of Horse-radish,	-	-	-	- ½ pint.
Cider Vinegar,	-	-	-	- 1 pint.

Mix, and apply night and morning.

Lice. These disgusting insects are easily transferred from one person to another, and give rise to much irritation and even eruption of the skin Sometimes they breed with inconceivable rapidity, especially on dirty or broken-down subjects

The treatment is simple. The clothes should be baked, for washing alone will not kill the insects; a warm bath should be taken, and the skin anointed with stavesacre ointment (one drachm to one ounce). Flannel should be discontinued next the skin until the eruption is well. A strong tea of tobacco, and mercurial ointment also destroy them.

The louse which frequents the hair differs from that found in the clothing. It can be destroyed by the same measures.

The crab-louse occurs in the hair on the trunk, under the arm-pits, etc. He is exceedingly tenacious of life. Mercurial ointment, or calomel rubbed in, is generally sufficient, however.

Mother's Marks. *Treatment.*—Ethylate of Sodium painted over the part, night and morning, for a time, will produce a scab, which must be removed, and the medicine re-applied until the mark has disappeared.

OTHER EXTERNAL DISEASES.

Hang-nails. The loose fragments of nails which bear this name should always be carefully removed, as they may, by their irritation, produce felons, and other inflammatory troubles.

Ingrown Nails. These usually are found on the toes, especiall the big toe. They are very painful and obstinate. To heal them, the middle of the nail should be scraped with a piece of glass until it is quite thin. Then the edge which has grown under the flesh should be gently but firmly pressed upward by inserting under it a roll of lint or soft cotton. The raw edge of the flesh should be fastened down, away from the nail, with narrow strips of adhesive plaster, and a loose shoe be worn. Perseverance in this plan will affect a cure.

Blood Blisters rise on the skin when one of the small skin blood-vessels is broken. If not the result of an accident, they are often the sign of bad health, and point to the necessity of a good diet, tonics, cod-liver oil, and change of air.

Boils or Furuncles. These are great pests to some people, and many ways are suggested to "backen" them. The best is to apply heat as soon as there is any sign of one. Hot water, long continued, will often succeed. When matter has once formed, warm poultices should be applied, and the boil opened with a sharp knife, so that the core can loosen.

When a person has crop after crop of boils, it is a sign of enfeebled health. Peruvian bark or quinine should be taken with iron.

For a *blind boil*, which is a dull, obstinate sore, opening with a knife, and poulticing, is the proper treatment.

Bunions are hard, tender swellings, which appear on the ball, the outer portion of the second joint of the big toe, or on the insteps. Tight or short shoes produce them. A loose, but well-fitting and flexible shoe must be worn, and the swelling painted, several times a week, with tincture of iodine or a weak solution of carbolic acid.

Carbuncle. This is a large and malignant boil, very painful, and even dangerous, as it indicates a low state of health. It is flat and firm, with a crust with several imperfect openings, from which the matter passes out. This, and its size, and the intense pain accompanying, distinguishes it from a common boil. At times it is infectious, induced by the reception of an animal poison (from cattle) into the system. This is called a "malignant pustule."

Treatment.—The only efficient plan of treating a carbuncle is to devide it crosswise, with a sharp surgical knife, early in its course, and then poultice it steadily. The system should be kept up with tonics, and strong, nourishing food, with ale or porter.

Corns. The cure of these common annoyances is easy, but trouble-some. The foot should be thoroughly soaked in warm water, and all of the corn removed with a dull penknife, but no pain should be produced. When this is felt, it is a sign that the knife is going too far. Then the little cavity should be surrounded with flat rings of wash leather, or felt, or corn plaster, leaving a hollow centre. A drop or two of sweet oil should be placed in this, and the foot clothed in a soft stocking, and a loose easy shoe. This process, repeated twice or three times a week, for a month or two, will cure a corn ; but it will return if tight, ill-fitting shoes are resumed.

Felon, or Whitlow This is a very painful inflammation of the finger, arising from a bruise, from the entrance of a needle or splinter, a hang-nail, or other irritant. When milder, it is called a " flesh felon ; " when severe, affecting the bone, a " bone felon." The former begins generally at one side of the root of the nail, or in the bulb of the finger end, with redness, swelling, and a throbbing, burning pain, shooting up the hand and arm.

These symptoms are very much increased in severity in the second variety. The patient holds his hand up, as the pain is more acute when it is dropped. The appetite suffers, and sleep is disturbed or prevented.

At the outset, a felon may be backened, at times, by holding the hand for a half-hour in water as hot as may be borne, and then wrapping it in a large hot poultice. After this is done, leeches may be applied, so as to draw blood freely from the part. These measures failing, the next measure, which should not be delayed, is to open freely the swelling, with a sharp knife, cutting, in a bone felon, fully to the bone. This, alone, will prevent the danger of having a stiff, mutilated, and useless finger for a lifetime. After the incision, poultices and warm water dressing will complete the cure.

A great German authority says :—Boils and felons may be often aborted by the free use of nitrate of mercury ointment, if suppuration have not commenced. It does not cause pain, but after about twelve hours, a drawing sensation is felt, after which all sensation ceases. The writer covers the entire finger with a coating of the ointment about ⅛ inch thick and covers with strong sticking plaster. The dressing is al-lowed to remain on for six hours, after which no further treatment is necessary.

Goitre, Derbyshire neck, or swelled throat, is a deformity common in mountainous countries. Its cause is not ascertained. The swelling is in front and at the base of the neck. It is painless, but is unsightly, and may become troublesome by pressure on the windpipe and large blood-vessels. When of many years' standing, it cannot be cured ; but when comparatively recent, rubbing with the following ointment, three times a week for several months, will cause it to disappear,

Take of Iodide of Cadmium, - - - 1 drachm.
Fresh Lard - - - - - 1 ounce.
Mix. Rub in a portion the size of a small pea.
Goitre is liable to return, however, unless a change of residence is adopted.

Sores and Ulcers. Cold sore, leg sore, or indolent ulcer. These are names given to those obstinate sores, or issues, that come on the legs of persons, especially those in advanced life, whose systems are below par, and who have suffered from swollen veins, rheumatism, frosted feet, and similar troubles. They are not very painful, but are foul, debilitating, and difficult to heal.

The treatment is local and internal. Good hygiene, great cleanliness, pure air, nourishing, simple diet, and iron or cod-liver oil, daily, are essential. Then locally apply the carbonate of ammonia ointment, given below, taking care to wash the sore by a stream of warm suds from castile soap. This done, cut long strips of sticking plaster about a half inch wide, and long enough to go once and a half round the limb. Applying the middle of one of these to the opposite side of the limb from the ulcer, the two ends are brought forward across this with a firm pressure, so as to bring together, as much as may be, the edges of the sore. A number of straps, so applied (leaving places for the matter to escape), completes the dressing. This should be repeated once or twice a week. This method of bandaging a leg is shown elsewhere.

Carbonate of Ammonia Ointment.—Take of
Carbonate of Ammonia, - - ½ a drachm.
Lard, - - - - ½ an ounce.
Mix. A useful application to sluggish, scrofulous sores.

Sweating, excessive. Persons of a stout habit occasionally suffer from excessive perspiration, which, though it may not be injurious, is disagreeable. It may be diminished by avoiding warm baths, changing the underclothing frequently, and sponging the body with a lotion consisting of two teaspoonfuls of dilute sulphuric acid, in a quart of water. The skin, when dried, should be powdered with starch or pulverized asbestos.

Fetid, or Foul Feet, from excessive sweating, are a great annoyance to some. Benefit in such cases will be derived from bathing the feet, night and morning, with a mixture of half an ounce of tannic acid in a pint of cologne water, drying, and powdering with starch or dry tannic acid.

Warts. These ugly excrescences are best dispersed by rubbing them night and morning, with a piece of muriate of ammonia (salammonic), moistened with water.

Ringworm. *Treatment.*—Ringworm on parts of the skin which are not covered with hair is usually cured very easily. The principle of treatment consists simply in the application of some material which destroys the vegetable parasite ; so soon as the plant is killed the irritation subsides, the crust is thrown off, the skin resumes its natural condition.

A considerable number of agents have been used for this purpose ; among household remedies may be mentioned kerosene oil and a solution of borax, as follows :

Borax, - - - - -	Half an ounce.
Water, - - - - -	Half a pint.

Some of the borax remains undissolved at the bottom of the vessel, but this is useful in order to keep the solution at full strength.

While these remedies are usually effectual, yet many people object to the application of kerosene, which is moreover irritating as well as unpleasant, and is apt to cause some swelling and pain of the skin around the diseased spot. For these reasons various other remedies are to be preferred in the treatment of ringworm. Among those most employed by physicians are the following :

Hyposulphite of soda, - - -	One ounce.
Dilute sulphurous acid, - - -	Two drachms.
Glycerine, - - - -	One ounce.
Water, - - - - -	Four ounces.

This should be applied to the diseased patches by means of a stiff brush or a coarse cloth, which is to be dipped in the lotion.

Saturated solution of sulphurous acid, -	One ounce.
Water, - - - - -	Three ounces.

The removal of the parasites is hastened by the use of a remedy in solid form, which can, therefore, be kept in contact with the skin constantly, and not simply applied at intervals, as is necessary when liquids are used. If the disease be located on the face, an ointment can be applied at night and allowed to remain till morning ; for this purpose any one of several ointments may be used, as follows :

Ammoniated mercury, - - -	Twenty grains.
Red oxide of mercury, powdered, -	Twenty grains.
Simple ointment, . - -	One ounce.

Mix thoroughly and apply directly to the skin, rubbing the salve vigorously into the pores of the skin.

The simple *ointment of mercury* will also be found efficient. In most cases, too, the ordinary sulphur ointment can be successfully used for destroying the parasite.

This affection is entirely a local disease of the skin, and requires no internal treatment.

Ringworm of the scalp is a far more troublesome and obstinate affection to cure, not because the parasite is any more difficult to destroy, but because it is extremely difficult to introduce the remedy into the follicles of the hair, which are a continual breeding place for these organisms. It will often happen that by a few applications of one of the remedies above given the disease will entirely disappear from the scalp, and the patient will consider himself cured ; but in a few days or weeks it becomes evident that the ringworm has started again in the same place.

It is, therefore, necessary to take especial pains to introduce the various agents used for killing the parasites into the hair follicles. For this purpose the first requisite is to shave the head closely, so as to remove all the hairs from the diseased patch and from its immediate vicinity. After this has been done, there may be rubbed into the scalp daily one of the following ointments :

Hyposulphite of sodium,	-	-	-	One drachm.
Vaseline,	-	-	-	-	One ounce.

Or,

Salycycle acid,	-	-	-	Twenty grains.
Vaseline,	-	-	-	One ounce.

In some cases good results may be obtained by brushing the surface thoroughly with strong tincture of iodine.

DIET IN DISEASES OF THE SKIN.

The physician is constantly asked by patients suffering from skin diseases, what they shall eat and what they shall not eat. Probably no other class of patients is so deeply impressed with the idea that their diseases are due to impurities of the blood, and that extreme care should be taken to avoid the use of certain articles of food. Most of these patients have theories and hobbies as to the diet which it is proper for them to take and to avoid ; and most of them seem to believe that dieting consists in the avoidance of food so far as possible.

It is true that the diet can be made to exercise considerable influence upon diseases of the skin as well as upon diseases of the internal organs ; but it is not especially necessary to regulate the food in diseases of the skin, with certain exceptions to be presently mentioned.

In every case it should be remembered that the plan of dieting does not mean to reduce the patient to the verge of starvation, but simply to grant him such articles of food and in such quantities as will, in the opinion of the physician, tend to restore his bodily functions to their natural condition. In most cases the patient needs to be *built up* rather than *torn down ;* for most diseases of the skin, even those of local origin, such as ringworm, indicate that the patient is in a more or less debilitated condition, since these diseases do not ordinarily occur in persons of the most robust habits.

There are certain affections of the skin which are provoked and aggra-vated by indulgence in particular articles of diet ; the patient soon learns to discriminate for himself upon this point ; he soon discovers what articles of food are especially apt to provoke the outbreak of his com-plaint. Thus the sufferer from nettle rash early ascertains that he has an attack of the disease whenever he eats strawberries, or oysters, or shellfish, or whatever his particular weakness may be.

Aside from these personal peculiarities, there are certain principles which apply to patients affected with chronic diseases of the skin. In most cases the appetite is a reliable guide, though it sometimes needs direction.

To begin with, it must be remarked that much of the difficulty from error of diet arises not so much from the nature of the substances eaten, as from the imperfect and careless way in which they are eaten. In our country especially, rapid eating and hurried chewing are prevalent habits, which are responsible for many difficulties of other organs than the stomach. For digestion really begins in the mouth ; here the food is not simply divided into small particles, so that it can be acted upon by the juices of the alimentary canal, but it is also mixed with the saliva, which effects certain changes in it. If the chewing be imperfectly per-formed, or if the saliva be but slightly mixed with the food, there will result first derangement of the stomach, and subsequently derangements of other organs. For the ill effects are not limited to the stomach alone. If this do its work but imperfectly, additional labor is required of other organs to piece out the work of the stomach ; while at the same time these other organs are supplied with imperfect blood, since the stomach does not digest and take up the food in a natural way. It is evident, therefore, that one of the first requisites for the diet of a patient affected with skin disease is that the food shall be easily digested. The patient's own sensations will usually indicate to him when he has indulged in indigestible food.

As to the quantity of food which should be taken, it may be said that but very few of the skin diseases are caused or aggravated by excessive indulgence in food. Patients with acne are perhaps the only ones whose complaint is aggravated by simple over-indulgence. Such patients should avoid hot drinks and soups, since these provoke flushing of the face and favor the development of the rash. It is well for them to avoid desserts, since these are usually just so much more than the individual requires or really desires.

A prevalent habit, which probably contributes largely to the preval-ence of indigestion, is the habitual use of large quantities of liquids with the meals. Aside from the injury which may result from alcohol or tea or coffee in excess, it is not desirable to fill the stomach with any liquid, however harmless, during the process of digestion ; since the stomach juices are thereby diluted and weakened and the process of digestion is, to say the least, retarded. This is especially true if the liquids taken be cold, since the effect of chilling the stomach is also to arrest the digestive process.

Patients with eczema are apt to dislike and avoid fatty food. It has been ascertained that the use of fats in the food generally exercises a good influence upon the course of the disease ; hence it is desirable that such patients take a moderate amount of fat with their food, even though they do not crave it. These patients with eczema are apt to eat vegetable food by preference, especially the starchy substances, such as rice, arrow-root, and oatmeal. It is well for them to bear in mind the popular idea that "oatmeal is heating," since there seems to be some foundation for this idea in fact.

One of the most frequent causes and one of the most constant means in prolonging the various diseases of the skin, is indulgence in fermented liquids. These generally exercise a decided influence in originating and in prolonging diseases of the skin. A patient suffering from such disease should, therefore, abstain from the use of beer, ale, wine, whisky, cider, etc., unless his condition is so debilitated as to require some of these liquids to increase and support his strength.

Especial care must be taken in the food of infants who are afflicted with diseases of the skin. The great bane of infancy among skin affections is eczema. This is doubtless often caused by a poor quality of food, whether this food be artificial nourishment or the milk from a debilitated mother. In every case of eczema in an infant, the attention should be directed to the matter and manner of nourishment of the child; the mother should scrutinize carefully her own condition ; should see that she eats only suitable food and avoids articles which she knows to be harmful, even though she have a craving for them. She should also endeavor to avoid any mental disturbances, excitement or emotion of any sort, undue sexual indulgence, and she should secure sufficient rest by retiring in due season at night.

Another habit which may assist in the development of eczema is the practice of giving the child the breast too frequently. The custom is almost universal of using the breast to soothe a crying infant ; this is usually a successful device, but it exerts a most injurious influence upon the child's digestion and promotes the development of several skin diseases, especially eczema. As will be observed under the appropriate heading, the child should not have the breast, even in the early weeks of life, oftener than once in two hours ; and as time passes this interval should be lengthened, in the interests of the child as well as the mother.

Doubtless another factor in the production of eczema in nursing infants is the use of fermented liquors by the mother. It is well ascertained that the use of such liquors promotes the occurrence of eczema in adults and in infants through the mother's milk ; hence, unless the mother's general condition is such as to absolutely require the support which can be given only by fermented liquors, it is advisable that she avoid these, in spite of the counsels and remonstrances of friends.

Eczema is especially frequent among artificially nourished or " bottle-fed " infants. It is often impossible to relieve an infant from eczema until its diet be radically changed. Directions for the feeding of infants will be found in the section on "diseases of women and children." It is scarcely necessary to remark, that the habit of feeding suckling infants with scraps from the table, pieces of cake "which don't do him any harm," sips of tea and coffee and the like, should be avoided even when the child is perfectly healthy, in the hope of keeping him so. It is all the more necessary when the infant is afflicted with a disease of the skin.

Much might be said also as to the hygiene of the skin during the existence of ailments affecting it. The popular idea that bathing is always desirable in all diseases of the skin, is a mistaken one ; some of these diseases, especially eczema, are greatly aggravated and prolonged by frequent contacts of the skin with water. Yet certain diseases, especially psoriasis, are certainly benefited by frequent bathing, especially at the sea shore. In fact everything which tends to increase the activity of the skin—muscular exercise in the open air, sunlight, fresh air in the bed-room, etc.—is highly desirable and important in the treatment of chronic diseases of the skin.

Dr. Fox gives the following directions as to diet in skin diseases :

First.—A distinction must be made between the diet of the private and the hospital patient. The latter often requires to be well fed and his disease then speedily goes ; the former, on the other hand, often needs to have a check put on the quantity and quality of his food.

Second.—In children, skin diseases may arise directly from defective alimentation, as in the case of eczema ; and it is frequently the case that the child who is the subject of eczema has not a sufficient supply of milk, either from excessive dilution or otherwise.

Third.—The regulation of the diet, setting aside the question of quantity or quality, is as a rule needed not so much to directly influence skin disease as certain states of the general health, which modify the particular disease present ; for instance, to meet especially dyspeptic, gouty and rheumatic conditions, but particularly the former.

In dyspepsia, in connection with eczema, acne, or congestion of the face, it is advisable, especially if the urine be very acid, to avoid sugar, tea, coffee, alcoholics, beer, raw vegetable matter, unripe or uncooked fruit, veal, pork, seasoned dishes, pastry and the coarser kinds of veg-etables, but especially all articles whose use is followed by heat or flush-ing of the face, and by flatulence and the like. Milk, the common meats, light kind of bread and some very light wine should be the diet of dyspeptic patients whose skins are at all in a state of irritation. In very many cases the stomach is at fault at the outset, and a careful regulation of the diet is of the utmost importance as an aid to the other means adopted to correct faults in other parts of the system.

In gouty subjects much the same line of treatment is to be pursued. As regards stimulants, hock, a good light claret, or whiskey and water are the best beverages.

In scrofulous patients the diet should consist of as much fatty matter as possible.

Fourth.—In children who suffer from ringworm it is desirable to give plenty of fatty food by means of milk, cream, eggs, and fat meat, if they can be persuaded to eat it.

Fifth.—In syphilis the greatest care should be taken to avoid everything beyond the most moderate use of stimulants ; their abuse in this disease is a source of the greatest aggravation.

Sixth.—In all cases in which the onset or early stage of a skin disease is accompanied by fever, however slight, stimulants should be avoided, and the plainest and simplest diet ordered. In marked cases of this kind a milk diet for a while is often found to be very beneficial.

Seventh.—In some cases in which the disease is accompanied by flushing of the skin, this condition is much increased by the consumption of food, especially if dyspepsia exist, in consequence of the sympathy existing between the stomach and the skin of the part affected. This state of things is especially marked in such diseases ; acne, congestion of the face, and sycosis (barber's itch). Stimulants must be avoided, unless they be diluted with some alkaline water. The use of a diet appropriate to the dyspepsia must be rigorously enforced.

Eighth.—In all cases where a skin disease has become chronic, and where there is debility, the patients hould be allowed a full unstimulating diet.

At the well-known hospital for diseases of the skin, Blackfiiars, London, the following directions are issued, to be observed by patients :

Remove flannel from next the skin affected, or line it with soft linen. Wash with warm water, and, as regards *diseased* skin, not more frequently than cleanliness requires.

Avoid using soap of any kind to the affected parts. To cleanse the diseased skin, substitute instead of soap a paste or gruel made of bran, oatmeal, linseed meal, arrowroot, or starch and warm water. Rinse off with warm water or warm milk and water ; and employ yolk of an egg and warm water to cleanse the scalp.

Dry the skin with soft linen, and smear it lightly with the ointments or liniments, or dress wounds with the same spread thin upon lint or linen. Afterward apply the bandages evenly should they be required. Bathe the affected part by means of a sponge or rag with the lotions or embrocations, or paint them over with a camel's hair brush, not more frequently than directed by the physician.

Rinse the mouth with water, and brush the teeth after taking the medicines, and observe that neither more nor less than the dose ordered is taken.

At the same institution the following rules of diet for patients are observed :

For breakfast—Bread and milk, rice, milk or porridge instead of much tea, coffee or cocoa—with or without eggs, and bread and butter, or a little animal food.

For dinner—Plain roast or boiled fresh meats, fish or poultry plainly cooked, egg or farinaceous (starchy) puddings, potatoes, and few other vegetables, plain boiled rice.

For supper—Milk and water, or gruel or other farinaceous food, with bread and butter, a little cream, cheese or poached eggs.

Beverages—Barley water, toast and water, thin gruel, beef tea, soda, potash or seltzer water.

N. B.—*To be avoided*—Salt meats, soups, sweets, acids, fruits, pastry and raw vegetables.

No malt liquors, wine or spirits are to be taken without the sanction of the medical officers of the hospital.

CHAPTER VI.

Small-Pox (*Variola*). This, like scarlet fever and measles, belongs to the class of eruptive fevers ; it attacks persons of all ages, but the young are the most liable to it. At no particular season of the year is it more prevalent than at any other, nor does climate appear to be influential in modifying its visitations.

Symptoms.—When it occurrs naturally, the premonitory symptoms are those of other fevers of its class ; there are usually cold chills, pains in the back and loins, loss of appetite, prostration of strength, nausea, and sometimes vomiting ; with young children, there are sometimes convulsions. About forty-eight hours after these symptoms set in, an eruption of hard red pimples begins to overspread the face and neck, gradually extending downward over the trunk and extremities. Each pimple is surrounded by the peculiar dull red margin termed areola, and has a central depression on the top, containing lymph ; at this period the eruption is decidedly vesicular, but it becomes afterward pustular ; this change takes place on about the fifth day of its appearance, when the central depression disappears, suppuration takes place, and the vessels are filled with matter, which shortly after oozes out and dries into a scab. In about ten days this falls off, and leaves a pale purple stain like a blotch, which gradually fades, unless the disease has penetrated so deeply as to destroy the true skin, in which case a pit, or, as it is usually called a " pock-mark," remains for life.

Small-pox may be either distinct or confluent. In the former case, the pustules are perfectly distinct from each other, in the latter, they run into each other ; this latter is the most dangerous form of the disease, the fever being more intense and rapid, and having no intermission ; it goes on increasing from the first, and frequently by its violence, in nine or ten days, so exhausts the system that coma, delirium, and death ensue, preceded by convulsions, hemorrhages, bloody stools, dysentery, and all the train of symptoms which indicate that a virulent and fatal poison has entered into the circulation.

Treatment.—As soon as the premonitory fever comes on, an emetic should be administered, and followed by a purgative of a tolerably active nature ; then keep on spare diet (certainly no meat), and give plenty of

63

warm diluent drinks; keep the bowels moderately open by means of saline aperients ; let the patient have plenty of fresh air, and sponge the skin with cool or tepid water, as may be most agreeable, to diminish the heat of the body. Sometimes there is not energy in the system to develop to pustules with sufficient rapidity; in this case nourishment and stimulants should be given in the form of broths, wine, whey, etc. ; warm or mustard foot-baths should also be resorted to ; and, to allay irritability, a ten-grain Dover's powder may be administered at bed time, or a ¼ of a grain of morphine, in camphor mixture. A good nourishing diet will be required in the secondary stage of the fever ; and if it assumes a typhoid character, the treatment should be the same as that of typhus fever. Frequently the face is much swelled, and the eyelids closed ; in this case rub the latter with olive oil, and bathe the whole with poppy fomentation. If the throat is sore, use a gargle of honey and vinegar, 1 tablespoonful of the former, 2 of the latter, added to a ½ pint of water or sage tea. If much affected, a blister should be applied to the neck. If there is much headache, cut the hair close, apply mustard poultices to the feet, and a spirit lotion to the head ; to reduce itching, apply to the eruptions a liniment composed of lime water and linseed oil, equal quantities, or smear the pustules with cold cream ; to check diarrhea, give chalk mixture, with 5 drops of laudanum in each dose ; if perspirations are too copious when the eruptive fever has subsided, take acidulated drinks. Smearing the eruption with mercurial ointment, or puncturing each pustule, and absorbing the pus with wool or cotton, has been recommended to prevent the deep pitting which is so great a disfigurement to the face. Painting the face once or twice a day with glycerine is said to effectually prevent pitting.

There is no disease more certainly and decidedly contagious than this; after imbibing the poison, a period of twelve days generally elapses before the commencement of the fever, and during this time no inconvenience may be experienced. Besides breathing the effluvia arising from a person attacked, small-pox may be communicated by inoculation with the matter of its pustules, and, the resulting disease being of a milder character, this method was formerly much practised to guard persons from a spontaneous attack ; since, however, the introduction of vaccination by Dr. Jenner this practice has been abandoned. This disease is frequently epidemic, and the statistics of its different visitations show that the mortality of those attacked who have not been vaccinated is one in four ; whilst of those who have, it is not one in four hundred and fifty ; a strong argument this for vaccination where the disease prevails.

A Certain Cure. Wm. Grandy, of Detroit, communicated the following item of Mr. Hines' to the Detroit *Tribune*, which he had seen in the Toronto *Weekly Globe*, with these remarks :

" Small-pox being so fatal and so much dreaded, an unfailing remedy like the following, so simple and so safe, once discovered, ought to be brought to the knowledge of the masses without hesitation or delay."

"I am willing," says Edward Hines, "to risk my reputation as a public man if the worst case of small pox cannot be cured in three days simply by cream of tartar. This is the sure and never-failing remedy : Cream of tartar, 1 oz., dissolved in boiling water, 1 pt. ; to be drank when cold, at short intervals. It can be taken at any time and is a preventative as well as a curative. It is known to have cured thousands of cases without fail. I have myself restored hundreds by this means. It never leaves a mark, never causes blindness, and always prevents tedious lingering.

Scarlet Fever or Scarlatina, Is a contagious febrile disease, almost always attended during a part of its course by a rash and by sore throat. Sometimes only one of these features is well marked, sometimes both. Though persons of all ages are susceptible of it, it is eminently a disease of children. Like small pox or measles, it rarely attacks a person more than once. Physicians distinguish three different varieties of scarlatina—namely, *scarlatina simplex*, in which there is a florid rash and little or no affection of the throat ; *scarlatina anginosa*, in which both the skin and throat are decidedly implicated ; and *scarlatina maligna*, in which the distress of the disease falls upon the throat.

Symptoms.—So plainly are the symptoms marked that it is scarcely possible to mistake this eruptive fever for any other ; almost invariably we have first sore throat, with shivering, headache, and loss of appetite ; probably there may be sickness and vomiting, with heat of skin, quick pulse, and great thirst. In about forty-eight hours from the commencement of the attack, we have an eruption of red spots upon the arms and chest ; these gradually become more thickly planted and widely spread, until they pervade the whole of the body, making the skin appear of one uniform scarlet tint, that is over the body generally ; in the extremities it is more in patches, the skin being perceptibly rough to the touch. On the second day, generally, the tongue presents the appearance of being covered with a white film, through which the papulæ project as bright red spots, as we see the seeds on a white strawberry ; then the white creamy-looking film comes away gradually and leaves the tongue preternaturally clean and red. On the fourth or fifth day the eruption begins to fade, and by the seventh or eighth has entirely disappeared, and with it, the febrile symptoms. Then commences the peeling off of the cuticle or scarf skin, which comes away in scales from the face and body, and in large flakes from the extremities. It is during this process that the greatest danger of contagion is to be apprehended, and, until it is com pleted, the patient should be kept apart from the rest of the family : it may be hastened by tepid bathing and rubbing. Sometimes with scarlet fever, there is little real illness ; the patient feels pretty well, and, in a few days would like to leave the sick chamber ; but it is always neces-sary to be cautious in gratifying such a wish, both for the sake of the invalid and of others ; after an attack of this fever, as after measles, **the**

E

system is peculiarly susceptible of morbific influences, and a chill taken at such a time may cause the most alarming results.

Sometimes we have a great aggravation of the symptoms above described ; the throat gives the first warning of the attack ; there is stiff neck, swelling of the glands, and the lining of the mouth and fauces becomes at once of an intense crimson color ; there are ash-colored spots about the tonsils ; the general eruption is of a deeper color, and spreads more rapidly than in the simple kind.

Then again we have the malignant form, with the rash in irregular patches of a dusky hue, which sometimes recedes and appears again. There is intense inflammation of the throat at the very outset, with general enlargement of the salivary glands ; the neck sometimes swells to a great size ; there is a sloughy ulceration of the throat, from which, and the nostrils—through which it is difficult to breathe—there comes an acrid discharge, causing excoriation of the nose and lips, and sometimes extending to the larynx and trachea, as well as to the intestinal canal, causing croup, vomiting and purging. The poisonous secretion enters into the circulation and vitiates the blood ; sometimes the sense of hearing, as well as of smelling, is entirely destroyed by the acrid matter coming in contact with and inflaming the mucous membrane. With this form of the disease it is extremely difficult to deal, and the patient often sinks beneath it in spite of the best medical advice and assistance. Scarlet fever may be distinguished from measles by the following characteristics :

In scarlet fever the eruption appears on the second day, accompanied with sore throat, but no running of the nose. In measles the eruption comes out on the third or fourth day, with running from the nose, and other catarrhal symptoms. The eruptions of measles are like flea-bites, slightly elevated from the surface, in patches the shape of a half moon ; whereas the rash of scarlet fever is smooth to the touch, spreads over the whole body, and is of a brighter color than the measles.

Treatment.—At first mild aperients only should be given, with diluted drinks, as flaxseed tea, and a spare diet ; the patient should have plenty of fresh air ; the head should be kept cool, the hair being cut close off or shaved. The following is a good febrifuge mixture :

Carbonate of Ammonia	-	-	1 drachm.
Solution of Acetate of Ammonia	-	-	2 ounces.
Water or Camphor Mixture	-	-	6 ounces.

A tablespoonful to be taken every four hours—that is for an adult ; a dessertspoonful will be sufficient for a child. The whole body should be sponged with cool water as often as it becomes hot and dry. If the throat swells much externally, and there are headaches, apply a blister or hot bran poultice, and soak the feet and hands in hot water, with a little mustard or Cayenne pepper stirred in. To gargle the throat, dissolve 1 drachm of common salt in ½ a pint of water ; with children who cannot gargle, this may be injected against the fauces or up the nostrils, by means of a syringe or elastic gum bottle. When the inflammatory action has ceased and the skin is peeling off, it is necessary to take good stimulant and nutritious food, with tonics such as iron and quinine, unless they cause bad head symptoms, in which case these must be discontinued and the diet chiefly depended on. When the system seems to be overwhelmed with the strength of the poison, a liberal administration of wine and bark will be required to sustain the flagging powers until the deadly agency has in some measure passed away. As gargles for the throat, a weak solution of chloride of soda, or of nitrate of silver is very useful. A solution of chloride of potash in water (a drachm to a pint) is recommended as a drink in this disease. The bowels also require to be carefully watched. It is of the utmost importance that the throat should be carefully treated. If neglected, the inflammation is liable to enter into the middle ear and cause life-long deafness, and perhaps ulceration of the ear, with discharges.

With regard to the more malignant form, but little is to be done ; the depressing effect of the contagious poison upon the whole body, and upon the nervous system especially, is so great as to defy all active treatment.

To assist the action of the skin, use the following :

Pulverized Gum Arabic,	- -	1 scruple.
Sweet Spirits of Nitre,	- -	½ an ounce.
Tincture of Veratum Viride,	-	20 drops.
Water, soft,	- - -	2 ounces.

Mix : give half a teaspoonful every half hour.

Measles is known also as rubeola, French measles, black measles, morbili, and may be confounded with reseola or rose rash. The latter has no fever or throat symptoms.

Symptoms.—The attack is ushered in by all the evidences of a cold. The nose runs, the eyes are red and watery, sneezing is frequent, and there is more or less cough. Along with this is more or less fever and debility. About the fourth day the eruption appears, first on the face, then on the chest and upper extremities, finally covering the whole body. It is of a dark red color, not fine pointed, and arranged in patches of a crescent shape. The fever soon declines, and the eruption fades after three days, being followed by slight scaling of the skin. There is also a soreness of the throat, though generally of a mild form.

French or German measles is also called "rötheln." It appears like a mixture of measles and scarlet fever. On one part, the eruption strongly resembles the one, while on another, it resembles the other. Some regard this as an entirely separate disease, and claim that it does not afford any safety against an attack of either measles or scarlet fever.

Black measles is that form when typhoid symptoms occur, and the attack is complicated with hemorrhages, etc., and generally proves fatal—*Treatment.*—The bowels should be acted on gently by some cooling saline, such as—

Cream of Tartar,	-	-	-	3 drachms.
Carbonate of Soda,	-	-	-	3 drachms.
Water,	-	-	-	1 tumblerful.

Put the whole in a stone jug or bottle and fasten the cork. To be taken in the morning before eating.

Then the solution of acetate of ammonia may be given freely with cooling drinks such as lemonade, etc. And if the cough and chest symtoms are severe ½ teaspoonful ipecac syrup as well. In very many cases nothing is required but good nursing and care. In debilitated cases quinine stimulants and nourishing food are required.

Table exhibiting the difference between small pox, scarlet fever and measles.

MEASLES.	SCARLET FEVER.	SMALL-POX.
The period which elapses between exposure to contagion and the beginning of the disease is usually *seven* to *fourteen* days.	The period between exposure to contagion and the beginning of the disease is variable, often *three* to *six* days, but may be several weeks.	The period between exposure to contagion and the beginning of the disease may vary from five to twenty days, and usually about *ten* days.
Fever is moderate ; it does not decrease, but often *increases* when the eruption appears.	Fever is intense; continues without interruption after the eruption appears.	The fever is usually high ; it subdues when the rash appears.
The eruption makes its appearance on the *fourth* day first on the *face* and neck ; it spreads *gradually* for two days over the rest of the body.	The rash makes its appearance on the *second day*, first on the *neck and chest;* spreads over the entire body *rapidly* in eight to ten hours.	The eruption makes its appearance on the *third* or *fourth* day, being first seen *around the mouth* and on the *forehead.*
The eruption appears as *crescent-shaped patches*, the intervening skin being healthy.	The rash spreads *uniformly* over the skin, without intervening patches of healthy skin.	The rash consists at first of *pimples* which become a day later watery *blisters.* Finally these blisters become white, and are drawn in at the centre —*umbilicated.*
The rash lasts *five* days, at the end of which time the skin peels off in very *fine scales.*	The eruption lasts *six* or *seven* days, when it begins to peal off in *large flakes.*	
The tongue is coated and red at the edges.	The tongue is covered with numerous fine red points, which give it the name "strawberry tongue."	The tongue is heavily coated and often swollen.
Running of the eyes and nose and *bronchitis* are usually *present.*	There is *rarely* any noticeable *bronchitis* or *running of the eyes and nose.*	There is no running at the eyes or nose, and not often bronchitis.

MEASLES.	SCARLET FEVER.	SMALL-POX.
Sore throat is very rare.	*Sore throat* is always *present.*	Sore throat is often present, but not so marked as in scarlet fever.
The *mind* is *not* affected.	The mind is usually affected : there may be *delirium* and *convulsions.*	The mind is often affected; delirium and convulsions may occur.
There is no *secondary fever ;* that is, after the first fever has subsided, which happens during the second or third day after the appearance of the rash, no further fever occurs.	There is no *secondary* fever.	*Secondary fever always* appears after the rash has been visible for several days.
Measles is often followed by chronic bronchitis, consumption and inflammation of the eyes.	Scarlet fever is often followed by *Bright's disease, dropsy,* inflammation of the eyes, deafness, and enlargement of the glands about the throat ; sometimes by paralysis.	Small-pox is not usually followed by other diseases, though the pocks may result in serious damage to the eye-sight, as well as cause unsightly scars on the skin.

Typhoid Fever. This disease is frequently spoken of as slow, nervous fever, and sometimes as common continued fever. It also bears the names of putrid fever, autumnal or fall fever, and night-soil fever.

Cause.—Exposure to foul air, as that of sewers, water closets, privy-wells, etc., is one of the most powerful causes of typhoid fever. Depressing influences, such as anxiety, home sickness, great fatigue, seem to occasion it in some instances. The discharges of the patient may, if not disinfected, impart the disease ; wells have been contaminated in this way, so that their water has become a source of wide-spread contagion. The age at which typhoid fever is most apt to occur is between fifteen and thirty years, rarely before ten, and still more rarely after fifty. The appearance of the disease does not seem to be influenced by climate or locality.

Symptoms.—Typhoid fever is usually preceded for several days by headache, loss of appetite, prostration of strength and great disinclination to make any physical or mental exertion. Bleeding from the nose is often an early symptom, so also is a slight cough. Increasing weakness and the coming on of fever force the patient to take to his bed. Considerable fever and thirst are then complained of. The nights are wakeful and delirious ; the days are passed in dozing and muttering. Soon, towards the close of the first week, the belly swells, and diarrhœa ordinarily sets in about the same time. The face takes on a dull look, and a dark purple flush. Hardness of hearing is not unusual about the middle of the second week. Towards the close of the second week, a few small rose-colored spots, which are peculiar to the disease, show themselves on the belly ; they disappear for a moment when pressed upon by the finger, but quickly return after the pressure is removed.

Recovery may begin at the end of the second week, but ordinarily not before the fourth week, while the attack, in rare cases, lasts two or three months. Great wasting of the body and troublesome bed-sores accompany protracted cases.

The patient gets well very slowly, and is liable, for a long time, to a relapse. About one case in twenty ends fatally, death taking place usually in the neighborhood of the eighteenth or twentieth day.

Treatment.—Good nursing is of more consequence than medicine in typhoid fever, as there is no remedy known which will cut the disease short. At the outset, if the bowels are costive, a teaspoonful of castor oil, or half a tumblerful of the effervescing solution of citrate of magnesia (to be obtained of any druggist) may be given; afterwards, however, no purgative is to be ever administered. When the diarrhœa becomes excessive, half a wineglassful of lime water mixed with an equal quantity of milk, is an excellent soothing drink. Headache is best relieved by cutting the hair short and applying iced cloths, or, in severe cases, pounded ice enclosed in a bladder or rubber bag. To allay heat of the skin, sponge the body with slightly warm whisky and water, or vinegar and water, care being taken to expose only one part at a time to the air. If much pain and tenderness of the bowels be complained of, apply a large hot mush poultice, mixed with one-fourth part of mustard. Great attention must be paid to preventing bed-sores, by keeping the bed-clothes always smooth, by frequently changing the position of the patient, and by bathing the parts most pressed upon with whisky, or with a mixture of spirits of camphor and sweet oil.

The *diet* is of the utmost importance in the treatment of typhoid fever. Neither during the sickness, nor for a long time after recovery, should solid food be given A neglect of this caution often causes the death of the patient, by occasioning a rupture of the sore places which always exist in the bowels in this disease. Although fluid, the diet must be supporting. The quantity given at a time must be small, not much more than a wineglassful, so as not to distend the stomach, but it is to be repeated frequently, every hour or two. An excellent diet is a wineglassful of milk, to which a tablespoonful of lime water is added, given every two hours; in the alternative hours, a wineglassful of beef tea. Great care must be taken not to let the patient sink, from the want of frequently repeated concentrated liquid food. Iced lemonade or ice water may be taken to quench the thirst, but not more than a wineglassful at one time.

As to medicines, few are required. A laxative may be given at the outset of the disease, as already mentioned. For a tonic, quinine is excellent. Few cases are treated without it at the present day. A good way of administering it is to procure it in powders of two grains each, from the druggist, and give one of these three times a day, in a spoon, surrounded with scraped apple, which will disguise the bitter taste.

Should diarrhœa cause trouble in typhoid fever, use—

Nitrate of Silver,	-	-	-	3 grains.
Powdered Opium,	-	-	-	6 grains.
Powdered Ipecac,	-	-	-	6 grains.

Mix and divide into 12 pills. Take one every 4 or 6 hours.

Dr. Bartholow.

Typhus Fever. Typhus Fever is a kind of continued fever, characterized by the ordinary symptoms of other fevers, accompanied with debility in the nervous and vascular systems, and a tendency in the fluids to putrefaction.

Causes.—Any of the ordinary causes of fever may give rise to typhus, but by far the most common cause of typhus is contagion, or febrile miasm, the activity of which is much increased by the crowding in close and ill-ventilated places, filth, insufficient nutriment, and other causes tending to depress the vital power. It is eminently contagious and infectious, and often prevails epidemically.

Symptoms.—The symptoms are great prostration of strength, heat intense, pungent, and more biting than in any other fever ; pulse hard, small, weak, and irregular ; nausea, vomiting, sometimes a greenish or blackish colored bile, countenance flushed, tongue parched and black furred, and thirst is excessive. In the worst cases black or purple spots appear, the urine is but little changed, and there is a peculiar fœtid smell, in cases of true typhus ; and sometimes there are discharges of blood. The duration of this fever is uncertain ; sometimes it terminates between the seventh and fourteenth day, and sometimes it is prolonged five or six weeks. Its duration depends greatly upon the constitution of the patient, and the manner of treating the disease. The most favorable symptoms are a gentle looseness, after the fourth or fifth day, with a warm sweat. These will continue some time, and carry off the fever. Hot scabby eruptions about the mouth and nose are good signs, as are also abscesses.

The unfavorable symptoms are excessive looseness, with hard swelled belly, black or livid blotches breaking out, sore mouth, cold clammy sweats, change of voice, inability to put out the tongue, a constant inclination to uncover the breast, difficulty of swallowing, sweat, and spittle tinged with blood, and the urine black, or depositing a black sediment, shows great danger.

Treatment.—In the early stages of this disease it is best not to interfere much with nature's operations. The principal aim ought to be to keep the patient alive until the fever-poison has expended itself. When seen early, however, it is often of advantage to administer an emetic or a purgative ; and the patient's uneasy sensations will be much soothed by sponging the surface of the body with cold or tepid water. Directly the powers of life begin to fail, a stimulating course of treatment should be commenced—such as strong beef or chicken tea, with wine or brandy frequently administered, taking care that it does not aggravate the febrile symptoms. When there is much irritability and sleeplessness, a dose of opium may be given. The patient should be in a large well-aired apartment, and the windows kept open as much as possible. As the patient begins to recover, a course of tonics will be necessary to expedite his restoration to health.

When sleeplessness and delirium are present the following is recommended by Drs. Graves and Murchison, who claim for it magical effects :

Sedative Liquor of Opium,   -   -   -   1 drachm.
Tartar Emetic,   -   -   -   -   1 grain
Camphor Water,   -   -   -   -   6 ounces.

Take a tablespoonful every hour till sleep is obtained.

Intermittent Fever (Fever and Ague) or Malarial Fever.

Symptoms.—The cold stage commences with a sense of languor and debility, and slowness of motion ; frequent stretching and yawning ; pain in the head and loins ; sometimes sickness and vomiting ; pulse small, frequent, and irregular ; urine pale ; to this succeeds a violent shivering and shaking, the patient feels very cold, and the breathing small, frequent and anxious, sensibility is much impaired. After a time these symptoms abate, and the second stage commences with an increase of heat and fever all over the body, redness of the face, dryness of the skin, thirst, pain in the head, throbbing temples, the tongue furred, the pulse becomes dry, hard, full, and regular ; when these have continued some time, a moisture breaks out on the forehead, which by degrees becomes a general sweat all over the body, the fever abates ; the water deposits a sediment ; the breathing and pulse are free, and the fit is over, but leaves the patient in a weak state.

Treatment.—In the cold stage, give warm diluent drinks, such as barley water, weak tea, or weak wine and water. Apply external warmth by means of extra clothing, hot bottles to the feet, mustard foot-baths, bags of heated bran, baked salt, &c. In this stage, an opiate is often beneficial ; give twenty-five to thirty drops of laudanum, with an equal quantity of ether, in a glass of water. During the hot stage, an opposite mode of treatment must be adopted. Sponge the surface with tepid or cold water, give cold diluent or iced drinks, and administer a full dose of laudanum. When the hot stage has subsided into the sweating stage, the action of the skin should be encouraged by tepid drinks ; and if the system is much exhausted, weak spirit and water in small quantities may be occasionally ventured on. During the intermissions, administer active aperients, as five grains calomel, with three grains of compound extract of colocynth ; followed by a mild purgative. Give bark to an extent as great as the stomach will bear, and combine with it wine and aromatics, accompanied by a generous but light diet, and moderate exercise. Quinine is a very powerful agent in ague; two or three grains of this medicine, administered twice or thrice daily, with such nourishing diet as the patient can take, will, in ordinary cases, put a speedy end to the disease.

The following tonic is recommended in Malaria:

Quinine	-	-	-	- 2 grains.
Arsenious Acid	-	-	-	- 1-30 grain
Powdered Capsicum	-	-	-	- 1 grain.
Extract of Taraxacum	-	-	-	- 5 grains.

Dr. Webster.

Should the disease become chronic use the following:

Pill of Carbonated Iron	-	-	-	- 1 drachm.
Arsenious Acid	-	-	-	- 1 grain.

Divide in 20 pills.
Take one three times daily.

Dr. Bartholow.

Bilious or Remittent Fever. When a fever is accompanied with a frequent or copious evacuation of bile, either by vomit or stool, the fever is denominated bilious, most frequent in the country at the latter end of summer or beginning of autumn.

Causes.—Exposure to damp or night air; frequently from intemperance when the body is disordered from cold or exposure, or similar to ague.

Symptoms.—Frequent flushings and shiverings, with vomiting, bilious phlegm, and sometimes purging, same as bilious diarrhœa.

Treatment.—Cleanse the stomach with the following: Emetic tartar, one grain; powdered ipecacuenha, fifteen grains; water, three tablespoonsful; mix and take; drinking warm camomile tea till it operates; and the bowels with this: Epsom salts, six drachms; glauber salts, three drachms; infusion of senna, seven ounces; tincture of jalap, half an ounce; compound tincture of cardamoms, one ounce; mix, and take two tablespoonsful every four hours, till it operates freely. Then take for a day or two the following: Subcarbonate of potash, four drachms; purified nitre, one drachm; syrup of saffron, six drachms; camphor mixture, twelve ounces; mix, take two tablespoonsful every four hours, with one of the following powders each time in the dose: citric acid or tartaric acid, half an ounce; divide into twelve powders, mix in the draught, and drink whilst effervescing.

When the fever has subsided take for a week or two the following pills: Sulphate of quinine, two drachms; extract of gentian, three drachms; mix well; divide into sixty pills, and take one every four hours. Then use the following excellent drink: Take well-crushed pale malt, three lbs.; dried wormwood, dried century, dried hoarhound, dried buckbean, dried betony, dried camomile, dried ground ivy, of each one ounce (but if fresh two ounces), gentian root, sliced, one ounce ; Virginia snake root, sliced, one ounce ; infuse all in two gallons of hot water, in a warm place, two hours, then boil together fifteen minutes, then strain off the herbs, etc., squeeze as dry as possible, put in two pounds of sugar, and boil again ten minutes ; when cool enough, put in some fresh yeast; work it well for two days, then bottle in sound bottles, putting two table-spoonsful of brandy to each quart. This is remarkably good for weak-ness, etc. Take three tablespoonsful three times a day, with a teaspoon-ful of the compound tincture of bark in each dose.

Rheumatism. Several distinct affections are popularly included under this term, rheumatism. First, an acute inflammatory affection of the joints, called in medicine *acute articular* rheumatism. Second, the disease, or perhaps series of diseases, called *chronic* rheumatism. And third, *muscular* rheumatism. Indeed this term is popularly applied with considerable license to almost any painful affection in which there are no local signs of disease.

Acute Articular Rheumatism. This disease affects the joints, as the name implies. It is indeed an inflammation of a smooth mem-brane which lines the joints, called the *synovial* membrane. At the same time it may extend to other parts of the body containing this same mem-brane, especially the heart. So long as the disease is confined to the joints it is not immediately dangerous ; the peril of life consists in the possibility that the inflammation may extend to the heart, in which case there often results serious difficulty, which may cause immediate death, or may result in permanent disease of the heart. Perhaps the majority of cases of so-called *organic* heart-disease originate in attacks of acute rheumatism.

Symptoms.—Acute rheumatism usually begins with a sudden attack, which may occur in the night. In some cases the manifestation of local difficulty—that is, pain in the joints—may be preceded for a few hours or days by more or less fever. In most instances, however, the fever and the local pain begin at about the same time ; it may indeed happen that soreness in the joints is felt for some time before the fever begins.

The commencement of the disease consists in a painful swelling in one or more joints, the skin around which is red and tender, the inten-sity of the pain varies, but in most cases it is quite severe, and becomes agonizing upon movements of the inflamed joints. Perfect quiet is therefore requisite as one of the essentials for diminishing pain. Pres-sure, too, over the joints is extremely painful, so that even the weight of the bed clothes is at times unsupportable. Those joints which are not covered by the muscles—the knee, wrist, elbow and ankle for example—exhibit considerable swelling ; while others, such as the shoulder and hip, may be equally affected and equally painful, but show only slight swelling. In some cases several joints become inflamed at the same time, or in rapid succession, in others the inflammation usually remains limited to one joint for some time before spreading to others; it rarely happens that the inflammation is permanently limited to a single joint, since several joints are successively attacked, and various ones may be suffering from the inflammation at the same time; in severe cases it seems as if nearly all the joints of the body were invaded by the disease during its course. It is a singular feature in the disease that the corresponding joints on the two sides of the body, both knees or both elbows for instance, are simultaneously affected. The relative liability of the different joints to the disease appears to be, first the knee, then the ankle, wrist, shoulder, elbow, hip and fingers.

Acute rheumatism is always associated with more or less fever, and is hence often called *rheumatic* fever. The appetite is impaired or lost, there is great thirst, the tongue is thickly coated, the bowels usually constipated. A special feature is the profuse sweating which occurs, especially at night; the perspiration evolves a sour odor. The mind is usually not impaired, unless the membranes of the brain become involved in the inflammation. The patient's strength is usually well preserved, his chief suffering consisting in the pain in the joints.

Although the local inflammation in the joints may result in some permanent stiffness or deformity, yet the chief danger from the disease occurs, as has been stated, from the possible complication in the heart, for the heart is lined with a membrane quite similar to that of the joints, and is covered over with another such membrane; either one or both of these may become the seat of the inflammatory process, a complication which may occur at any time in the course of the disease, though most frequently in its first half.

Another unpleasant feature in the disease is its liability to recommence when apparently about concluded; that is to say, the patient will sometimes, after weeks of illness, become quite free from fever and pain, the joints are no longer swollen nor tender, and complete recovery seems to be at hand, when suddenly the disease begins again affecting perhaps the same joints as before, and manifesting the same intensity.

Treatment.—Until a few years ago, the treatment of acute rheumatism was not entirely satisfactory, as was proven by the fact that many methods were in use. The best results have been obtained by the use of *alkalies*, with or without *colchicum*. A formula frequently used was the following:

> Carbonate of potassium, ⎱ Each two and a-half drachms.
> Nitrate of potassium, ⎰
> Water, eight ounces.

Dissolve and take a tablespoonful three times a day.

Much value has been obtained also from the use of lemon juice in water, say a tablespoonful every three hours. Yet, since 1876, there has been but little resort to these measures, because means have been found by which the disease can be promptly and effectually checked. Under the use of the alkalies and lemon-juice, the patient was usually ill for two or three weeks at least, and ran the risk of complications in the heart, which might prolong the disease indefinitely. At present, however, we are enabled to cut short acute rheumatism usually within three days, sometimes within twenty-four hours ; and not the least valuable feature of this treatment is the avoidance of the heart complications, which often prove the most serious feature of the entire illness. The measure employed for this purpose is the use of *salicylic* acid, or some of its compounds. The best form for general use will be the compound of the acid known as the *salicylate of sodium*, which is less disagreeable and more easily administered than the acid itself. Perhaps the best way to take it is in powder, ten grains of which may be taken every two hours until six doses have been swallowed. It may then be desirable to discontinue the drug for six hours. If, at the end of this time, the symptoms of the disease have not materially subsided, the powder may be administered in the same way and quantity for another twelve hours. In the majority of instances the remedy works like a charm, especially if it be administered early in the disease, before complications have arisen in other structures than the joints. The fever subsides, the joints are less sore and not at all painful, the appetite returns, and not infrequently the patient who a day previously was writhing in agony upon the slightest movement, flushed and feverish, arises from his bed and walks without pain. It must be said that this result cannot always be depended upon. If the case has already lasted one or two weeks, the drug does not always act so promptly nor so efficiently, though even then it is usually the best treatment that can be employed. Then again, there are cases in which the disease is not very acute nor painful, in which the swelling of the joints is moderate and the fever slight. These cases are apt to be especially obstinate, and to resist the action of salicylic acid. It is impossible to say in advance which of the cases will yield, and which will resist this drug, although one may feel sure in the cases in which the fever is high, and be somewhat doubtful in regard to those in which the fever is very moderate. Yet, in every case we should begin the treatment—and the earlier the better—with the salicylic acid. If this drug be efficient in controlling the case, the fact will be evident within three days, at the outside, in the rapid diminution of the pain, swelling and fever. If, at the end of three days, there be no evidence of improvement, it will be wise to resort to the alkalies, as in the formula above given, and the lemon juice.

Local treatment may also be used for the swollen joints ; these may be enveloped in flannel or cotton wool, which may be surrounded with oiled silk.   Or the swollen joints may be wrapped up in cloth saturated with chloroform liniment, as it is obtained in the drug store.   Much benefit is derived in some cases by gentle friction of the joints with the dry hand or with chloroform liniment.   The choice of measures may be left largely to the selection of the patient, since some individuals will prefer one and some another of these local applications.   In employing the friction, the pressure must be at first, of course, extremely light, to avoid giving pain ; though many times the force employed can be gradually increased with comfort to the patient until the attendant may use as much pressure as he can conveniently employ.   A method which has been recently much employed consists in the application of fly blisters to the affected joints permitting them to remain until some blistering occurs.   This plan doubtless relieves temporarily the pain in the particular joint which is blistered, but seems to have no effect upon the progress or course of the disease.   On the whole it is an undesirable measure, since the blisters may subsequently give trouble.

The following is much used in Bellevue Hospital, New York.

Salicylic Acid,	-	-	-	- 160 grains.
Acetate of Potash,	-	-	-	- 320 grains.
Glycerine,	-	-	-	- 1 ounce.
Water, to make	-	-	-	- 4 ounces.

Take a teaspoonful every 3 or 4 hours.

Chronic Rheumatism.   This name should, according to all medical usage, represent a continuation of an acute rheumatism in a less violent and painful form, and such cases are actually found under the name chronic rheumatism.   Yet this name, as ordinarily employed, designates several affections all of which are characterized by pains in the joints or in the muscles, which have a tendency to persist indefinitely.   There is a form of chronic rheumatism which affects the patient like the acute disease, except that the symptoms are less marked : there may be no fever, the pain and soreness are less intense, the tenderness on pressure is comparatively slight, and the swelling of the joints may be scarcely noticeable.   As in the acute variety, various joints are affected successively ; the disease may finally become concentrated and remain fixed in a single joint.   In this disease there is but little disturbance of the general health, insufficient, indeed, to disturb the patient's avocation.   Yet there are instances in which movements of the affected part cause considerable pain, and patients may be even confined to the bed.   After long continuance of the disease the affected joints may present irregular enlargements and stiffenings, while the muscles of these limbs become small from lack of use.

In many cases of acute rheumatism the severity of the pain varies extremely with the weather ; so that such individuals are usually able to foretell, by a few hours, the occurrence of cold and moist weather.   There is a variety of rheumatism, so-called, in which the pain is felt chiefly along the leg bones, the "shins," and occurs especially at night. This affection is often due to a syphilitic taint, and will be discussed in speaking of syphilis.

Treatment.—One of the most important features of treatment of chronic rheumatism, is care in wearing flannel next to the skin throughout the year. The administration of drugs is by no means certain to produce beneficial results. Some cases are materially benefited by the regular employment of the hot air, or hot vapor bath, the Turkish bath, etc. The fact is, that the treatment of each case of chronic rheumatism is largely an experiment which can be successfully accomplished after considerable time has been spent in trials of drugs and remedial measures. Among the medicines which are most frequently useful, are the iodide of potassium, guiac, and cod liver oil. The following formula may be given :

Iodide of potassium,	-	-	-	Five drachms.
Tincture of guiac,	-	-	-	Two ounces.
Water,	-	-	-	Two ounces.

Mix, and take a teaspoonful four times a day.

Other cases will be benefited by using colchicum with the alkalies. An example of such mixture is the following :

Wine of colchicum root,	-	-	-	One drachm
Bicarbonate of potassium.	-	-	-	Three drachms.
Rochelle salts,	-	-	-	Three drachms.
Peppermint water,	-	-	-	Four ounces.

Take a tablespoonful three times a day.

Muscular Rheumatism. Although this affection is designated rheumatism, there is every reason for believing that its cause is quite different from that of the disease just described. It seems to be of the same nature as neuralgia, and might properly be so described, though it is convenient to follow the usual designation as already given.

Symptoms.— The disease is usually developed gradually. A dull pain is felt in certain muscles, gradually increasing until it becomes quite severe. The pain is usually increased upon movement, sometimes becoming convulsive and cramp-like, causing the patient to groan, or even cry aloud. These movements, and the consequent pain,, may occur during sleep and awaken the patient. The muscles thus affected are somewhat tender upon pressure, but present no other signs of disease, such as swelling or redness. The constitutional condition is otherwise good, and no interference occurs in the bodily functions except such as are consequent upon the painful movements.

The duration of this disease may vary from a day to a week or may last, in less severe form, indefinitely. The muscles most apt to be affected are those of the forehead and of the temples. Another familiar example is in the neck, resulting in what is popularly termed " wry-neck." The pain may also be located in the muscles of the back and loins, constituting the affection known as *lumbago* It occurs not infrequently in the muscles of the chest, where it may for days render deep breathing impossible without pain.

Wherever it may occur, the characteristic feature of muscular rheumatism is the occurrence of cramp-like pain, aggravated by the movement of the part. In this particular it is distinguished from neuralgia, which is a constant pain, affecting perhaps the same localities. This form of rheumatism is undoubtedly promoted by exposure to cold, and seems often to result from getting the feet wet. The treatment should consist in improving the general condition by such measures as will invigorate the health. In the acute cases benefit can also be derived from the use of liniments and of gentle friction.

Gout. Gout is closely related to articular rheumatism ; indeed, by some the two diseases are regarded as essentially identical. However, they are so distinct in their symptoms and course as to be generally recognized and described as distinct diseases. They may be regarded as allied, just as typhoid and typhus fever are allied. Gout occurs both as an acute and a chronic disease.

Symptoms.—In describing the symptoms, it becomes necessary to distinguish between the transient acute cases, and those of long duration, called chronic. The acute attack usually occurs suddenly, often during the night. In many cases the patient has retired in his usual health, though in others there may be certain premonitory symptoms, such as pain in the region of the heart, belching of gas from the stomach, and mental depression or irritability. These symptoms so frequently precede an attack of gout that many who have already experienced one attack, recognize in the symptoms the onset of another. The attack itself begins by extreme pain in one of the joints, usually that between the great toe and the foot This pain is variously described by patients as like that of the gnawing of an animal, the dislocation of the bone, or like tearing the flesh with pincers. A feeling of throbbing accompanies the pain, and there is usually considerable fever, sometimes preceded by a chill. These features continue for several hours before subsiding ; at the end of this time the patient is reasonably comfortable, and usually falls asleep.

This attack is usually followed on succeeding nights by repetitions of a similar nature ; either during these repetitions, or perhaps during the first attack, the painful joint becomes swollen, reddened and exquisitely tender. The local affection may be limited to the joint originally affected, but often attacks also the corresponding joint of the other foot, or spreads to the instep and hollow of the foot. Indeed, in individuals who have repeatedly suffered from this disease, numerous joints in different parts of the body, even of the hands, may become similarly affected. In many cases the pain lasts but a few hours, and returns again on the following night. There are instances, however, in which the pain merely decreases in intensity during the day without entirely ceasing. Cramp may affect the muscles of the legs, the thigh, and even other parts of the body, though this is a somewhat unusual symptom. These attacks may return nightly for a week or for several weeks. These long continued attacks are usually less severe than the brief ones; the result may be entire recovery or continuation as the chronic form of the disease. Chronic gout may be the result of a series of acute attacks of gout, or may begin in a very mild manner. In either case the symptoms which characterize the acute affection are present in but slight degree. The disorder is rather a constitutional one, that is to say, the symptoms manifested in the foot are so slight that attention is directed to the constitutional derangement. There are usually disorders of indigestion—dyspepsia, pain in the stomach, derangement of the bowels. In many cases these derangements have caused serious impairment of the general health.

Sooner or later there occurs in the chronic cases of gout a condition quite characteristic of this affection : masses resembling chalk are formed in and around the small joints, and, indeed, some of the large ones ; the joint of the great toe is especially apt to exhibit this appearance. As a result, the joints become deformed, stiffened and even incapable of any movement. Sometimes collections of this chalk-like matter can be felt just under the skin, and if opened this matter can be pressed out. Sometimes matter forms around these masses and makes its way through the skin, resulting in the formation of openings, called *fistulæ*. The result of all these changes is that both hands and feet become strangely deformed, presenting appearances which have been likened to a bunch of parsnips. Similar formations of chalk-like matter have been known to occur in various parts of the skin, on the eyelids and in the ear.

There has long been an impression among medical men, which has become a popular belief among the people, that gout may be manifested, not merely in the joints, but also in the internal organs, such as the stomach, heart and brain. Disorders which affect these organs during an attack of gout are supposed to be due to the gouty influence ; the gout has "struck in." This expression is particularly applicable to those attacks of gout during which the affection of the joints suddenly ceases, and symptoms indicating some derangement of some internal organ follow. There is, doubtless, some connection between the local attack and the affection of the internal organs. This is amply shown in the fact that remedies addressed to the gout relieve these symptoms.

Those who have long been subject to gout usually present certain disorders which constitute the dangers of the disease ; for the local affection of the feet, or of the hands, however painful and unsightly, contains no element of danger to the patient. One of the most serious complications occurring in gouty persons is a disordered action of the heart. This disorder occasions not only palpitation, shortness of breath, sense of suffocation and pain in the region of the heart, but may even cause sudden death, in consequence of the sudden stoppage of the heart's action. Various affections of the lungs also are popularly, and doubtless properly, attributed to the gout. Among these are persistent cough, bronchitis and asthma. The explanation of many an intractable case of asthma has been found in the gouty condition of the patient, and relief from the asthma has followed remedies addressed to the gout. So, too, various pains designated as neuralgia,—particularly neuralgia of the face and of the hip (sciatica)—periodical headache and even hysteria, seem to be promoted by the gouty condition of the patient.

Another most serious complication of the gout is a disease of the kidney, one form of the so-called "Bright's disease." This disease is so common among those affected with gout that the small, hard kidneys characteristic of this affection are called "gouty kidneys." This form of Bright's disease is not indicated by the symptoms characteristic of the ordinary acute Bright's disease; there is no dropsy, no fever, in fact none of the usual signs. The most characteristic feature, at least the one most readily recognized by the non-professional observer, is the fact that the patient passes an unusually large quantity of urine. These patients are almost alw: · ··nced in years.

Treatment.—One of ⸱ a. t important measures in the treatment of gout consists in the regulation of the diet ; the avoidance of excesses at the table and in the use of liquors is an evident necessity. Among the remedies used during the attack, reliance is chiefly placed upon *colchicum* and *alkalies*. Instead of the latter, various mineral waters have been highly recommended by different authors : perhaps the most noted and universally used of these is the so-called "Lithia Water," which may be obtained at the druggist's. This is in reality, not a natural mineral water, but an artificial solution made by dissolving the citrate of lithia. If this be not procurable, a solution of magnesia and colchicum may be employed as follows :

Wine of colchicum root,	- - -	One drachm.
Magnesia,	- - - - -	One drachm.
Peppermint water,	- - -	Four ounces.

Mix ; take a tablespoonful three times a day.

It may be necessary also to protect the inflamed joint ; this may be accomplished by the means prescribed in treating of rheumatism—by enveloping the joint in cotton, or by wrapping it with cloths which have been saturated with the tincture of belladonna or with the tincture of aconite. The chloroform liniment will also be found advantageous for

F

local use in this affection.   During the paroxysms it may be necessary
to relieve pain by the use of opiates; for this purpose twenty drops of
laudanum, or ten grains of Dover's powder may be employed.   If the
opiate cause unpleasant effects, such as nausea and headache, we may
substitute ten drops of the tincture of belladonna, which may be repeated
in two or three hours.   In the intervals between the attacks the diet
must also be restricted—should embrace no wines or malt liquors, and
but little, if any, meat; a simple, unstimulating diet of milk, eggs, fish
and vegetables will materially assist in preventing the return of the attack.   So, too, the remedies useful in treating the paroxysm—especially
the colchicum—may be continued during the intervals between the
attacks.   If the indications, such as dyspeptic symptoms, render the patient suspicious of an approaching attack, it will be well to employ the
colchicum and the alkalies in advance.   For there is reason to believe
that impending attacks can be in this . ay warded off.

The local affections of the joints may sometimes require attention
after the severity of the paroxysm has subsided, for, as already said, the
joints are often rendered permanently stiff and deformed by the deposit
in and around them of chalk-like masses.   This stiffness can be sometimes partially relieved by systematic gentle friction, and by the arrangement of the shoes so as to afford the parts ample room.   The swelling
can be sometimes diminished before the deposit of chalk is apparent by
the use of repeated blisters, not allowed to remain long enough to produce excessive blistering.   After this chalk-like matter has been de-
posited in abundance no measures can be successfully applied for its
removal.

With reference to the use of mineral waters, Dr. Garrod, the most
eminent authority on this subject, says : "They should be altogether
prohibited when there is considerable structural disease in any important
organs, especially in the heart or kidneys ; and even when the organic
mischief is slight, the greatest caution is necessary in their use.   They
should be avoided when acute attack is either present or threatening.

"The waters should be selected according to the nature of the case.
When the patient is robust and of a full habit, the alkaline-saline springs ;
when torpidity of the bowels predominates, the purgative waters ; when
there is a want of vascular action, the saline water ; when the skin is inactive, the sulphur springs ; lastly, when debility prevails, then the more
simple warm waters should be chosen.   In all cases the use of the
waters should be cautiously commenced, and care taken not to oppress
the stomach by giving too much liquid, nor to induce debility or any
other injurious effects by allowing a too long sojourn in the bath."

Sciatica.   This is a painful rheumatic affection, confined to the hip-
joint and lower extremities, and affecting the large nerve (called the sci-
atic nerve) of the leg.

Treatment.—Apply a small blister on the spine at the bottom of the loins, and when it is removed sprinkle the surface with one-third of a grain of acetate of morphia, mixed in a little starch-powder. Or, apply to the part affected a bran poultice, to be followed twice or three times a day by an embrocation composed of one part of turpentine, and two parts of soap and opium liniment. A couple of drachms of this should be rubbed in for ten minutes at a time. Meanwhile, cleanse the bowels by a purgative, and if there is no tendency to fever, take drachm doses of carbonate of iron, three times in twenty-four hours. When the pain is very severe, accompanied with general fever, leeches should be applied, and cooling purgatives taken. It will also be advisable to employ the hot bath at a temperature of 105 degrees, and to remain in it from fifteen to twenty-five minutes. This should be repeated two or three times a week.

CHAPTER VII.

Falling Sickness. This is known as epilepsy. It is an attack of convulsions, with unconsciousness, and which recurs from time to time, without any regularity.

Cause.—This may, and often *does*, result from parentage, intemperance, or excesses of any kind ; injuries to the head, or fright.

Symptoms.—In many instances an attack is ushered in by a headache, vertigo, a sudden feeling of terror, or what is called the aura ; a feeling as though something was creeping up a limb, or a wind blowing on it. This extends slowly upward, and then the patient screams, starts, and falls suddenly, convulsed, foaming, grinding the teeth ; the face is flushed, the eyes roll wildly, respiration is performed with difficulty, and there may be vomiting and involuntary passages This lasts usually five or ten minutes, the fit passes off, the patient goes into a sleep, or a drowsy state, or may arise bewildered, and stagger on his way, with more or less headache, weakness, or even delirium and frenzy, which impels to attack those around. A fresh attack may occur in a few minutes, or hours ; or even months may elapse before a renewal.

Treatment.—During the fit but little can be usefully attempted. The patient should be cared for, and placed so that no injury may result from striking his limbs or head as he tosses about. The clothing should be loosened, and he should be allowed plenty of fresh air. If prolonged, ether, chloroform, or nitrate of amyl may be given by inhalation, to arrest its continuance. The special effort will be to prevent a recurrence, and for this purpose, many and diverse remedies have been proposed. Perhaps the best results have been obtained from the valerianate of zinc, one grain, two or three times a day ; the bromide of potassium, in full doses, and continued for a long time, say 15 to 30 grains, three times a day, for months.

How Prevented.—This indication will be met by the careful avoidance of the causes ; by living temperately, yet using nutritious diet, keeping away from all excesses, avoiding all bad habits, and taking regular exercise in the open air. Tobacco, especially, should be avoided, as highly detrimental. The bowels should be kept as regular as possible.

The following is the famous Brown-Séquard mixture which, if taken faithfully for six or nine months, is most successful in this disease :

Bromide of Soda,	- - -	3 drachms.
Bromide of Potash,	- - -	3 drachms.
Bromide of Ammonia,	- -	3 drachms.
Iodide of Potash,	- - -	1½ drachms.
Iodide of Ammonia,	- - -	1½ drachms.
Carbonate of Ammonia,	- -	1 drachm.
Tincture of Calumba,	- -	1½ ounces.
Water to make 8 ounces.		

The full dose is 1½ teaspoonfuls before each meal and 3 teaspoonfuls at bedtime.

Dr. Brown-Séquard.

Hysterics (*Hysteria*).—This is a nervous affection seen sometimes in males, but usually in females suffering from uterine irregularities. Single women, and the married who do not bear children, are the most subject to it, although it sometimes occurs at the early period of pregnancy and immediately after childbirth. Persons of studious and sedentary habits, and of scrofulous and weakly constitutions, are especially likely to be the subjects of hysteria, as are indolent and plethoric persons, and those debilitated by disease or excesses of any kind. It may be excited by excessive evacuations, suppression of the natural secretions, strong mental emotions, or sympathy with others so affected. It is a curious circumstance connected with this affection that it simulates almost every disease to which humanity is liable.

Symptoms.—An attack generally comes on with a sensation of choking. It seems as if a ball were rising in the throat, and threatening to stop the passage of the air ; then the trunk and limbs become convulsed, so much so that an apparently feeble woman will require three or four strong persons to restrain her from injuring herself ; then follows the hysterical sobbing and crying, with alternate fits of laughter. Generally the head is thrown back, the face is flushed, the eyelids closed and tremuluus ; the nostrils distended, and the mouth firmly shut. There is a strong movement in the   t, which is projected forward, and a wild throwing about of the arms and hands, with sometimes a tearing of the hair, rending of the clothes, catching at the throat, and attempts to bite those who impose a necessary restraint.

A fit of hysteria may last for a few minutes only, or for several hours, or even days ; persons have died under such an affliction. It may generally be distinguished from epilepsy by the absence of foaming at the mouth, which is nearly always present in that disease, and also by the peculiar twinkling of the eyelids, which is a distinguishing symptom of great value, and a sign of safety. In epilepsy, too, there is complete insensibility, not so in hysteria ; the patient retains partial consciousness ;

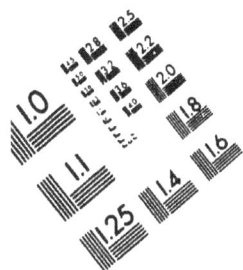

IMAGE EVALUATION
TEST TARGET (MT-3)

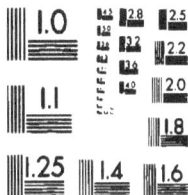

hence it behoves those about her to be cautious what they say.   If any
remedies are suggested of which she is likely to have a dread, her recovery may be greatly retarded thereby.   In epilepsy there is laborious or
suspended respiration, a dark livid complexion, a protruding and bleeding tongue ; rolling or staring and projected eyeballs, and a frightful
expression of the countenance.   Not so in hysteria ; the cheeks are
usually red, and the eyes, if not hidden by the closed eyelids, are bright
and at rest ; the sobbing, sighing, short cries, and laughter, too, are
characteristic of the latter affection.   We point out these distinctions
that no unnecessary alarm may be felt during a fit of hysteria, which is
seldom attended with ultimate danger either to mind or body, although
the symptoms are sufficiently distressing to cause some anxiety.

Treatment.—First, prevent the patient, if violent, from injuring herself.
Confine her hands, by wrapping tightly round her a sheet or blanket.
The dress should be loosened, especially round the throat ; and the face
freely exposed to fresh air, and both that and the head well washed with
cold water.   If she can and will swallow, an ounce of camphor-mixture,
with a teaspoonful of ether, sal volatile, tincture of assafœtida, or valerian,
may be administered.   Strong liquid ammonia may be applied to the
nostrils ; and if the fit is of long duration, an enema injected, cohsisting
of spirits of turpentine, castor-oil, and tincture of assafœtida, of each half
an ounce, in half a pint of gruel.   What is required is a strong stimulus
to the nervous system ; therefore, dashing cold water on the face, and
hot applications to the spine, are likely to be of service.   Carlisle recommends that a polished piece of steel, held in boiling water for a minute
or two, be passed down the back over a silk handkerchief.   This has
been found to prevent the recurrence of the paroxysm, which has before
been periodic,—by which it would seem that the patient has some power
of controlling the symptoms, when a sufficiently strong stimulus is applied, to enable or induce her to exercise it.

The patient's mind, during the intermission of the attack, should be
kept as tranquil as possible, and a tendency to all irregular habits or excesses held in check.   If plethoric, there should be spare diet ; if scrofulous and weakly, good nourishing food and tonic medicines particularly
some form of iron, the shower-bath, regular exercise, and cheerful company.   Antispasmodics, and remedies which have a gently stimulating
effect, will frequently relieve the sleeplessness complained of by hysterical patients better than opiates and other narcotics.   In such cases Dr.
Graves recommends pills composed of a grain of musk and two or three
grains of assafœtida, to be taken two or three times a day.

St. Vitus' Dance—(*Chorea*).   The disease known as St. Vitus'
dance, is characterized by irregular and aimless contractions of different
muscles, without the agency of the will, in fact often in opposition to the
desire of the individual.   The muscles first affected are commonly located either in the arm and hand or in the face.   The irregularity of

movement may remain limited to a single extremity for a long time, or may rapidly progress so as to involve all of the extremities and even the trunk itself. The appearances presented by the subject of this affection are most ludicrous ; the first impression derived by the spectator is that it is a voluntary performance designed for the amusement of the by-standers.

The constant activity of the muscles causes exhaustion, so that the patient may be unable to perform the acts necessary to supply his bodily wants. He may even be unable to walk, not because his legs are too weak, but simply because they do not obey his will. In most cases the contortions become more violent as the patient's efforts to control his muscles become more vigorous.

St. Vitus' dance may occur at any period of life, but is most frequent during the years preceding puberty ; that is from ten to fifteen. Girls are more frequently affected than boys, in the proportion of three to one.

The affection lasts ordinarily from two to four months ; it may terminate earlier than this period, or may, on the other hand, last for many years.

Treatment.—In the majority of cases chorea terminates in recovering spontaneously. Hence it has happened that a vast number of remedies have acquired a reputation as specific cures for the disease, for the patient recovers no matter what remedy, provided it be harmless, be administered to him ; hence every remedy which is thus used seems to cure the disease. It would be quite useless to name all the medicines which have been given for the treatment of St. Vitus' dance with apparent success. Those most frequently used are the bromide of potassium and the valerianate of zinc. In most cases it becomes necessary to administer tonic medicines, for the children are evidently in need of some blood-making remedies. They are pallid and become more so as the disease progresses. It is therefore desirable to administer iron and cod-liver oil. For this purpose one of the following prescriptions may be given :

Syrup of the iodide of iron, - - - Two ounces.

Take ten to fifteen drops in water after meals.

Citrate of iron and quinine, - - One drachm.
Cod liver oil, - - - -
Glycerine, - - - - - } Each two ounces.

Mix and take a teaspoonful after meals. This dose may be gradually increased if the stomach bear the oil well.

Arsenic, - - - - - One-quarter of a grain.
Reduced iron, - - - - Ten grains.
Extract of nux vomica, - - Two grains and a half.
Sulphate of quinine, - - - Ten grains.
Mix and make twenty pills. Take one before meals.

In other cases Fowler's solution has been used with advantage, two or three drops being given before meals three times a day, and the dose gradually increased.

If the child is compelled to remain at home, the cold bath should be employed every morning; if possible a course of sea bathing will be found very beneficial.

Another point is that the child should receive the sympathy and attention of parents and friends, and should be protected from the thoughtless ridicule which is naturally excited among children by the grotesque movements that cannot be controlled.

The following are the prescriptions of two very eminent men for this disease :

> Bromide of zinc, -   - One drachm.
> Simple syrup, -   -   -   -   - One ounce.

Take ten drops three times daily, increased as rapidly as the stomach will bear it.

Dr. William A. Hammond.

> Valerianate of zinc, -   -   -   -   - Two scruples.
> Sulphate of cinchona, -   -   -   - One scruple.

Mix and make into twenty pills.
Take one pill three times daily.

Dr. Da Costa.

Neuralgia. This disease is truly a nerve pain, and may affect any portion of the body. From its location it derives a variety of names, as *hemicrania*, when it affects one side or half of the head ; *tic douloureux*, when it attacks the face, sometimes called facial neuralgia ; *sciatica*, when it affects the great nerve of the hip ; *pleurodynia*, attacking the side, or pleura ; *gastrodynia*, the stomach, *angina pectoris*, the heart, etc.

Cause.—It is generally caused by a sudden exposure of the part to cold ; an injury of the nerve ; weakness, as want of red blood ; the actions of some poisons, as that of lead ; or by malarious influence, when it takes the form of intermittent neuralgia.

Symptoms.—This form of pain is not associated with an inflammation, but seems limited entirely to the particular nerve affected. The pain is acute, shooting, sometimes expressed as a stabbing pain, with great tenderness on pressure over the affected part. Particularly is this shown when the pressure is made with the points of the fingers, when the patient will shrink or cry out. But if the pressure be more firm, the pain diminishes or disappears entirely. The pain is generally intermittent, that is, not always recurring at the same time, as in the form of *intermittent neuralgia ;* but it comes and goes ; there are paroxysms of acute pain, and in the intervals even the tenderness on pressure may disapper. Again, certain movements increase it, as a fall, or sudden concussion, coughing, laughing, sneezing, the application of cold or heat to the seat of the pain.

It may, and often does, cause contraction of the muscles of the part, cramps, etc. No fever is ever present, though neighboring glands may become enlarged and painful.

Treatment.—Neuralgia is essentially *pain*, and as such is merely the symptom of a disease. In every case, therefore, treatment is to be preceded by an attempt to ascertain the seat of the difficulty. The promiscuous application of liniments and plasters to all parts of the body for pain is not a rational way of treating the disease.

In many cases neuralgia is easily curable. If the patient be living in a malarial district, it is quite probable that the pain is of a malarial origin, and that three grains of quinine, administered four times a day, will cure the neuralgia. If the patient be pale and bloodless, and evidently in poor health, tonic medicines are required; for this purpose the following prescription may be given:

> Tincture of the chloride of iron, - - One ounce.
> Sulphate of quinine, - - - - One drachm.
> Syrup of orange peel, - - - Half an ounce.
> Water to make four ounces.

Mix and take a teaspoonful in water before meals. Such individuals should of course have good food and plenty of air, sunshine, and exercise.

Neuralgia may be the result of some constitutional taint which has been inherited or acquired. Thus syphilis frequently causes intense pain, particularly in the legs and in the body; the various minerals, especially lead, may also cause severe neuralgia, as one of the symptoms of poisoning. In all these cases the treatment consists, in part, in the effort to remove the cause.

In every case the treatment must also aim at relieving the pain. For this purpose various measures have been employed, all of them with advantage in certain cases. The tincture of aconite may be rubbed upon the skin every hour (and this simple measure sometimes suffices to relieve the pain), or the following ointment may be used:

> Veratria, - - - - - - Fifteen grains.
> Pure Lard, - - - - - - One ounce.

Mix and apply to the skin. If the pain be severe, relief can be obtained immediately by the inhalation of chloroform; meanwhile a quarter of a grain of morphine may be given, either dropped dry upon the tongue or dissolved in a little water, or the following combination may be employed:

> Tincture of gelsemium, - - - }
> Tincture of belladonna, - - - } Each two ounces.

Take fifteen drops every two hours, increasing the dose gradually to thirty drops, if required; or the following may be found useful:

Chloroform, - - -	-	-	Four drachms.
Muriate of morphia, -	-	-	Five grains.
Ether, - - -	-	-	Two drachms.
Oil of peppermint, -	-	-	Eight drops.
Dilute hydrocyanic acid, -	-	Two drachms.
Tincture of capsicum,	-	-	Six drachms.
Gum arabic, -	-	-	Two drachms.
Water and molasses, -	-	-	To make five ounces.

Mix and take a teaspoonful every two hours.

In some cases of neuralgia relief has been obtained by the application of blisters along the course of the nerve, and in severe cases a grain of morphine may be applied to the raw surface after the blister is removed.

After other remedies have failed, relief can often be obtained by the simple application of cloths wrung out in hot water, or by a hot bath. Electricity, when properly applied, is a valuable agent in many cases of neuralgia, and seems, indeed, sometimes to exercise a curative influence.

Nearly all cases of obstinate neuralgia are benefited by the use of iron. In females particularly the following prescription can be employed with advantage :

Carbonate of iron,	-	-	-	-	Forty grains.
Sulphate of quinine,	-	-	-	-	Thirty grains.
Extract of belladonna,	-	-	-	-	Five grains.

Mix and make twenty pills. Take one before eating.

Many cases of obstinate neuralgia *in the face* are relieved by *croton chloral hydrate ;* five grains of the drug may be given in a teaspoonful of syrup, and repeated until five doses have been taken or relief afforded.

Headache. Headache is the symptom of numerous affections ; in most instances the cause of the difficulty is to be found not in the head, but in various organs of the body. If, for any reason, the different functions of the body are not properly performed, so that the waste materials are not carried off as they should be, there is apt to occur, among other symptoms, a headache. Thus in Bright's disease of the kidneys, headache is often a prominent symptom ; in habitual constipation of the bowels, the same sympton is apt to occur ; irregular menstruation may be also accompanied by it ; most of the acute diseases are ushered in by headache among other symptoms. In fact, almost any derangement of the body or its functions may be accompanied by a pain in the head.

There are, therefore, almost as many causes for headache as there are diseased conditions of the body ; it is impossible to refer to them all in detail, and we shall be content with describing two conditions accompanied with headache, which are especially common, and therefore especially important. These are *sick headache* and *nervous headache.*

By sick headache is understood that frequent form of pain in the head accompanied with nausea. **In many cases this sick headache can be** traced directly to a derangement of the digestive organs, and is then usually one of the symptons of *biliousness*

There is usually impairment of the appetite, an uneasy feeling in the region of the stomach, a bitter taste in the mouth ; the tongue becomes coated and the breath may be offensive. There is pain in front of the head especially, and a general indisposition for exertion.

In other cases sick headache appears to be a form of neuralgia ; it occurs without perceptible cause, and seems to run in families.

In some cases the attack occurs, as has been said, without any perceptible cause. At other times it is the direct result of excessive emotion, bodily fatigue, exhaustion, the consumption of indigestible food, exposure to cold and over-heating.

Treatment.—Until the cause and seat of the difficulty can be accurately located, attempts at treatment must be necessarily somewhat experimental in character. There are numerous remedies which have been used to relieve this affection, and it is quite certain that many individuals can be completely cured of the difficulty. Yet it is impossible to say in advance just what plan of treatment will be adapted to a particular case. We shall, therefore, mention several plans, which may be tried in succession.

In many instances, particularly those associated with biliousness, a mercurial laxative will secure relief. For this purpose, take :—

> Calomel, - - - - - - 5 grains.
> Bicarbonate of sodium, - - - 10 grains.

Where there is no evident disturbance of digestion to account for the difficulty, and where the individual is " nervous," the following prescription may be given :—

> Extract of guarana, - - - - 40 grains.
> Extract of cannabis indica, - - - 30 grains.
> Citrate of caffeine, - - - - 60 grains.

Mix, and make forty pills ; take one pill, and repeat the dose after two hours, if not relieved.

In many cases, thirty or forty grains of the bromide of potassium, taken in half a glass of water, will secure relief.

In other cases, three grains of the *monobromated camphor*, in the shape of a pill, will be efficient.

One to two teaspoonfuls of the *fluid extract of guarana* have occasionally relieved sick headache after other measures had failed.

One or two grains of the *citrate of caffeine* may be placed upon the tongue dry, and swallowed. This has proven efficient in many cases where the ordinary remedies had been used without success.

Lock-Jaw (*Tetanus*.—This is a spasmodic seizure of a dreadful and generally fatal character. By this disease, not only are the muscles of the jaws, but those also of the whole body, thrown more or less into spasm, often so violent as to break the teeth or bones.

Causes.—The cause of tetanus is frequently exposure to cold and damp, or it may be some local injury, such as a cut, puncture, or laceration. It more commonly results from either of these in warm climates, although intense cold alone has not unfrequently produced it. It often affects a large number of the wounded on a field of battle, who are exposed to the vicissitudes of the weather. Lock-jaw, which is produced by a wound, will sometimes show itself in four days; sometimes not for two or three weeks after the wound has been received.

Symptoms.—The most common form of this fearful malady is that in which the muscles of the neck and throat are chiefly affected. It generally comes on in a gradual manner: there is a slight stiffness in the back of the neck, which extends to the root of the tongue, causing great difficulty in swallowing; then the whole muscles of the face probably become implicated; there is soon tightness of the chest, and the spasmodic pain extends to the back; while the teeth become so closely and firmly set together that no food of any kind can pass them. If the spasm extends further than this, the muscles of the trunk, and lastly of the extremities, become involved, contracting and drawing the body to the side, or backward, as the case may be, so as to form an arch, resting on the head and heels. The suffering caused by the tetanic spasm is frightful to contemplate. The face is pale, the bones contracted, the skin covering the forehead wrinkled, the eyes fixed and prominent, sometimes suffused with tears, the nostrils dilated, the corners of the mouth drawn in, the teeth exposed, and the features fixed in a sort of grin. The breathing is performed with difficulty and anguish; there is great thirst, and the sufferings are greatly increased by attempts to swallow; the pulse is feeble and frequent; the skin is covered with perspiration; and yet, with all this torture, the intellect remains clear and unaffected. Death at length closes the scene, being due partly to suffocation, and partly to exhaustion.

Treatment.—It seems to be well established that tetanus can often be controlled and cured by the use of *calabar bean*, given in frequently repeated doses, if treatment be begun sufficiently early. Good results have also been obtained by the use of *curare*. But these remedies are too powerful and too dangerous to be used by any one except a physician. Other measures to be applied are opium, a grain of which may be given every two hours; chloral, twenty grains of which have been employed in the same way; bromide of potassium, and stimulants, especially brandy and whisky.

If the services of a physician cannot be at once procured, the following mode of treatment should be adopted until a medical man can arrive:

Ice should be applied to the spine, wrapped up in soft, thin cloths.

A grain of opium may be given in a tablespoonful of brandy or whisky, mixed with the same quantity of milk, every two hours.

The violence of the spasms may be controlled by permitting the patient to inhale chloroform.

The great difficulty attending success in the treatment of tetanus arises from the fact that the disease is not recognized until some time after its appearance, because the early symptoms of lock-jaw are not distinguished from the stiffness and soreness of the neck, such as occurs after simply exposure to cold. It should be remembered that if the patient has been wounded or injured in any way, if even his skin has been bruised, the possibility of tetanus must not be forgotten ; and the appearance of stiff ness in the jaws and of difficulty in swallowing must be regarded as the possible evidences of the onset of this disease.

Wakefulness. Known as insomnia, morbid vigilance.

Cause.—This symptom, for it cannot be regarded as a disease, marks the presence of some form of disease of the brain ; or is the result of strong mental excitement, or mental labor ; or is caused by excessive pain, or the inordinate use of articles, as strong tea, coffee, tobacco, etc.

Symptoms.—When not the result of acute pain, the patient finds himself, at the hour for retiring, wide awake, and is unable by any of his usual methods to compose himself to slumber. The history of the case will enable the physician to decide as to whether it is the result of excitement, over-work, the use of stimulants, etc., or a symptom of hidden brain affection. When no other cause can be assigned, the gravest fears may be felt lest it result in insanity, inflammation of the brain, or soft ening, etc.

Treatment.—The treatment will depend solely upon the cause. Whatever this may be, it must be removed, if removable. The patient should be made fully to understand the danger he is in, and to lend his entire concurrence in the efforts for his relief. Mental labor should be given up ; overwork of any kind must be abandoned : articles liable to keep up the trouble must be forbidden, such as tobacco, coffee and tea. If general debility be present, the system must be brought to a natural standard by tonics and a proper diet. If there are symptoms of a fulness of the brain, this must be relieved by cups, or leeches to the nape of the neck, the temples, etc., or a blister to the same spot, dressed, when requisite, with morphia. Occasionally, this fulness may be relieved by proper physical exercise, as a long walk, calisthenics, etc. All excitement of any form must be avoided just prior to the hour for sleep, hence, light exercise at this time will be b   as diverting the attention from business, etc. Instances are common where the patient suffers from an empty stomach, and a quantity of easily digested food will satisfy the craving and promote a sound, refreshing slumber. Again, cold to the head, or hot foot baths, with or without mustard, or a warm bath, will relieve the patient promptly. The patient should examine his surroundings, as to whether any cause exists, such as improper position in bed; the best is always where the head and shoulders are above the level. For the relief of this symptom, except where it is the result of pain, medicines should be employed with caution.   It is always best to obtain sleep with the mildest means first.   A glass of beer at bed-time, a hop pillow, or the preparations of hops, hyoscyamus, lactucarium, bromide of potassium or of sodium, will generally relieve, and preferably in the order given.   Some observers have found that opium or morphia, added to the bromides, appear to correct the unpleasant action of the opium, and aid its effects.   The best remedy of all is the chloral.   This should be employed in positive doses, say fifteen or twenty grains, repeated, when necessary, in one, two or three hours.   A variety of other remedies have been proposed, but those quoted will answer the purposes.

Dr. Brown-Sequard recommends the following :

Bromide of potash,	-	-	-	6 drachms.
Water,	-	-	-	5 ounces.

Take 3 teaspoonsful before dinner, and 4 at bedtime, with a little wine and water.

Sunstroke.   This is a common accident in the intense heat of American summer.   It does not always need the sun to bring it on ; a great heat in the shade may cause it.

Before it comes there is pain in the head, wandering of the thoughts, irritable temper, nervousness and general illfeeling.   "Sun-headache" and "sun-pain" show that there is unusual danger from this source.

How prevented.—During the hot weather all alcoholic drinks must be avoided.   A bath should be taken daily.   Abundant cool, but not ice-cold, water should be taken.   The head should be well protected by a tall straw hat, with a damp newspaper or sponge, a wet handkerchief or a handful of green leaves, in the crown.   When the heat is felt unpleasantly, exposure to it should cease.

Treatment.—The person attacked should at once be carried to a cool, airy spot, in the shadow of a wall, or to a large room in a house with a bare floor ; or, what is often better, if there is no sun, he should be placed in a back yard, on the pavement.

The clothing should be at once gently removed, and the patient placed on his back, with the head raised a couple of inches by a folded garment.   Then the entire body, particularly the head and chest, dashed with cold water in profusion.   While preparations are being made for this, a messenger should be despatched for a good supply of ice.   A large fragment should be placed in a towel, and struck a few times against the side of the house, to reduce it rapidly to small pieces. These pieces, mixed by the hand into a bucket of water, will promptly supply ice-water.   Two buckets can be used, each half full of the small ice, and as soon as the water of one is used for dashing against the patient, another will be ready for the same purpose. The ice-water must not be *sprinkled* over the person, but *dashed* against him in large bowlfuls, particularly against the head and chest. While one person makes the ice water, and another uses it, a third should, in the same manner, with a towel, break some ice in fragments not larger than almonds. A double handful, at least, of these bits should be placed in a thin, coarse towel, the ends gathered up and fastened with a string, as you would a pudding. Then holding to the tied portion of the collection of ice, the entire surface of the body should be rapidly rubbed.

These measures are to reduce the heat of the body. When the decline in the heat is noticed, the cold applications should be abandoned, the patient carefully removed to a dry spot, and the entire surface of the body dried off with towels. Should a tendency to a return of the high temperature be seen, as sometimes happens, even after consciousness is restored, it must be met by a renewal of the cold application.

Artificial respiration, until the natural returns, must be resorted to as soon as the heated condition of the body is overcome. The dashing of cold water over the chest and face is a useful means of encouraging a return of the life force.

This plan of treatment is only to be used in those cases of sunstroke where the skin is hot, face flushed, pulse strong and bounding. There is another form of sunstroke where the face is pale, skin cool and pulse feeble.

In these cases the clothing should be loosened, especially from the neck ; the patient's head should be kept low, the air allowed to circulate around the body ; half a tablespoonful of whisky or brandy may be administered every fifteen minutes until six doses have been taken ; hartshorn should be applied to the nostrils. If there be any vomiting, the whisky or brandy may be administered as an injection into the rectum. In this case, an ounce (two tablespoonsful) may be given for each dose.

It is highly important that such a patient be not moved nor agitated ; he should not, therefore, be taken home nor to a hospital, unless the distance be short, but should be treated at once at the nearest convenient place.

Whooping-cough. This affection is by many classed among nervous diseases, notwithstanding the apparently infectious nature of the complaint. The symptoms of whooping-cough are so familiar that no description is necessary.

Treatment.—The paroxysms can usually be shortened by the use of emetics, which not only provoke vomiting, but also loosen the phlegm. For this purpose, we may give a full dose of ipecac or squills.

Aside from this measure, but little treatment seems beneficial. In fact the mild cases do best without medicine, if care be taken to avoid exposure to the cold and to wrap the body well in flannel underclothing.

To cut short the disease, a great many remedies have been administered ; the fact that these remedies are so numerous, indicates that no
one of them can be relied upon for all cases. Good results have been
reported from the use of belladonna. The following mixture will be
found of service :

Extract of belladonna,	-	-	-	- One grain.
Mucilage of gum arabic,	-	-	-	- Two ounces.

Give twenty or thirty drops of this every three hours. We may also
use to advantage the following prescription :

Fluid extract of hyoscyamus,	-	-	Half a drachm.
Orange flower water,	-	-	- Four ounces.

Mix and give a tablespoonful every three hours. This dose is suited
to a child of 12 years, and must be correspondingly reduced for a younger
child.

Apoplexy. This is the deprivation of life or motion by a sudden
stroke or blow ; it is one of the most awful and appaling modes of sudden death ; in an instant a healthful and vigorous man is smitten down
—one who has exhibited no signs of decay or disease—who has perhaps
received no premonitory warning, lies before us motionless and stark.

Apoplexy may be either cerebral— proceeding from congestion or rupture of the brain—or pulmonary—proceeding from hemorrhage into the
parenchyma of the lungs. The first is the more common form, and this
may be spoken of under four heads ; first, when it is sudden and violent
at once ; second, when it is comparatively slight at the commencement,
and progressively increases in severity ; third, when it commences in
apoplexy and terminates in paralysis ; fourth, when it commences in the
latter, and terminates in the former.

Causes.—The causes of apoplexy are either predisposing or exciting ;
among the first may be named, first : Sex—men are more liable to it
than women, because they are more subject to its exciting causes, of which
which we shall presently speak ; second : Age—it is very rare in childhood, rare also in youth, most common between the ages of forty and
seventy—rare much beyond the latter age ; third : Bodily Conformation
—the man of sanguine and plethoric temperament, with large head, short
neck, and full chest, is most liable to its attack, although one of the opposite state and condition of system is sometimes smitten down with it ;
fourth : Mode of Life—persons of sedentary habits, who live luxuriously,
are its frequent victims ; fifth : Suppression of evacuations or eruptions—
as the piles, perspiration, healing of the seton or wound ; sixth : Mental
Anxiety—such as a long continuance of harassing fears, business per-
plexities, grief, or any violent emotion or passions. All these are predisposing causes of apoplexy, to which it has been said that the studious
are more liable than others ; but this is an error, as the history of lawyers,

FIG. 21.

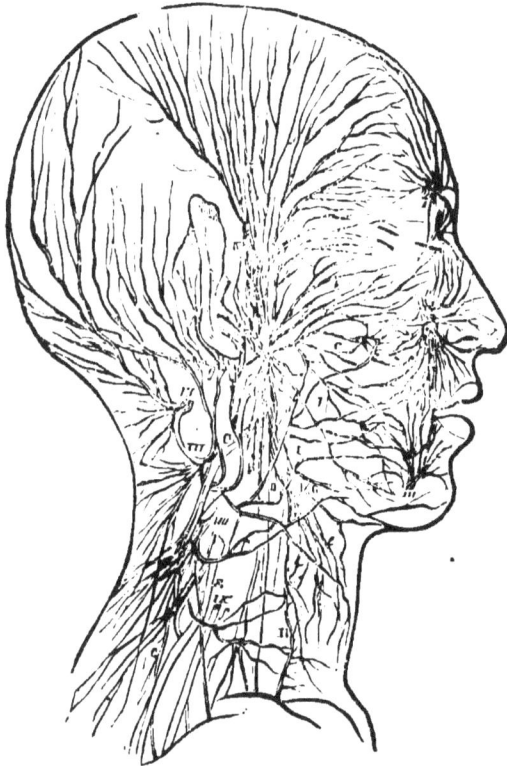

DESCRIPTION OF FIG. 21.—I. Frontal branch of the fifth nerve of the brain which bestows sensation alone. II. Superior maxillary, or that branch of the fifth nerve which supplies the upper jaw, and which, like the last, arising from the sensitive root, bestows sensation alone. III. Mental or inferior maxillary branch of the fifth nerve. This also comes from the sensitive root. It is called mental, because it is involved in that expression which indicates the emotions of the mind. IV. Temporal branches of the same fifth nerve. They are distributed on the temples, and are for sensation. V. The only branch of the fifth nerve which arises from the smaller or motor roots, and assists in the motion of those muscles which are employed in mastication or chewing. VI, VII, VIII, IX. These are spinal nerves—the first of the series which come out between the vertebra, in the whole length of the spine, to supply the body generally with motion and sensation. A. The facial nerve. It is situated in the front of the ear, and is the motor nerve of the features. It sends branches (a) to the muscles of the forehead and eyebrows. Branches (b) to the eyelids. Branches (c) to the muscles which move the nostrils and the upper lip. Branches (d) to the lower lip. Branches (e) going down to the side of the neck. Connections (f) with the

G

spinal nerves of the neck. A nerve (g) to a portion of the muscle that is in the back of the head, and to muscles of the ear. B. The nervus vagus, or the wandering nerve, so named from its extensive distribution. This is the grand respiratory nerve. C. The spinal accessory nerve. D. The ninth nerve, which is the motor nerve of the tongue. E. The nerve which supplies the diaphragm. F. Branch of the sympathetic nerve. G. A branch of the nervus vagus which goes to the superior portion of the larynx or windpipe. H. Another branch of the vagus, which goes to the inferior portion of the larynx. I. The nerve which goes to the tongue and upper part of the gullet called the pharynx.

judges and philosophers, ancient and modern, is sufficient to show. Persons of advanced age, who take rich and stimulating diet in more than sufficient quantity, and whose intellectual faculties are exercised but little, are those most frequently carried off by this embodiment of the Greek idea of the "skeleton at a feast." The most powerful exciting causes of apoplexy, then, are intemperance, whether in eating or drinking, as well as violent exertions of the mind and body—whatever, in short, tends to determine the blood with an undue impetus to the brain or impedes its return from it, is an invitation to this dreadful destroyer to step in and arrest the vital current in its flow, as the breath of frost stays the water of the river.

Symptoms.—Apoplexy may be known by the patient falling down in a state of insensibility or stupor, out of which it is impossible to rouse him by any of the ordinary means; the face is generally flushed, the breathing difficult and stertorous; the upper lip-margin is projected at each expiration; the veins of the head and temp.... protrude as though overfilled, the skin is covered with perspira.... and the eyes are fixed and blood-shot; sometimes, however, the face .... ale, with a look of misery and dejection; and the pulse, instead of .... ing full and hard, is weak and intermitting.

Treatment.—This, of course, must vary considerably in accordance with the pathological condition of the brain of the person attacked, and with other circumstances which only those accustomed to the treatment of disease can judge of. The immediate measures to be adopted when a fit of apoplexy comes on are the following: Place the patient in a sitting position, with the legs depending; remove everything about his neck, and let the air be freely admitted; apply cold wet cloths to the head and neck, and mustard plasters to the soles of the feet; if the patient be old and the pulse weak and feeble, the skin bloodless, and the countenance pinched, warm flannel and hot bricks should be used, and cold water should be dashed in the face, strong spirits of ammonia applied to the nostrils, the feet put into a warm bath with a little mustard, and every means taken to arouse the patient from his state of lethargy. As soon as this is so far effected that he can swallow, give ½ drachm of aromatic spirits of ammonia in 1½ ounces of camphor mixture, as a stimulant draught, but it is only when the pulse is feeble and fluttering that the stimulant may be administered; that is the exceptional case in apoplexy—most commonly the symptoms are those first described.

Purgatives must be got down as soon as possible ; 10 grains of calomel placed on the tongue, and washed down with a black draught, or 2 or 3 drops of croton oil may be rubbed on the back of the tongue, and an injection composed of 2 tablespoonsful of common salt, with a little oil or butter, and a pint of warm water ; or a tablespoonful of soft soap mixed with the same quantity of water ; or an ounce of spirits of turpentine, rubbed down with the yoke of an egg, and a pint of thin gruel ; one of these should be repeated every two hours until some decided effect is produced. Other means of relieving the system may be taken should these fail, such as blisters behind the ears, to the nape of the neck, or calves of the legs ; should the head be very hot let it be shaved, and a cold lotion be applied to it—water and vinegar or acid water will do best. Should the attack be soon after a full meal, administer an emetic—a scruple of sulphate of zinc with a grain or two of tartar emetic; something like this should always be given when apoplexy arises from the effects of opium or spirits. In all cases after the crisis of the disease is over, and when the patient has become convalescent, it behooves him to be very careful, as a slight indiscretion may bring on a fresh attack.

We have said that apoplexy comes without warning, but this is not strictly true. However sudden the attack itself may be, there are certain premonitory symptoms which no prudent man will disregard. Among those may be named a sense of fulness in the veins of the head, and a feeling of pressure in the head itself, with occasional darting pains, giddiness, vertigo, partial loss of memory, and the powers of vision and of speech ; numbness of the extremities, drowsiness, and a dread of falling down ; irregularity in the action of the bowels, and involuntary passage of the urine. These all indicate that some internal mischief is going on, and if their warning is attended to the threatened attack may, perhaps, be avoided. Persons whose full habit of body and modes of life predispose them to this disease should, when such warnings reach them, live sparingly, avoid stimulants, especially fermented and spirituous liquors, take regular and moderate exercise, sleep on a firm pillow with the head elevated, and nothing round the neck to impede the act of breathing ; the mind should be in a cheerful condition, and free from excitement ; sexual indulgence should rarely be resorted to ; late suppers must be avoided, and a hard hair mattrass used for sleeping on. Keep the bowels regulated by an occasional dose of saline purgatives. Those of a spare habit should take light, although nourishing diet, a little beer or wine, if they have been accustomed to it, and it does not affect the head ; spirituous liquors and hot spices should be avoided, and great bodily fatigue or nervous excitement of any kind.

Convulsions, or Fits Involuntary contractions of the muscles of a part or the whole of the body—generally with corresponding relaxations, but sometimes with rigidity and tension ; in the former case they are called clonic spasms, as hysteria ; in the latter tonic spasms, as lock-

jaw ; when the convulsions are slight and rapid they are called tremors.
They are universal, affecting all the limbs more or less, and the muscles
of the face and those of respiration, as in epilepsy, and the convulsions of children ; and partial, when they only affect some of the muscles
irregularly, as in chorea or St. Vitus' dance.

Treatment.—In adults, convulsions may be apoplectic, epileptic,
hysterical, or puerperal, as the case may be. Some narcotic poisons produce them, such as opium, prussic acid, some kind of fungi, ardent spirits, and indigestible substances. In all these cases, emetics would be
the first remedies, and the stomach-pump ; then volatiles and stimulants
—as ammonia, valerian, and a stream of cold water poured upon the
head from a considerable height. Convulsions may be caused by excessive mental emotion, and sometimes by long continued diseases, such
as dropsy, jaundice and fever.

When a person is taken with a fit, proceed thus : Loosen any part of
the dress which may appear tight, especially about the neck and chest ;
if a female, cut the stay-lace, as tight lacing often causes fits ; sprinkle
cold water on the face, and apply volatile stimulants to the nostrils ; rub
the temples with eau de cologne, ether, or strong spirits of some kind,
and blow upon them ; and as the patient can swallow, give 30 drops of
sal volatile in water, or the same of ether, or, if neither are at hand, a
little cold brandy and water.

When the fit is over, a gentle aperient should be taken, to be followed
by cold bathing, exercise, and, if possible, by a change of air.

Dizziness. Many persons are subject to a fulness and rush of
blood to the head, either with or without any excitement. It is a symptom of a deranged system, and it may be a symptom of a tendency to
apoplexy.

Causes.—This condition may be caused by heart disease, by debility
arising from hemorrhages, indigestion, constipation or excessive mental
labor.

Treatment.—What has been said on congestion of the brain applies to
this affection ; a dose of some gentle purgative should be taken, as castor oil, salts, or salts and senna, at night, and the following in the morning :

Rochelle salts,   -   -   •   -   2 drachms.
Bicarbonate of Soda,   •   •   -   2 scruples.
Water,   -   -   -   •   -   ½ pint.

Mix. To this mixture add 35 grains of tartaric acid. Take the whole
while foaming.

Delirium Tremens. Delirium ebriositatis, or mania-a-potu, is a
disease of the brain, usually caused by an abuse of spirituous liquors, but
sometimes also by great mental anxiety and loss of sleep ; or it may result from bodily injuries or accidents, loss of blood, etc. Delirium some-

times makes its appearance in consequence of a single debauch ; but more frequently it is the result of protracted or long-continued intemperance. It usually supervenes on a fit of intoxication ; but it not unfrequently occurs, also, when the habitual drunkard omits his accustomed draught.

Symptoms.—The approach of an attack is almost invariably preceded by the patient being remarkably irritable, with fretfulness of mind and mobility of body. He becomes very nervous and uneasy ; is startled by a sudden noise, the opening of a door or the entrance of a visitor ; is restless ; the hands and tongue are tremulous ; he complains of inability to sleep, and if he dozes for a moment, he is awakened by frightful dreams. Soon delirium manifests itself ; if questioned, the patient often answers rightly enough ; but if left to himself, he begins to talk or mutter ; he is surrounded by frightful or loathsome animals ; is pursued by some one who has a design upon his life ; has terrible and ghastly visions. Though most commonly of a frightful or terrifying character, the delirium is not always so ; occasionally the appearances are dull and ludicrous, and the patient seems amused by them ; at other times it turns on some matter of business, as settling of accounts or telling of money, and the patient is in perpetual bustle, and his hands are constantly full of business. The predominant emotion with the delirious patient is fear, and in his efforts to escape from an imaginary enemy, he may be guilty of murderous assault, or, as is more frequently the case, may take his own life ; and hence he requires to be very carefully watched. "The strong features of this complaint," says Watson, "are sleeplessness, a busy but not angry delirium, constant chattering, a trembling of the hands, and an eager and fidgety employment of them. The tongue is moist and creamy ; the pulse, though frequent, is soft; the skin is perspiring, and most commonly the patient is drenched in sweat." The delirium continues until the patient sinks into a sleep, from which he awakes comparatively rational, or dies from exhaustion. In such cases death is often sudden, the patient rising for some trivial purpose, and falling in a faint, from which he never recovers ; or at length, after passing many nights without sleep, he sinks into a state of coma, which terminates in death. This disease, however, is rarely fatal, unless where the strength of the patient has been seriously impaired by long-continued excesses.

Treatment.—The chief object in the treatment of delirium tremens is *to induce sleep.* For this purpose he must be carefully protected from the inquisitive gaze and questioning of friends and acquaintances. It is also necessary that he be confined in such a way as to prevent him from harming himself. In some cases it will be necessary only for a friend to stay in the room with him, and to judiciously soothe and quiet him during the more violent parts of his delirium ; in other cases it will be necessary to put him in a straight jacket, or in a padded room. In general no more violence should be used than is absolutely necessary to control him.

The following prescription may be ordered :

Bromide of potassium, -	- - - -	2 ounces.
Hydrate of chloral,	- - - -	1 ounce.
Syrup of orange peel, -	- - -	2 ounces.
Water, -	- - - - -	2 ounces.

Give a teaspoonful in water every two hours, until four doses have been taken, unless the patient becomes quiet. In using this mixture certain caution must be observed, especially after three or four doses have been given.

It will be advisable not to give opium or any of its preparations, which are apt to aggravate the patient's mental condition. So soon as the violence of the attack is over, measures should be taken to tone up the patient's nervous system. For this purpose the following mixture may be used :

Tincture of nux vomica,	• - - -	6 drachms.
Tincture of digitalis	- - - -	6 drachms.
Tincture of gentian, -	- .. - -	6 drachms.
Wine of pepsin to make four ounces.		

Mix, and take a teaspoonful before meals.

Much good has been observed from the application of a small mustard plaster over the pit of the stomach, especially those cases in which obstinate vomiting occurs.

One of the most important items in the treatment of delirium tremens is the employment of nutritious food in an easily digestible form. For this purpose, milk and eggs are the staple articles ; they may be supplemented by soups and beef tea. These should be given in small quantities, at intervals of two or three hours.

It must be remembered that delirium tremens is a disease of exhaustion ; notwithstanding the patient's frenzy and frequent exhibition of strength, his nervous system is profoundly prostrated ; hence all measures employed in the treatment of the disease should tend to build up the patient's exhausted powers.

Paralysis. By paralysis we ordinarily understand a loss of the power of movement. The term, however, is used in medicine also to embrace a loss of the sensibility of a part. In this discussion we shall employ the word in the popular sense, namely, as designating an impairment in the power of motion.

In order to understand the conditions which cause paralysis, we must remember the conditions which must exist in order that a part of the body can be moved at will. Movement is, of course, performed by the contraction of muscle ; but this muscle does not contract of itself. Under natural conditions a contraction of the voluntary muscles occurs only under the influence of nervous force. This originates in the nerve centres—especially in the brain and spinal cord —and is conducted along the nerve, just as electricity passes along the wire. When this nervous force reaches the muscle, contraction occurs and the part is moved. In order, therefore, that a voluntary movement shall occur, it is necessary that the nervous force shall be manufactured ; that is, that the brain or spinal cord must be in a healthy condition. Secondly, it is essential that the nerve leading from the brain to the muscle shall be sound ; if this be injured in any way, the force which is produced in the brain is interrupted in its passage along the nerve, just as the electric current is interrupted if the conducting wire be cut ; and finally, the muscle itself must be in a condition to respond to the influence of this nervous force.

It is evident, therefore, that paralysis - -that is, loss of motion—may result from any one of three causes : first, disease or injury of the brain or spinal cord ; second, disease or injury of the nerve ; third, disease or injury of the muscle.

In various diseases we have illustrations of these different causes of paralysis ; thus, in apoplexy a portion of the brain is destroyed and a portion of the patient's body is paralyzed, though the muscles and the nerves of the paralyzed part remain uninjured. In certain cases a nerve of the arm for instance is cut or injured by a wound, paralysis of the muscles to which this nerve runs is a consequence, although the brain and the muscle itself are uninjured. Then, again, the muscle itself sometimes becomes incapable of contracting, as in the disease known as wasting palsy. In this case there is paralysis, although the brain and the nerve remain intact.

Whenever, therefore, a patient is paralyzed, it becomes necessary to ascertain what part of the apparatus is at fault ; whether the paralysis results from disease of the brain, so that no nervous force is generated ; or whether the nerve going to the part is injured, so that the nervous force cannot be conveyed to the muscle ; or whether finally the muscle itself is diseased, so that it fails to respond to the nervous influence. Paralysis therefore is a *Symptom* of a disease rather than a disease itself.

As to the paralysis of *Sensation*, it will not be necessary to enter into any detailed discussion. Sensation, like motion, may be lost in any one of three ways : injury to the brain, injury to the nerve, injury to the skin of the part affected. In many cases sensation and motion are transmitted by different nerves ; that is to say, there may be paralysis of motion in consequence of injury to a given nerve, while the sensibility of the part remains unimpaired ; in the same way the sensibility may be lost while the part can be moved without difficulty.

Paralysis receives different names according to the part of the body which is paralyzed and according to the nerve which is injured. It

sometimes happens that an entire half of the body—one side of the face
one arm, one side of the body, and one leg—will be paralysed, while the
other side remains intact.

Writers' Cramp is a form of paralysis usually limited to certain
muscles of the hand.   As the name indicates, the affection is especially
common among those whose occupation compels them to hold the pen
many hours a day.   It may be indicated by actual paralysis, so that the
finger and thumb cannot be brought together with the usual power ;
in other cases, the muscles controlling the fingers are firmly contracted,
so that the thumb and fingers cannot be moved or are moved irregularly.
Unlike most of the forms of paralysis, this affection can usually be cured
by rest and treatment.

Persons engaged in other occupations than writing, who are compelled
to use the same muscles constantly for many hours daily, are often similarly affected. Taylors and sewing-girls, for example, may lose the power
of holding and guiding the needle ; and women who are compelled to
work a sewing machine may have a similar affection of the feet and legs.

Treatment of Paralysis.—In every case the first object is to ascertain
the *cause.*   In many cases careful investigation will show that the cause
can be removed and the paralysis relieved.   Thus paralysis affecting
various parts of the body, even an entire half, as in hemiplegia, may be
due to *syphilis*, for an individual who has had this disease is liable to
inflammations in the brain which may paralyze his muscles.   These are
the most favourable cases for treatment, since, if taken early, they may
be readily cured by the following prescription :

Iodine,	Eight grains.
Iodide of potassium,	Ten drachms.
Syrup of sarsaparilla,	Eight ounces.

Mix, and take a teaspoonful after meals ; the dose may be gradually increased to two or even three teaspoonsful.

In other cases paralysis results from slow poisoning of some of the
metals, such as lead and mercury.   These forms of paralysis are, of
course, found with especial frequency in those who are compelled to
handle and work with these metals.   Lead poisoning may occur, too,
among women who apply cosmetics containing the article, and from the
use of drinking-waters which pass through imperfectly constructed pipes.
In these cases relief may be obtained by the use of the following in connection with the measures to be presently mentioned :

Iodide of potassium,	Five drachms.
Water,	Four ounces.

Take a teaspoonful four times a day.

In addition to this the sulphate of magnesia may be given in doses
sufficient to keep the bowels active ; for this purpose it may be necessary
to give from a teaspoonful to a tablespoonful of this laxative every day.

It would be impossible to follow out in detail all the different measures which may be at times useful in the treatment of paralysis; for every case must be studied and treated separately; it has been already stated that paralysis is a symptom and not a disease.

Yet there are certain measures which will be found useful in almost all cases, and which may be therefore mentioned here. Prominent among these is *electricity*. Physicians have come to rely upon this agent as furnishing excellent results, though it must not be expected that a cure can always be effected. When, for example, paralysis results from an inflammation or hemorrhage in the brain, the application of electricity to the arm or to the leg can be of no service.

Another most valuable agent is *Massage*. This process, which is now extensively employed by physicians, is performed as follows: the patient is stripped, or at least as much of the body is laid bare as is required for treatment; an attendant then kneads, pinches, pulls and rubs the flesh until a gentle glow and feeling of warmth are excited. This process may seem at first somewhat rough, and may leave a slight soreness, but in a short time these symptoms no longer occur and evident benefit results. Considerable practice is required for the skilful performance of massage, but much benefit can be conferred even by an inexperienced person who will persevere in the effort.

It is highly important that those afflicted with paralysis should have the benefit of fresh air and of such exercise as they are capable of taking; for this purpose it may be necessary to furnish them with the assistance of perambulators, easy chairs, and other mechanical contrivances.

Among the remedies which may be employed with advantage in certain cases of paralysis, are strychnine and phosphorus. The former may be given in the following prescription:

Sulphate of strychnina,	- - - Half a grain.
Reduced iron,	- - - Thirty grains.
Extract of belladonna,	- - - Eight grains.

Mix and make thirty pills. Take one morning and night.

Phosphorus can be best given dissolved in almond oil; one-fourth of a grain of phosphorus may be dissolved in two ounces of the oil, and a teaspoonful of this may be taken morning and night.

Nervous Exhaustion. This term designates a condition which is known by physicians as *neurasthenia*. It may be defined in short as a lack of nervous force. It often exists in pallid, bloodless people, and disappears when the patient's general condition is so improved that the blood-producing organs again perform their functions properly, and the individual acquires again the ruddy glow of health. Yet it often happens that nervous exhaustion exists in individuals whose general appearance would not lead any one to suspect any serious disease; the person may be stout and of full habit, may have a good appetite and digest the food well, and yet may be and feel quite incapable of performing those duties which he had previously fulfilled without difficulty.

The affection seems usually to proceed from an improper degree of activity of some part of the nervous system, more especially in the exercise of the mental faculties. It seems also to be subject to certain hereditary influences ; the children of parents who have suffered from chronic diseases of the nervous system, such as epilepsy, hysteria and insanity, are especially prone to the manifestation of nervous exhaustion.

Physicians, especially those who practice in large cities, are often consulted by individuals who, although manifesting no well-defined disease, are evidently not in good health. It is possible that these cases do not receive as much attention from friends of the person, or even from the physician himself, as they deserve ; for the tendency to complain, to exaggerate slight indisposition, is so common, that unless there is some definite and tangible derangement of the body, the tendency is to ignore and make light of the symptoms presented. In nervous exhaustion, moreover, the indications of the difficulty are of a subjective rather than of an objective character ; that is, they are symptoms which the patient can himself feel, but which no one else can perceive.

The subjects of nervous exhaustion complain of lassitude, a want of buoyant feeling, an indisposition for exertion, mental depression, and sometimes wandering pains and aches felt in various parts of the body. Such individuals are wakeful at night, and arise with a sense of fatigue and a feeling that their sleep has not refreshed them. When stimulated by some unusual excitement they are capable of the usual exertion, but when the excitement has subsided they feel exhausted. Such patients usually fancy that they have some serious disease, and often become melancholy at the thought that their powers are being undermined and that they are "in a decline."

Spinal Irritation is a manifestation of nervous exhaustion which afflicts many of those engaged in active mental effort, and is especially common among women who are subject to diseases of the womb. In this condition there is extreme tenderness along the spine : there are usually flying pains, especially in the chest and abdomen ; and the occurrence of hysteria as well as of convulsive spasms of the limbs is a frequent symptom.

This condition of spinal irritation is usually periodical, and is especially apt to occur after over-exertion or excessive emotion. In nervous women it occurs particularly during the period of menstruation.

There are also conditions which seem essentially the same as spinal irritation, though there is no tenderness on pressure along the back bone. The condition is manifested by unpleasant and annoying sensations in different parts of the body. Some individuals suffer from neuralgic pains in the limbs ; others have throbbing sensations in the chest and in the head. Another symptom is itching, which may occur in any part of the body without apparent cause, and may be quite intense and persistent.

One of the most distressing symptoms is the wakefulness of such patients. They lie awake and toss about for hours and perhaps fall into a heavy sleep toward morning, from which they awake without feeling much refreshed. This condition is obstinate and may not yield even to the bromide of potassiumorto chloral, unless taken in excessive doses.

Another manifestation of nervous exhaustion, is dyspepsia, which is rarely so distressing as those forms of dyspepsia which result from organic disease of the stomach, but is nevertheless a source of much annoyance and uneasiness to the patient.

The special senses are also liable to derangements. Among the most common of these are specks before the eyes, which appear especially when the individual feels exhausted. Another occurrence is noise in the ears, which sometimes takes the form of a continual humming, and sometimes appears as sudden and loud noise.

One of the symptoms of nervous exhaustion, which is brought to the notice of the physician with especial frequency, is derangement of the sexual functions. This may take the form of *impotence*, partial or complete. This is often manifested by a loss of sexual power before the appetite disappears. Under these circumstances the patient is extremely depressed and despondent, as a result of which the symptoms are aggravated. Sometimes this sexual weakness takes the form of seminal emissions. These are of course natural and in perfect accord with health and those who are continent ; but in conditions of nervous exhaustion the emissions are apt to occur with far more frequency than in health. These emissions when excessive are of themselves somewhat exhausting, but they are especially important as indications of nervous prostration. The popular idea ascribes to seminal losses the symptoms which occur in the individual at the time ; in other words the emissions are assumed to be the *cause* of the patient's prostration. As a matter of fact they are a *result* rather than the cause of the condition, and the patient's despondency should be relieved by the assurance that when his general health shall be improved, this symptom will disappear, provided there be no organic disease of sexual organs.

In females nervous exhaustion is manifested by pain and unusual prostration at the time of the menstrual epochs. Here also the menstrual disorders are the result rather than the cause of the nervous prostration accompanying them.

There may be in various parts of the body derangements of motion and of sensation which are to be explained simply by the general condition of the patient, and not by any local disease. Thus it may happen that certain portions of the skin become quite numb, and remain so for hours and days at a time ; in other cases certain parts, such as a finger or toe, an arm or a leg, become extremely sensitive both to pain and to changes of temperature.   At times too there may occur what seems to be a genuine paralysis ; the patient loses control of fingers, of thumbs, or even of the entire hand or fore-arm.   In other instances twitchings of the muscles are constantly observed ; this is especially frequent in the muscles of the eyelids.   Such patients are annoyed by the consciousness that they are constantly winking, and yet they are unable to control the eyelid.

Occasionally such patients are troubled also with unusual diffidence, and even timidity, which sometimes manifests itself by an aversion to society ; this is particularly apt to occur in those whose nervous exhaustion takes the form of sexual incapacity.

Treatment.—Nervous exhaustion usually requires, first of all, complete relief from care, anxiety and exertion.   It is not desirable that the patient should entirely relinquish his occupation ; but a respite for a certain period seems absolutely necessary.

Not less important is the avoidance of errors in the habits of life. The inordinate use of stimulants, excesses of any kind, etc., are of course to be avoided. The best sanitary regulations also should be observed, and one of the most efficient remedies that can be employed is a course of sea bathing.   If this cannot be procured, the cold bath in the morning at least, or morning and night if the patient can bear it, is a good substitute.

The diet should be generous and varied, even though the patient may already seem to have an abundance of flesh.   Among the remedies to be employed, two are especially valuable—electricity and massage.   Exercise should be provided for, but not taken in excess, since exercise of the body requires exertion on the part of the nervous system.   In some cases severe measures have been employed to relieve spinal irritation; small blisters and even the white-hot iron have been applied along the spine. Such measures must, of course, be used only under the advice of the physician, since in every case it is the *patient* and not the *disease* that is under treatment,   The drugs that are to be used vary in different cases. In most cases strychnine, arsenic and quinine, with or without iron, will be useful.   These may be given in the following prescription :

Sulphate of quinine,	-	- Forty grains.
Arsenious acid,	-	- One third of a grain.
Reduced iron,	-	- Twenty grains.
Extract of nux vomica,	-	- Four grains.
Extract of cannabis indica,		- Five grains.

Mix, and make twenty pills.   Take one before meals.

If there be symptoms of dyspepsia it will be advisable to use in addition to the above pepsin and extract of malt, as in the following prescription :

Extract of malt, - - - Three ounces.
Wine of pepsin, - - - Three ounces.
Mix, and take a teaspoonful after meals.

If the patient be troubled with sleeplessness, the following may be administered at night :

Bromide of potassium, - - - Two ounces.
Hydrate of choral, - - - - One ounce
Syrup of orange peel, - - - Two ounces.
Water, - - - - - Four ounces.

Mix, and take a teaspoonful before retiring. The dose may be repeated in an hour if needed.

CHAPTER VIII.

How to Bandage. There is not a more important art connected with household surgery than that of bandaging. To do it well requires much practice and no little judgment. The material employed in bandaging is usually stout unbleached cotton, from two or three to nine or ten inches wide, and from six to twelve yards long : the former length and breadth will do best for the leg. If commenced at the ball of the foot, and evenly applied so that each fold overlaps the other about one-third, it will reach to the knee. Fig 98 will best show the mode of application. The bandage having been first tightly rolled up, is taken in the right hand of the operator ; the end is passed under the

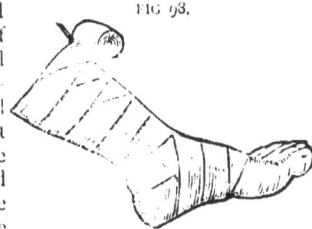

FIG 98.

foot, and held there by the left hand until it is secured by one turn of the bandage over it ; an upward direction is then taken, so that a couple of folds bring the bandage up to the front of the leg, over the instep ; the next turn will naturally pass above the heel behind ; and then, if proper care be observed, it will go on, fold above fold, each over-lapping the other slightly, all up the leg. The bandage is passed from the right to the left hand each time it goes round the leg, and great care should be taken to hold it firmly, and equalize the pressure, as well as to smooth out any wrinkles that may occur in the process of binding. A firm and even support is thus afforded to the limb, which is not likely to crease, or get displaced by the motion which may be afterwards necessary ; it may be made fast above the calf by a couple of pins, or a needle and thread. Great care should be taken in this, as in all similar operations, **to get the bandage rolled up tightly and smoothly, before commencing ;**

it may thus be grasped in the hand, and kept well under the command of the operator, who should on no account let go his hold of the bandage, so as to relax the pressure.

The arm does not require so long or broad a bandage as the leg ; about two inches, by three or four yards, being the average size : this limb is rather more difficult to manage, half turns being necessary to effect a proper envelopment. How this is effected may be seen in Fig. 99. The bandage is folded back upon itself, so as to take a different direction, and cover the space which would be left exposed by the ordinary method of folding ; these half turns, unless they are done tightly and evenly, will be apt to slip and derange the whole binding. Some operators avoid half turns, by letting the roller take its natural course, and then coming back to cover the exposed parts ; but this method, besides requiring a larger bandage, does not effect the required purpose so neatly and efficiently. One mode of fastening a bandage is to split it up a short distance, so as to leave two ends, which can be passed round the limb, and tied. It should always be borne in mind that the chief art in applying bandages is to give firm and uniform support, without undue pressure upon any part ; and to effect this properly, the strain in winding should be upon the whole roll held in the hand, and not upon the unrolled portion of it. This strain should not be relaxed during the progress of the operation.

Fig. 100 represents the mode of applying what is called a many-tailed bandage,—useful to apply over a wound, or wherever it requires frequent changing, or in cases in which it is desirable not to exhaust the patient by much movement of the limb. This is a strip of cotton somewhat longer than the limb to be enveloped : on it are sewn, at right angles, other strips, about one half longer than the circumference of the limb, each overlapping the other about one-third of its breadth, so that when drawn tightly over in regular succession, each secures the other. The end of the strip passes under the heel, and coming up on the other side, is made fast to the bandage there, and so all is kept firm.

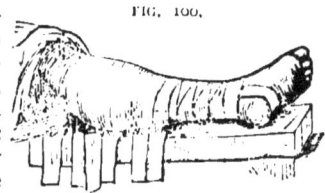

For keeping poultices on the lower part of the back, or in the groin, a cross bandage is used, the fashion of which is this : make a cotton band large enough to pass around the loins, and tie a buckle in front ;

to this is attached another piece which proceeds from the centre of the back to the anus, where it divides in two, which pass under the thighs, up on either side, and are fastened to the band in front. The bandage used to close a vein after bleeding is made thus : lay the tape obliquely across the wound, pass it round the arm above the elbow, and bring it back again over the same spot ; then let it go round the arm below the elbow, and returning, let the two ends be tied in a secure bow, in the bend of the arm, with the free movement of which the bandage should not be tight enough to interfere, although it must be sufficient to retain its position. This mode of bandaging is called the figure of 8, from its resemblance to that figure. Fig 101 will probably make our explanation clearer.

FIG. 101.

For a sprained ankle, place the end of the bandage upon the instep, then carry it round and bring it over the same part again, and from thence round the foot two or three times, finishing off with a turn or two round the leg above the ankle.

For a sprained wrist, begin by passing the bandage round the hand, across and across, like the figure **S** ; exclude the thumb, and finish with a turn or two round the wrist.

For a cut finger, pass the bandage (a narrow one) round the finger several times, winding from the top, and splitting the end ; fasten by tying round the thick part above the cut : or, if it be high up, tie round the wrist.

The best bandage for the eyes is an old silk handkerchief passed over

FIG. 102.   FIG. 103.

the forehead, and tied at the back of the head.   For the head itself, it is best to have a cross bandage, or rather two bandages,—one passing across the forehead, and round the back of the head, and the other over the top of the head, and below the chin, as in Fig. 102.   Or, better than this is, perhaps, a large handkerchief, which will extend all over the forehead and crown, two ends of it passing to the back, and after crossing from thence round the neck, then tying the other two beneath the chin, as in Fig. 103.

For a bandage to support a pad of poultice under the arm-pit, a handkerchief may be used, put on as in Fig. 104 ; or a broad piece of cotton arranged in the same way.

For fracture of the ribs, bandages should be about nine inches wide, and drawn round the body very tightly.   In this case, as in that of any other fracture or dislocation, only a properly qualified person should attempt their application.

FIG. 104.

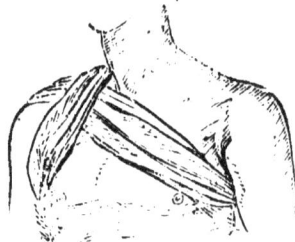

We have not yet spoken of the **T** bandage, which is simply a broad band to pass round the body or elsewhere, having attached to it one of the same width, or narrower, like the upright part of the letter after which it is named ; or, there may be two stems—if they can be so called—in which case it is a double T bandage, as in Fig. 105.

Starch bandages are those in which the roller, before it is put on, is saturated in a strong solution of starch.   Sometimes a covering of brown paper is put over this, and another dry bandage is applied.   This makes a firm and compact case for the limb.   It is useful in cases of fracture, especially if the patient has to be removed to a distance.   Sometimes, when it is not desirable to make the covering so thick and durable, the displacement of the bandage is guarded against by brushing a weak solution of starch or gum over the folds.

Bandaging should be performed, in nearly all cases, from the extremities upwards, or inwards to the heart, except where the injury is seated above the seat of vital action.   If they give much pain, there is reason to suspect inflammatory swelling beneath ; and they should be loosened, if moistening with cold water does not relieve the pain.   Flannel for bandages is used where warmth as well as support is required.

Burns and Scalds.   There are no more frequent, distressing, and dangerous accidents than those which result in the above.   They cause great pain, often amounting to agony ; local injuries of the most serious character, and permanent constitutional derangement, even if death does not immediately or quickly ensue.   The first rule to be observed in the

H

event of the clothes catching fire, is to avoid running away for assistance, as the motion will only fan the flame, and increase the evil. Presence

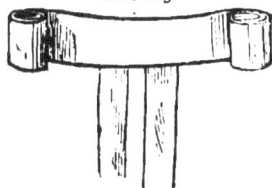

FIG. 105.

of mind in the sufferer is rare on such an occasion, but the best plan is to lie down and roll on the floor, screaming, of course, for assistance. Whoever answers the call should snatch up a rug, or piece of carpet, or other woollen article, and completely envelop the person in it. This will be sure to extinguish the flame. Then cut the clothing away from the burnt parts, taking care to use no violence where it adheres, nor to break any blisters which may be raised. The great object is now to exclude the air from the blistered or raw surfaces, and it is a usual plan to cover them with flour, and then wrap them in wadding, or cotton-wool. A good application is either of the above substances saturated in lime-water and linseed-oil, equal parts mixed ; this is extremely cool and soothing, and greatly assists the healing operation. It should not be disturbed for some days, unless the discharge should be great, and the wounds painful, in which case a fresh application of the same should be prepared, and put on immediately on the removal of the other. The wadding or cotton-wool covering is sometimes applied quite dry, with good effect ; and where the tissues are not deeply or extensively injured, a lotion composed of an ounce and a half of vinegar to a pint of water is a good application, as is also a saturated solution of carbonate of soda. The flour dredging is that which is the most readily available, and it is as good as any. It should be applied immediately and repeated as often as moisture is perceived issuing through the crust which it forms over the burnt parts ; if these have fresh sweet oil brushed over them with a feather, previous to the application of the flour, it will adhere better.

That which is most to be apprehended in severe burns is the great constitutional depression which often follows the excitement and severe pain; especially is this the case with children, and when the seat of this injury is the chest or abdomen, or other vital part. Hence the effects should be closely watched, and stimulants administered, if there are such symptoms as shivering, pallor of the countenance, sinking of the pulse, or coldness of the extremities. Ammonia wine, or spirits, must then be given in doses sufficient to rouse the failing powers, without too much exciting the brain. If there is excessive pain, a slight opiate should be administered to allay the irritation of the nervous system, which, however, frequently receives so severe a shock as to lose its sensibility for a time ; and when this is the case there is great reason to apprehend a fatal result. A burn, if properly treated, and unless very severe, will generally do well, and require little after-dressing ; but if the blisters are suffered to break, and the true skin beneath becomes inflamed by exposure, matter will be secreted, and troublesome ulcerations formed. Bread and-water poulticing will be the best treatment in this case, with Goulard lotion, if there is much inflammation, or an ointment composed of extract of Goulard, one drachm, mixed with one ounce of fresh lard. This should be applied spread on soft linen.

When the burn is deep, after the flour has been on for some days poultices as above should be applied until the coating of flour all comes away, and the wound looks clean and clear ; after which the simple water dressing will be best, and when nearly healed, the Goulard ointment as above.

When parts immediately contiguous are involved in the burn, care must be taken to interpose dressings, or they may become permanently united.

After the more immediate constitutional effects of a severe burn have passed off, it will be necessary to be careful as to the patient's diet, which should be sufficiently nourishing and stimulative, especially while discharge is going on,—taking care, however, to reduce it if febrile symptoms should set in. So constantly are these painful accidents occurring, and so frequently does it happen that the care of a medical man can not be obtained for them, that it behoves all heads of families to make themselves acquainted with the best remedial measures. It should be borne in mind that the principal aims in the treatment of such cases are, first, the protection of the injured parts from atmospheric influence ; secondly, to keep down inflammatory action, both local and constitutional ; thirdly, to soothe the nervous irritation which may arise, and to sustain the system should too great depression take place.

Cancer of the Tongue. This is a frequent and serious affection. It begins as a sore usually on the side of the tongue, which remains open for a considerable time and does not yield to the treatment that ordinarily suffices to heal such ulcers. After a time the sore becomes deeper and its edges sharp and everted. The ulcer gives rise to considerable pain and may be the source of hemorrhages. Before it has attained large size it emits a peculiarly fetid odor which, with its appearance may suffice to arouse a suspicion of the nature of the disease. In a few months the glands at the angle of the jaw become enlarged and hard.

If allowed to progress, the cancer finally destroys a large part of the tongue, and, if the patient live long enough, may spread into the throat.

The only hope of relief lies in the early extirpation of the ulcer and of the adjacent part of the tongue. If this be done sufficiently early, the patient may escape with his life ; but if the cancer return, as it usually does when the operation is too long deferred, the most that can be hoped for is a relief from suffering for a few months.

Enlarged Tonsils. This, if neglected, may lead to deafness, therefore it should be treated at once.

Treatment.—In most cases the child's general health needs attention. He should be provided with the best of food and allowed plenty of re-creation ; in short, all those measures which evidently conduce to the improvement of the health should be employed. In addition, it may be well to administer tonic medicines. The following formula may be given :

Syrup of the iodide of iron,	- - -	One ounce.
Glycerine, -	- - - - -	One ounce.
Water,	- - - - - -	Two ounces.

A teaspoonful of this may be taken after meals.

If the child be evidently scrofulous, as indicated by the pallor, enlarge-ment of the glands in the neck, and the other usual symptoms, cod-liver oil should be administered.

Local treatment in the throat can sometimes be made effectual in re-ducing the size of the tonsils, or at least in preventing the occurrence of unpleasant symptoms. This treatment may consist in the use of astrin-gent gargles, and in the application of remedies directly to the enlarged tonsils by means of a camel's-hair brush. The following gargle may be employed :

Alum, -	- - - -	One drachm.
Glycerine,	- - - -	One ounce.
Tincture of myrrh,	- - -	Three drachms.
Water -	- - - -	Four ounces,

The local applications should be made by brushing the tonsils once or twice a day with the following solution :

Tannin,	- - - - -	Twenty grains.
Brandy,	- - - - -	One ounce.
Camphor water, -	- - - -	Five ounces.

This may be used to *swab* the throat. For this purpose a piece of sponge as large as a hickory nut may be tied firmly unto the end of a piece of wood or whalebone. The sponge is moistened with the solution, and then rubbed thoroughly over the surface of the tonsils and the neighbor-ing part of the throat. It will generally be necessary to hold the tongue down with the handle of a spoon during this process. Care must be always taken to fasten the sponge firmly upon the handle, in order to prevent the possibility of its slipping off into the throat.

Ulcers of the Leg. One of the most common and troublesome affections, among poor people especially, is ulcer of the leg, particularly on the part of the leg just above the ankle.

In most cases of severe ulceration of the leg in which the individual has not had syphilis, the veins of the leg and thigh will be found to be enlarged, constituting the condition known as "varicose veins." These ulcers occur almost always in middle or advanced life, though they may be found in children who are poorly nourished.

Treatment.—The healing of the ulcers will be promoted by improvement of the general health. In most cases, however, the sufferers are unable to enjoy the recreation, air and exercise which form such important elements in improving the health. Yet what can be done in the way of food and personal care should not be neglected, since such measures will have a marked effect in hastening the healing of the ulcers.

The treatment consists chiefly of local applications. Sometimes the ulcers can be healed by the constant application of astringent ointments, of which the following is a good example :

> Diachylon ointment, - - - - - One ounce.
> Vaseline, - - - - - - - One ounce.

Mix. Apply the ointment spread upon soft cloths, which should be bound over the ulcer by means of a bandage.

The healing of the ulcer will be promoted by measures which tend to keep the blood out of the leg. For this purpose the leg may be enclosed in a bandage of soft flannel which is applied from the foot to the knee. Muslin bandages should be avoided, since it requires considerable practice and skill to apply these evenly and firmly. As they are ordinarily put upon the leg, they do injury rather than good ; for they are generally arranged so as to make deep impressions in the skin, and even to cut or abrade the surface.

The healing of the ulcers will be hastened by keeping the foot elevated as many hours in the day as possible. This can be best accomplished by having the patient lie down, or at least by supporting the foot upon a chair. Yet, as a matter of fact, it is practically impossible to persuade a person to remain in bed or on a chair all day and night, even though he have the opportunity ; and for most individuals the opportunity is lacking.

It was, therefore, a godsend for persons afflicted with ulcers of the leg when Dr. Martin introduced to the profession the rubber bandages, which he had himself employed in his private practice for twenty-five years. These bandages are simply made of pure rubber, of varying widths and lengths, according to the needs of the patient. The bandage is applied directly to the skin without interposing any dressings or ointments over the ulcer. It should be put on in the morning before the patient leaves the bed, or even puts his foot out of the bed. It is applied to the foot first, and then wound snugly around the ankle and leg some distance above the site of the ulcer. The patient can then rise and attend to his usual duties. The bandage is quite warm, and causes profuse perspiration of the limb ; there is apt to be also an increased discharge from the surface of the ulcer. Yet these elements do not inter-

fere at all with the beneficial effect of the bandage ; in fact the benefit
seems to depend largely upon the moisture and warmth secured by the
bandage as well as by the support to the veins of the skin.

At night the bandage is removed and cleansed with warm water, after
which it may be hung up to dry until morning. The limb should be
also bathed and cleansed, and the ointment above mentioned may be
applied during the night.

The success obtained in the treatment of ulcers of the leg by the use
of this bandage astonished every physician who employed it. The most
obstinate ulcers, even those which had resisted ordinary measures for
years, were healed in a few months or even a few weeks by the constant
use of this bandage ; and the patient had moreover the pleasure and
profit of pursuing his usual avocation instead of being compelled to sit
or recline during the day.

Choking.   This accident, caused by substances getting into the gullet, or stopped between the mouth and the stomach, is extremely dangerous, and generally the effect of carelessness.

Treatment.—Slap the back smartly, but not too heavily, and in the
meantime let the person swallow some crumbs of bread, and drink a
draught of water.   Or, press a finger immediately down the throat as far
as possible.   Or, take large draughts of water, and make great efforts to
swallow.   The quantity of water distends the gullet above the lodged
food, alters its position, and both water and food pass into the stomach
with a sudden jerk.   If the foregoing efforts fail, make a hook with a
strong iron wire or a thin and narrow flat piece of iron, sufficiently long
not to slip out of the operator's hand.   The hook should be covered by
sewing over it a piece of wash-leather or tape.   This is to be introduced
into the throat, and by that means the obstruction removed.   A strong
emetic will sometimes effect the purpose when other means fail ; mustard mixed with warm water is as efficacious as any.

Frost Bite.   Lengthened exposure to the cold is apt to render parts
of the body numb and inanimate.   The fingers, toes, lips, nose and ears,
are especially liable to be affected.

Treatment.—To restore the natural warmth of the part gradually must
be the main object ; and on no account must a considerable degree of
heat be applied suddenly, as it would either kill the part outright, or
cause violent inflammation to result.   Friction with snow or cold water
merely should be used, until a circulation is somewhat restored, and
equal parts of brandy or some other spirit mixed with cold water may be
applied, until the restoration is completed.   Frost-bites are apt to leave
troublesome sores, which are difficult to heal.   The red precipitate ointment is the best application ; and, if much inflamed, they should be
poulticed.

Goitre, or Big Neck. *Treatment.*—An important item of treatment is the removal of the patient from those influences, whatever they may be, which induce the disease ; hence, a change of residence is almost essential. No medicines can be relied upon to check the growth of the tumor, though much good seems to have resulted in many cases from the use of iodine. This should be applied to the skin in the following form :

Tincture of iodine,	-	-	-	- One ounce.
Glycerine,	-	-	-	- Two ounces.

This may be painted over the enlarged gland every day or two ; if the skin show much evidence of irritation, the painting process may be performed less frequently.

At the same time the patient may take iodine internally in the form of iodide of potassium. The following prescription may be administered :

Iodine,	-	-	-	- Four grains.
Iodide of potassium,	-	-	-	- Four drachms.
Syrup of sarsaparilla,	-	-	-	- Four ounces.

A teaspoonful of this may be taken three times a day after meals.

Housemaid's Knee. This affection consists of a swelling on the front of the knee, or rather on the upper part of the leg just below the knee. It consists of an enlargement of a little sac which naturally exists over the knee-pan. This sac becomes filled and dilated with watery fluid, constituting a soft fluctuating tumor. The swelling may vary in size from that of a hazel-nut to the dimensions of a walnut. The swelling is at first painless, and remains so until irritated by mechanical violence ; it may then become acutely inflamed and occasion much pain.

This affection is termed housemaid's knee, because it occurs with especial frequency in servant girls, presumably in consequence of kneeling upon hard, damp floors. So long as it remains painless it need not be interferred with, unless it attains such a size as to inconvenience the patient. In this case it may be punctured with a fine needle, and the fluid allowed to escape. If it becomes inflamed, the patient suffers great pain and high fever ; the knee swells so that walking is impossible.

Treatment.—During an inflammation of such a tumor the patient should lie quietly in bed. Hot cloths must be wound around the knee and frequently changed, in order to keep up a constant warmth and moisture. In two or three days the pain and swelling usually subside, and the patient's condition remains as before. In other cases matter forms, and it becomes necessary to open the swelling with the knife.

The tumor can usually be made to disappear by passing a seton through it : this consists in inserting a needle armed with clean silk into and through the sac, the silk being permitted to remain. This causes some inflammation, as a result of which the sac gradually dries up. Several other plans of treatment are in use, but can be practised only by the surgeon.

Wounds. Wounds are either incised, lacerated, contused, or punctured. They are called incised wounds when they are made with a sharp cutting instrument, as when a shoemaker cuts himself with his knife, or a carpenter with his chisel. They are called lacerated when the flesh is torn, either by machinery, hooks, or other blunt instruments. Wounds are said to be contused when there is an irregular breach of surface, accompanied by injury and a bruised condition of the surrounding parts ; they are generally produced by falls or blows of dull instruments. Punctured wounds are produced by the forcible entry of sharp instruments, such as bayonets, swords, scissors, hooks, or the pointed ends of broken bones.

Incised Wounds. It has been observed before, that incised wounds consist of a mechanical division of the parts by a cutting instrument ; all, therefore, that is necessary to be done, is to bring the edges of the wound nicely together, and maintain them in that position until union takes place. This is effected, if the wound be trifling, by means of straps of sticking-plaster, which should be so applied as to preserve the edges of the wound in apposition. Collodion answers admirably. If the wound be of considerable extent, and bleeds freely, the first thing to be done will be to arrest the hemorrhage ; this will be effected, if the bleeding vessels be small, by making pressure with a sponge for some considerable time. All extraneous matter should be cleared off, and the lips brought together ; a piece of lint should be dipped in the blood and placed over its edges. This is found to be an excellent application, as the blood in drying, in consequence of its adhesive qualities, seems to maintain the union of the edges of the wound. In the course of four or five days, the parts will be found to be united, unless some accidental circumstance, such as too great a degree of inflammation or an untimely meddling with the dressings, should occur. The strappings or dressings should on no account be disturbed before the fourth, fifth, or sixth day, unless the parts should be in great pain or much swollen. If the incision takes place about the cheeks or lips, or other parts which are unsupported, and where sticking plaster could not be applied, it will be necessary to put in two or three sutures, according to the extent of the wound. Should the parts swell, a cooling lotion may be applied, such as Goulard-water, and the bowels should be kept in a free state. Frequently the edges of the wound must be maintained in apposition by means of sutures.

Sutures are for the purpose of holding together the edges of a wound in soft, fleshy parts that are loose and movable, where sticking-plaster would not, of itself, hold sufficiently secure. They FIG. 128. consist of stitches, from half to three-quarters of an inch apart, between which strips of plaster are placed, and are not drawn out for several days if they do not irritate the part much; but if they do, then they must only be continued one or two days. The needle should be threaded with silk or hemp thread well waxed and flattened. It should always be borne in mind that the edges of wounds are never to be drawn together with any degree of strain or force to the parts, as then the process of healing will not take place. In what is technically called the interrupted suture, a stitch is taken straight through the edges of the wound, as in ordinary sewing, and then knotted. In the twisted suture fine steel needles, with flattened points, are passed through the edges of the wound ; then silk is twisted in the figure of eight around them, as a boy twists his kite twine on a stick. It is better not to cut the silk, but continue each end down to the next needle, and so on ; secure the ends with a small knob of wax. Fig. 128 shows the mode of making and tying sutures.

Lacerated Wounds. Lacerated wounds, in consequence of the great injury done to the parts, and from the fact of their not bleeding much, are very subject to active inflammation. If the wound be considerable and the parts much injured, the patient should enjoy perfect rest ; the parts should be covered with cooling lotions, all dirt and extraneous substances being previously washed off; the bowels should be opened by the common black draught. If inflammation run high, leeches should be applied, and the bleeding encouraged by the application of hot water ; the cold lotion should now give way to fomentations and poultices ; the patient should live low. When the inflammation has subsided, the wound may be dressed with basilicon or Turner's cerate. Erysipelas frequently follows lacerated wounds of the scalp ; in this case the parts should be freely fomented with hot water, and the patient should take a fever mixture. Tetanus, lock-jaw, and spasm, often arise from lacerated wounds : in such cases opium should be administered in doses suited to the age and circumstances of the patient. One grain might be given every three or four hours until relived.

Contused Wounds. Contused wounds will require the same treatment as that already described. Cold applications in the first instance, and if inflammation sets in, leeches and hot fomentations. They generally terminate in suppuration and sloughing, or mortification of the parts, according to the extent of the injury. In order to expedite these processes, poultices of bread and water, or linseed-meal, should be applied three or four times a day, and when the abscess opens or the slough is thrown off, they are to be treated as common ulcers with basilicon or some other stimulating ointment, for the purpose of promoting healthy granulations, and thus healing them. During the active stage of inflammation, the patient should live sparingly ; but tonics, such as quinine, and a generous diet should be allowed under the stage of suppuration or sloughing.

Punctured Wounds. Punctured wounds are extremely dangerous—much more so than the others already described. A punctured wound from a nail, hook, or any other pointed instrument, gives rise to inflammation of the absorbents (a set of vessels running from the wou. d into the neighboring glands), and is manifested by red lines taking the course of these vessels. Abscesses of the glands, and of other parts of the body, in their course, frequently ensue ; and if the matter be deep seated, such a degree of irritative fever is produced as to cause death.

Lock-jaw (tetanus) and frightful convulsions are often the result of tendons or sinews receiving punctured wounds. In the first instance the puncture should be laid open with the lancet, cold lotions should then be applied, and if inflammation sets in, the parts should be covered with leeches according to the age and strength of the patient ; the diet should be sparing, fomentations and poultices should be constantly applied, and the limb should be supported on an inclined plane, in order to favor the gravitation of the blood towards the body. All stimulating drink should be cut off. The bowels should be kept freely open, and the patient should observe perfect rest. As soon as matter has formed, it should be let out by free incisions with the lancet, after which the parts should be poulticed three or four times a day. In order to allay irritation and pain, and to procure sleep, great advantage will be derived from the administration of ten grains of Dover's powder, at bedtime.

Bleeding, or Hemorrhage. This always accompanies wounds, and is generally most alarming to bystanders. There is no occasion for fright, as people do not bleed to death very quickly. Retain your presence of mind, and remember that three things are to be done, all of which you can do at once. Take time to notice the color of the blood. Blood from the arteries is a bright red color, and bursts out in spurts, while venous blood is a purple red, and flows in a steady stream.

The three steps you can take to stop the blood are :

1. *Pressure.*—Should an artery or branch have been divided (indicated by a spurting of a spray of bright blood at each beat of the heart), the firm pressure of the finger for some time, to the point of division, should be used, to diminish the size of the vessel at that point, until a clot is formed there.

Sometimes, pressure to the supposed seat of the injured vessel does not reach the artery. In such a case the pressure must be used to some known trunk between the original supply of the blood and the injured branch. Thus, if the finger or the toe is the seat of the arterial hemorrhage, firm pressure applied each side of the finger, close to the hand, or toe, close to the foot, compresses the arteries passing along to be distributed to the extremity. If the hand or foot is the seat of the injury, pressure on the wrist, over the point where the artery is felt for the "pulse," or at the inside of the ankle, will materially retard the passage of the blood beyond these points.

2. *Position.*—The part from which the blood comes should be raised above the rest of the body, and if the patient become faint he should not be roused immediately, since faintness acts as nature's remedy by lessening the force and activity of the flow of blood.

3. *The Application of Cold.*—This plan answers best when the bleeding is from several points scattered over a large surface ; it is conveniently applied by letting cold water drip from a sponge upon the bleeding points, or by the application of ice in a rubber bag, or bladder.

When these immediate measures have been used, there is time enough to use what physicians call hemostatics, to stop the blood. Gallic acid is a cheap and convenient one. Still more handy is alum. Either may be dusted on the part in powder, or poured over in solution. Or the wound may be touched with nitrate of silver, or tincture of iron. But these measures are needless in ordinary cases. Sometimes, when a tooth is drawn, and the blood will not cease running, a piece of cotton, dampened with alum water, or sprinkled with alum powder, and applied, will check it promptly.

Bleeding from the Nose. This is seldom serious, and may generally be controlled by the application of a little cold water. The patient should be kept upright with his head thrown back, and his hands raised above his head. Ice to the back of the neck is useful. If very obstinate, throw into the nose with a syringe, ice water or a solution of Gallic acid, or pure vinegar. If these fail, the nostrils should be plugged with lint dipped in a strong solution of sulphate of copper, or the lint first moistened and then dipped in finely powdered charcoal.

When the bleeding has stopped there should be no haste to remove the clotted blood from the nostrils. Let it come away of itself ; do not blow the nose violently, nor take stimulants, unless there be excessive faintness, in which case a little cold brandy and water may be taken. Where there is full habit of body, cooling medicines and low diet may be safely advised.

Bleeding from the Bowels. When blood escapes from the bowels the patient is usually afflicted with hemorrhoids, or "piles." Bleeding from this source need occasion no alarm ; indeed the patient's sufferings are usually alleviated by it.

In other cases an escape of blood from the bowels is a symptom of disease higher up in the intestine. The affection which is most frequently accompanied by hemorrhage from the bowels is typhoid fever. In this disease severe bleeding sometimes occurs ; and in some cases but little blood escapes from the body, so that the patient may even die from unsuspected loss of blood into the bowel.

We can usually distinguish blood which escapes from some point high up in the bowel from that which comes from piles by the color ; blood which issues from piles is usually of a bright red color, while that which proceeds from the upper part of the intestine is generally very dark, or even black ; its true nature may in fact escape detection, since it looks very much like pitch.

Treatment.—Bleeding from the bowels should be treated by giving half a teaspoonful of the spirits of turpentine in a tablespoonful of milk and by the application of cold cloths to the abdomen. If these measures do not suffice, ice-water may be injected into the rectum, or pieces of ice wrapped in soft cloth may be inserted into the bowel. In these cases the tincture of ergot is a valuable remedy ; half a teaspoonful of this may be given, and a similar amount taken at the expiration of fifteen or twenty minutes. The patient should of course lie perfectly quiet, and resist, so far as possible, the inclination to evacuate the bowel.

Fainting. (*Syncope.*)—This is a state of total or partial unconsciousness caused by diminished action of the heart causing less rapid circulation of blood through the brain.

Causes.—The causes of it are various and sometimes very peculiar, such as a particular smell ; that of a rose, for instance, has been known to cause it ; certain objects presented to the sight ; surprise, joy, fear, or any sudden emotion ; loss of blood or anything which tends to debilitate the system by diminishing the vital energy.

Symptoms.—The first sensation of fainting to the patient himself is generally a singing in the ears ; then the sight becomes confused, and all the senses deadened ; a clammy sweat breaks out over the person, the countenance becomes deadly pale, and the limbs refuse to support the weight of the body, which sinks to the earth as helpless and motionless as a corpse : indeed the condition so closely resembles that of death, that it is difficult to distinguish it therefrom. This is a complete faint ; frequently the fits are only partial, and very limited in duration.

Treatment.- Place the patient in a horizontal position ; free the face, neck, and upper part of the chest from all incumbrances ; let the fresh air play freely upon them, and sprinkle the former with cold water ; holding to the nostrils from time to time some volatile stimulant, such as hartshorn or ammonia ; as soon as swallowing can be accomplished, administer about thirty drops of wine, or sal volatile in water. The after-treatment will of course depend on the cause.

As the first stage of some forms of apoplexy and paralysis is one of faintness, a little discrimination should be used in the administration of stimulants. Where the seizure, too, is in consequence of loss of blood, no violent efforts at restoration should for a time be made, as this state is necessary for the patient's safety.

Persons subject to fainting should be careful in frequenting crowded rooms, or going anywhere where the air is bad. Tight dresses should be avoided ; and no excitement be allowed. A well-regulated diet, cold bathing, and vegetable tonics, will usually cure this distressing infirmity.

Bones, Broken or Fractured. Splints is the name surgeons give to the apparatus used to place along the broken fragments, to keep them in position. They are made of wood, rubber, leather, binders' board, tin, and many other materials. For general use those manufactured of stiff felt are the best. They come in sets, and are readily moulded, when hot, to any limb, and are firm and soft when cool.

In instances of suspected fracture or dislocation of the thigh or leg, the injured parts should be placed in a comfortable position, and as well supported as possible, to prevent the *twitchings* of the leg from the spasmodic action of the muscles of the injured extremity. If necessary to remove the patient to his home or the hospital, from the spot where the accident happened, the arrangement of the limb should be made after he has been placed on the stretcher or substitute.

If found necessary to carry the injured person some distance, and a litter for the purpose cannot be had, the arrangement of the fractured limb against the other, and kept there by handkerchiefs, is often of great comfort to the sufferer.

By a little ingenuity a comfortable litter can be made by fastening four stout poles together, and tying a blanket securely to them.

We shall now describe particular fractures which may be either Simple, Compound, or Comminuted.

Simple.—When a bone is broken in one place without any external wound.

Compound.—When a bone is broken in one place, and there is an external wound leading down to the broken bone.

Comminuted.—When a bone is broken in two or more places, as when a splinter of bone is broken off.

Fracture of Skull.—Put the patient in bed, and let his head be shaved for some distance round the seat of injury, and wet lint with gutta percha tissue applied over it.

Fracture of Lower Jaw.—In this fracture the parts of the bone should be replaced in their natural position, the mouth closed, and the face bandaged so as to retain the fragments in place. The patient should be fed by a tube, which can be inserted where he has lost a tooth. Broth and milk must be his diet until the bones knit. A dentist can easily make a splint of rubber to fit inside the mouth, and thus hold the parts in position at less discomfort.

Fracture of the Finger.—After employing extension, and thus bringing the ends of the bone together, place a small smooth piece of wood, or of gutta-percha, on the under, and another on the upper side, and proceed to bandage somewhat tightly, so as to keep the finger extended ; put the arm in a sling, and keep it so for a month. If the injured part swells and becomes painful, the bandage must be loosened, and a cold lotion applied ; this is generally by no means a difficult case to treat.

Fracture of the Bones of the Hand or Finger.—These bones, which intervene between the wrist and the fingers, should be treated in the following manner : place in the palm of the hand, a soft, but firm, spherical body, and clasping the fingers and thumb over it, in a grasping position, keep them so with a bandage ; by this means the natural arch is preserved, which it will not be if flat splints are applied. In this case, too, the arm had better be slung, and from a month to five weeks will be the time required to effect a union

Fracture of the Forearm.—Fracture of the forearm may be either of the ulna or the radius, or of both : the former is the inner and thicker bone of the two (see Fig. 116), and the fracture of this does not much disturb the general outline of the arm ; it may be broken at any part of its length or at the elbow process, called *olecranon* (3), or at (4). In the first place the plan will be to bend the elbow, and bring the hand into such a position that the thumb points upwards ; use extension until no unevenness can be discovered in the course of the bone, and then apply two splints, the inner one reaching from the elbow to the tips of the fingers, and the outer from a little beyond the elbow to the middle of the back of the hand, which should be raised well towards the chest so as to make a sharp angle and draw the ulna from the radius. When the fracture is in this latter bone (2) the same method must be adopted, only that the hand must be depressed instead of raised, in order to keep the two bones apart. When these are both fractured, the setting is, of course, more difficult, and much time has often to be spent in extension and manipulation, before the four broken ends can be brought properly together. The splints should be put on as above directed, bandaging the hand firmly to the longer one, and placing it so that it is neither raised nor depressed, but in a right line with the axis of the arm. When there is a fracture of the *olecranon* there is little or no power of extension in the elbow behind which a bony lump may be felt. A true osseous union in this case is scarcely to be looked for ; but the injury will probably be repaired by a band of ligament. There is commonly inflammation and swelling, which must be reduced before pressure can be applied ; the arm should

FIG. 116.

The Fore-Arm.

be kept straight, and wet with cold lotion ; and apply a splint as soon as it can be borne; let it be a long one, reaching on the inside from the shoulder to the hand.   Bandage the arm in a straight position, beginning from the top, and making, as you go, extension downwards, so as to get the broken bone in its place ; it is long ere the limb is in a serviceable condition after a fracture like this.   When the coronoid process is broken, the matter is more easily managed.   The forearm must be bandaged in a bent position and kept so.   In about a month, slight exertion of the limb may be allowed, and there must be great care taken that it is not too violent.

Fracture of the Humerus.   Fracture of the humerus or upper arm-bone, very commonly takes place in the shaft, or any part of it within an inch and a half of the extremities.   It is easily detected by the mobility of the limb at the seat of the injury, and the patient's incapability of raising the elbow ; the broken ends of the bone, too, may readily be felt, and the crepitation heard, when they are rubbed together.   In this case two

Fig. 117.

wooden splints will be required—one to go before and the other behind ; or, if the arm is very muscular, four may be necessary to embrace it properly ; they should be padded with tow, wadding, or lint, as here represented, and furnished with tapes to buckle or tie, as may be most convenient.   The padding should be placed upon a soft piece of cotton or linen, a little longer than the splint at each end, and three times as broad ; turn in the ends and sides, so that the pad may be a little larger than the splint in every way, and about half an inch thick, and make all fast by tacking ; place the turned-in ends of the cotton next the wood, so that there is

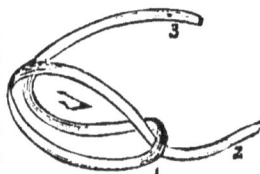

Fig. 118.

a smooth surface presented to the skin.   The tapes, three in number, are put on to the splints double, so that there is a loop at one end (1), through which, after it has encircled the limb, the other end is passed (2), then drawn tight and tied to the remaining end (3) with a bow-knot, as shown in Fig. 118.   A bandage, very easily loosened, may be made in this way of a strip of cotton or broad tape.   The setting of the bone is not difficult in this case ; the ends are easily brought together, and being so, the splints may be placed, and made firm by means of loose tapes ; these should not, at first, be drawn tighter than is required to keep the splints right, and prevent movement of the arm.   After the first few days, when the swelling has subsided, a more permanent investment of the limb may be made.   Place two splints, one on each side, of stout pasteboard, gutta percha or leather, cut so that they will come down and

FIG. 119.

cover part of the forearm, as represented by the dotted lines in Fig. 119. The splints should have been previously shaped, or moulded, to the sound arm, and should be well fixed by more bandage, which, as it is rolled, should be brushed over with starch to prevent it slipping. Sometimes, where there is not much muscle, the starch bandage alone is used ; but in this case, the whole of it must be well saturated with strong starch, paste, gum, or white of egg, with strips of brown paper stuck down across the folds here and there. Care must be taken not to move the arm until all this is dry and firmly set. The hand and wrist must be supported with a sling, but the elbow had better hang free, as its weight will tend to keep the bone straight and the muscles extended.

Fracture of the Neck of the Humerus.

FIG. 120.

Fracture of the neck of the humerus is that which takes place when the upper extremity, or head is broken off. The symptoms here are very much like those which attend dislocation of the shoulder, and the treatment must be much the same. Draw down the shaft of the bone, and push up the head by means of a pad in the armpit ; then bringing the arm close to the body, with the lower part at right angles with the upper, fix it to the chest by a splint on its outside, and a long bandage encircling it and the whole body, as shown in Fig. 120.

Fracture of the Condyles.

This is when the lower part of the humerus is the seat of the injury, the condyles being the rounded eminences which fit into the socket-like hollows at the head of the ulna to form the elbow joint.

Treatment.— Bend the elbow to a considerable angle, and keep it so by means of bent splints of gutta-percha, or millboard, moulded to the shape, the first being softened by heat, the last by moisture. Bandage, and keep all quiet until adhesion of the bone takes place, then put the arm in a sling, and let it remain thus supported for a month or six weeks.

Fracture of the Shoulder-Blade.

This commonly happens near the neck, and is very likely to be confounded with dislocation of the shoulder, or fracture of the neck of the humerus, like which it should be treated, only that the arm, instead of being drawn down, must be supported.

Fracture of the Collar-Bone. This is, perhaps, one of the commonest accidents of the kind that can happen, and one of the most
easily detected. It is generally occasioned by a blow on the shoulder,
which falls forward, pushing the ends of the broken bone one over the
other. The main object in the treatment
must therefore be, to keep the shoulder back
until the bone has united, and become sufficiently firm to do this without artificial aid.
This end is accomplished by various means,
but the following plan is the most simple and
successful for unprofessional adoption (see
Fig. 121). A wedge-shaped pad of any soft
material—a pair of old stockings, for instance
—is made, and put in the middle of a small
shawl or a large handkerchief; it is then
placed well under the arm, but on the injured side (1); the ends of the envelope are
brought back and front over the opposite
shoulder, then crossed, and tied beneath the
sound arm (2); another broad bandage of some kind is then passed
several times round the body and injured arm (3), so as to bind the latter closely to the former in such a manner that the pad beneath the armpit acts as a fulcrum, and allows the outer end of the broken collar-bone
to be pulled backward and outward during the process of binding, which,
when completed, sets it fast in the right position. We have then only
to envelope the whole of the fore-arm in a sling, and the apparatus is
complete; it should be worn a month at least.

FIG. 121.

Sometimes the collar-bone is broken externally, near the point of at-
tachment to the coracid process. In this case there is scarcely any displacement of the fractured ends, and little need be done beyond keeping the patient quiet, and slinging the arm. In any fracture withinside of
this point, the arm falls down, and is drawn inwards, and the above
should be the plan of treatment.

Fractures of the Ribs. Fractures of the ribs are not of unfrequent
occurrence. They commonly result from a fall or blow, and may be
complete or only partial, involving one or more of the bones.

Symptoms.—A sharp pain is felt at the injured spot, especially in
breathing and coughing; irregularity to the touch; and distinct crepitation.

Treatment.—The chief risk involved is injury to the lungs, from the
sharp ends of the bone and consequent inflammation. Leeches are sometimes applied to the seat of pain, and hot bran bags. A band of stout
cotton or flannel, from eight to ten inches wide, should be passed round
the chest several times, beginning close under the armpits and going
down to the end of the ribs. It should be drawn so tightly as to keep

I

the ribs from falling in the act of respiration. If inflammation follow, the pulse must be quieted by five to eight drops of tincture of viratrum viride, every one or two hours. The patient should be kept perfectly quiet, and on low diet, for a fortnight at least, assuming the position which is found most easy, which will probably be a half-sitting one supported by pillows.

Fracture of the Leg between the Knee and the Ankle-Joints.

The leg is composed of two bones, an inner larger (the tibia), an outer smaller bone (the fibula). One or both may be broken. The tibia is more frequently broken about two-thirds of the way down. It is detected without much difficulty by passing the hand down the line of the shin, although the displacement may not be great. It is not easy to detect a fracture of the fibula, nor is it of great importance to do so.

FIG. 122.

Treatment. — Extension must be made as in the other fractures, and sufficiently so to bring the broken surfaces together. The old-fashioned straight splint would probably be the most readily available in domestic treatment. It is shaped as shown in Fig. 122, and should be sufficiently long to extend from a little above the knee to four inches beyond the sole of the foot. It may be quickly made out of half-inch board, planed smooth ; the breadth should be about four inches ; this must be padded throughout its whole length, except the matched end, which is to project beyond the

FIG. 123.

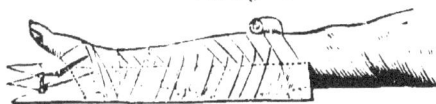

foot, with tow, lint, or other soft material, taking care to have the pad thicker at the lower part, to suit the diminution in the size of the leg. This splint must be carefully placed against that side of the limb from which the foot exhibits a tendency to turn. We will suppose that a stout cotton bandage, about two and a half inches wide, and twelve feet long, has been provided. With this, beginning at the foot, and bringing it down from the instep between the notches at the bottom of the splint, envelope the limb evenly, fold over fold, up nearly to the knee, just below which a broad piece of tape should be passed, with the ends through the holes in the top of the splint, which ends are to be firmly tied at the moment when extension of

FIG. 124.

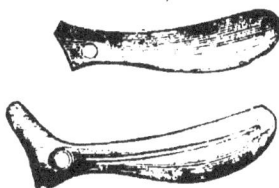

the limb is made by an assistant ; the bandage is then to be carried on over the head of the splint, and made secure. In Fig. 123 we see the limb, before this process is completed. When both bones are broken, it is generally necessary to apply the angular splint adapted to the ankle, of which Fig. 124 exhibits the outer and inner sides.

Fracture of the Knee-Pan. Fracture of the knee-pan sometimes happens from the more muscular exertion of kicking or throwing out the leg violently. It may be at once detected by the depression of the bony plate and separation of the broken fragments ; these cannot be kept in close apposition, and the injury is made good by a ligamentous band, which connects them. To facilitate this process, the leg should be kept in a straight position, above the level of the hip, so that the muscles of the thigh, which are attached to the upper edge of the knee-pan, may be relaxed. A long splint, bound beneath the leg from the thigh to the foot, will effect this object. Over the broken patella a piece of cotton is bound, and the knee is bandaged tightly above and below this, so as to bring the broken pieces as closely together as possible, and to keep them so. The bandage will have to remain on probably for two months, as a fracture of this kind unites very slowly. The knee is generally weak after, and it is best to support it with an elastic knee-cap.

Fracture of the Thigh. This is a very serious accident. The bone may be broken just above the knee, in the shaft, or near the neck. In the first of these cases the nature of the injury is sufficiently obvious, as the broken bone can be felt beneath the skin. This also is the case with the second, in which, as in the third, there is shortening of the limb, and generally turning out of the foot. This accident may be readily distinguished from dislocation of the hip, by the mobility of the hip-joint.

Treatment.—There is always much difficulty in keeping the ends of the bone in apposition here, in consequence of the power exerted by the muscles of the thigh, which are constantly pulling lengthways and causing the ends to overlap, or, as we say, " ride " upon each other ; this is especially the case if the fracture is oblique. It is best to use the straight splint first in either of these cases, and put it on with a light bandaging, gradually tightening it, to accustom the limb to the pressure. The splint must be made in the same way as that shown in Fig. 122, but much longer, reaching from the hip to beyond the toes. When inflammation has subsided, and the pressure can be borne, the case had better be treated in this way : let the patient lie on a hard mattress, with the leg extended and uncovered ; then place the splint, previously well padded, in its place, and make it fast with rollers to the foot, ankle, and leg, taking care that the former is in the position which it is to occupy— that is, pointing straight upwards ; next, take a silk handkerchief, in the middle of which some wool has been rolled up, to make it a considerable thickness, and pass it between the legs, bringing one end up behind, and one before ; these ends pass through the holes at the top of the long splint, and tie them as tightly as possible, without displacing the fracture. Then after confining the splint to the waist, with a bandage, insert a short stick between the loop of the handkerchief, and give two or three turns; this will have the effect of shortening the handkerchief, and pulling down the splint, which will carry with it the part of the limb attached to it below, producing the necessary extension. Keep on at this until you find that the injured leg is as long as the sound one; and when this is the case, lay a well padded short splint along the inside of the thigh, and bandage tightly and smoothly, from the knee to the hip. When it is

FIG. 125.

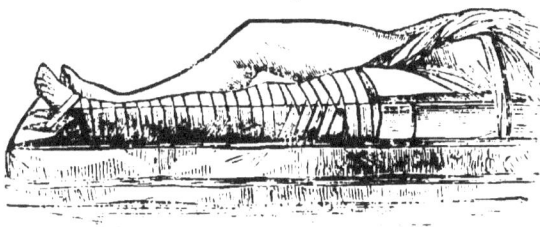

completed, the patient will appear as in Fig. 125. The extension must be kept up for about six weeks, at the end of which time the fracture may be sufficiently united to bear the strain of the muscles upon it.

Fracture of the Pelvis.—Fracture of the pelvis sometimes occurs in falls from great heights, or in being run over, or having some crushing weight thrown on the body. When it occurs there is generally serious injury to the viscera of the abdomen and pelvis, indicated by the passage of blood from the bladder and bowels. The nature of the mischief in this case is not easily detected, and little can be done beyond enjoining perfect rest and a lowering diet, unless there are symptoms of collapse, in which case stimulants must be given.

Compound Fractures.—The term *compound* is applied to a fracture in which the skin is broken or torn, the wound of which communicates with the broken bone. This case is more serious and much more tedious than of common fractures, especially when the wound is large; but be it ever so small, it proves great violence and injury done to the soft parts, muscles, &c. Some weeks after the injury elapse before the bone begins to unite, in consequence of the large formation of matter that generally takes place, and the process of union of the bone does not begin until this action has ceased. During this stage the patient often becomes seriously ill, and his vital powers are exhausted by the large quantity of matter poured out around and among the injured parts. In young and healthy persons, in whom the injury to the muscles, &c., is not very great, these stages are not very strongly marked, and the cure proceeds more rapidly.

Treatment.—The wound must be healed, if possible, the edges being brought together by adhesive plaster. The splints, whatever bone may be affected, should be applied as in simple fracture, care being taken, however, not to press on the wound if possible, and this may be avoided by dividing the pad that lies over the wound, into two parts, leaving a space for the wound, which should be untouched ; the pad should be very thick. A better method than this, however, is to divide the splint and to connect the two parts by means of an arch of iron, so that the wound may be dressed without difficulty. This is called an interrupted splint. If the wound does not heal at once, the plaster may be removed, and a linseed-meal poultice substituted.

During the stage of formation of matter (suppuration), the patient will require tonic medicines, as bark, porter, &c., and small doses of opium at night, and nourishing diet, if the stomach will bear it ; but this treatment must not be carried too far. Strong purgatives are injurious. The case will continue to progress very slowly for some weeks ; abscesses may form ; and should matter collect under the skin so as to be felt on examination, or the skin become red and thin, the part should be punctured, and great relief will be afforded by its escape.

The splints should be removed as often as the matter renders the pads foul, or the wound appears to suffer from their presence ; perhaps this may be required every other day, or even oftener. When the suppurative stage has passed, which may occupy from one month to two, the wound will look florid and healthy ; and as soon as it begins to heal, the bone will begin to unite, but not till then. A month or five weeks will still be required before the union is complete, and two or three weeks yet longer before the patient is enabled to use the limb. The above periods refer to compound fracture of the thigh-bone. Compound fractures of other bones pass through these stages more readily.

Injuries to Achilles' Tendon. This is the great tendon which passes from the muscles of the calf down to the heel, upon which it acts with the whole force of those muscles. It sometimes happens that by a sudden jerk, or violent exertion, the tendon gets torn across, or ruptured, and great pain or lameness is the consequence.

Treatment.—On the first occurrence of the accident, if swelling and inflammation ensues, apply three or four leeches, and encourage the bleeding for some time with warm fomentations, or a linseed poultice. Afterwards resort to cooling lotions, such as the following : Liquor of acetate of lead and tincture of opium, of each two drachms; common vinegar, one ounce ; distilled water, fifteen ounces ; keep lint or linen rag wet with this lotion constantly applied. When the inflammation has subsided, if there be still swelling and stiffness, rub in, night and morning, this liniment : strong liquor of ammonia and tincture of opium, of each one drachm; spirits of turpentine and soap liniment, of each one ounce. If it is merely a *strain* of the tendon, a little rest and the above remedial measures will soon afford a cure; but if a positive *rupture*, there may be much difficulty in getting the parts to unite. To accomplish this end, it is best to use a slipper with a strap attached to the heel, which passing up and encircling the thigh, may be drawn tight and kept so, as in Fig. 125. During the process of uniting, if the patient walk at all, it should be with a crutch; and after the cure has been effected, a high-heeled laced boot should be worn to protect the part.

FIG. 125.

Dislocations. By this term, we understand a displacement by violence, of one part of a joint from its natural connection to the other. By a knowledge of the structure of the joint, we are enabled to lay down rules by which the displaced bone may be returned or reduced. The ligaments which have been torn asunder re-unite, and the joint regains its healthy structure. The sooner this is done the better, and the easier will it be effected; but the attempt may be made even after the expiration of three or four weeks, if in the larger joints. After this period, the displaced bone adheres to the part it is in contact with, and the attempt should not be made but by an experienced surgeon. After the reduction, inflammation of a mild character may follow, which the application of a few leeches will suffice to remove. The joint may be bound up lightly with a wet band, and cold water, or vinegar and water, applied.

Dislocation of the Lower Jaw. *Symptoms.*—The mouth, is fixed open, pain in front of the ear, and extending up to the temples. This state of jaw occurs suddenly, while gaping, eating or talking, while the jaw is in motion, and is apt to recur.

Treatment.—Place the patient on a low seat, cover the two thumbs with a silk pocket-handkerchief, pass the thumbs into the mouth, and press with force, slowly applied, on the last four lower teeth, and at the same time raise the chin, pushing the jaw backwards. Considerable pressure is required by the thumbs; two pieces of wood may be employed as a substitute for the thumbs.

Dislocation of the End of the Collar-Bone. Either end of the collar-bone may be dislocated by a blow or a fall, indicated by a swelling over the joints which the bone forms either with the breast-bone or shoulder-blade, and by the suddenness of its occurrence. The treatment is very much like that of the fracture of the collar-bone, to which reference must be made. A pad of lint should be put on the swelling, and the arm raised high in a sling. This accident will require three weeks' rest.

Dislocation of the Shoulder. *Symptoms.*—Flatness of the shoulder, compared with the roundness of the sound side; inability to move the arm; the elbow placed at from two to three inches from the side; the attempt to press it to the side occasioning pain in the shoulder. If the fingers be passed up under the arm to the armpit, the head of the bone will be felt out of the socket, and may be revolved to make it perceptible.

Treatment.—A round or Jack towel, through which the arm should be drawn; the towel carried up to the armpit and twisted over the shoulder, and the two ends thus twisted passed over the back of the neck, and fixed

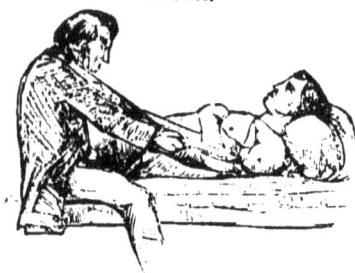

FIG. 111.

into a staple by a rope, or otherwise. Wash-leather, or other soft material to be wound around the arm, just above the elbow; a close hitch-knot of good quarter-inch line made upon it. The patient to be placed in a chair and held firm, or to lie down on a bed, and fixed. The arm may be drawn slowly and steadily, at an angle half way between the horizontal and vertical, and the extension to be continued for ten minutes to a quarter of an hour,—during which, frequently, the surgeon or superintendent should raise the arm, near the upper or dislocated end, upwards, with his two hands, with some force; the head will return into the socket with a sound, or slight shock. If the head of the bone be thrown forward on the chest, the extension to be carried a little backward; if backwards, a little forward. After reduction, a sling and three weeks' to a month's rest. The reduction may also be effected by laying the patient on the ground on his back, while the operator places his left heel in the left armpit (as in Fig. 111), if the dislocation occur on the left side, and his right heel in the right armpit, if it occur on the right side, and makes a powerful extension on the affected arm by both hands.

A person who has repeatedly dislocated his shoulder, may, if he have courage to bear a little pain for a few minutes, even manage himself to reduce it, if the accident should happen while he is out in the fields, and if there be a five-barred gate at hand. All he has to do is to get his arm over the top rail and then, having grasped the lowest rail he can reach, hold fast, and let the whole weight of his body hang on the other side of the gate; and then if he make some little attempt to change the position of his body, still, however, letting its weight rest on the top of the gate, the bone will probably slip into its place.

Dislocation of the Elbow Joint. The elbow-joint consists of three bones—the bone of the arm spreading out across the joint, and the radius outside, the ulna inside. The most common dislocation is when both radius and ulna are thrown backwards.

FIG. 112.

The Elbow-Joint.

1. The Humerus, or upper bone of the arm. 2. The Ulna. 3. The Radius:— these two being the lower bones, they are all held together by ligaments connected with both extremities of the bones, and with the shaft. 4 marks the insertion of the external lateral ligament, which passes beneath into the orbicular ligament 5, of which the hinder part (6) is spread out at its insertion into the Ulna. 7 marks the situation of the anterior ligament, scarcely seen in this view; and 8 is the posterior ligament, thrown into folds by the extension of the joints. There are other ligaments not shown here; nor are the muscles by which the complicated movements of the joint are effected.

Symptoms.—The joint motionless, a little bent; skin tight in front of the joint; a projection behind formed by the elbow, which, with its tendon is pushed back. The joint can neither be bent nor straightened.

Treatment.—Two men will be sufficient generally, unless the patient be very muscular. Extension to be made in a straight direction by both. The force required is not generally very great, and the reduction takes place commonly with a snap. Both bones may be forced forward—when this accident occurs, the elbow (*olecranon*) is broken. The imperfect line of the joint will be readily observed when a comparison is made with the opposite site joint.

Reduction.—Simple extension, as before, and, when reduced, the joint should be placed straight, and bound on to a splint. This accident will require from five to six weeks. Other accidents of this kind occur to the elbow joint, but they may all be treated on the same principle, namely, forcing the bones back to their natural position, which may be ascertained on comparison with the opposite sound limb, or the limb of another person.

Dislocation of the fingers and Toes. Dislocation of the fingers and toes is of rare occurrence; and, when it does happen, it is generally between the first and second joint. They may be easily known by the projection of the dislocated bones, and reduced without much difficulty, if done soon after the accident.

FIG. 113.

Treatment.—Fig. 113 will show the method of reduction; the clove-hitch, made with a piece of stout tape, may be used if there is much difficulty; the wrist during the operation should have a slight forward inclination given to it. This will relax the flexor muscles.

Dislocation of the Wrist-Joint. The hand may be forced backwards or forwards, but this accident is very uncommon. The nature of the case will be apparent to the slightest observation.

Treatment.—The hand should be grasped firmly by a powerful man, and drawn straight. If the hand slips, a bandage may be applied around it to aid the application of the extending force : but all that is required is full extension, by which the hand may be drawn straight. The same observation will apply to dislocation of the fingers.

Dislocation of the Hip-Joint. These dislocations are very important and very numerous being not less than four in number. The hip-joint consists of the head of thigh-bone and the socket formed by the pelvis, or continuation of the haunch-bone, towards the middle of the body. These accidents generally arise from a fall from a height, or a very severe blow, and are attended with severe injury to the structure of the joint and surrounding parts, although the consequences are not generally so severe as fracture of the neck of the thigh, detailed above.

The head may be thrown from the socket in four directions : First— upwards and backwards. Second—backwards. Third—downwards and inwards. Fourth—upwards and inwards. The most frequent is the first—upwards and backwards.

Symptoms.—Shortening of the leg to the extent of about two inches. The foot is turned in, and lies over the opposite foot ; the ball of the great toe towards the opposite instep ; the leg can not be turned out, nor the attempt made without pain. On examining the side of the buttock where the head is thrown, it will be felt on the bone, with the great projection formed by the end of the shaft of the bone placed in front of it. If the leg is rotated, the head and the great process, or prominence (trochanter), will be felt to revolve also. The line of the thigh is altogether too far outwards.

Treatment.—A round or jack towel should be applied, as in the case of the dislocation of the shoulder, and drawn up around the thigh as high as possible, and twisted over the hip-bone somewhat tightly, and fixed behind into a staple. Wash-leather, or a soft towel, to be wound around the thigh, above the knee, and around this the cord or line with two clove-hitches, one on each side of the thigh. The aid of six men will be required, who must draw very slowly and very cautiously. The patient should be placed nearly on the sound side, and the limb should be drawn a little across the other limb ; and after it has begun to descend, yet a little more across the opposite leg. When the thigh is fully extended it will generally reduce itself, and may be heard to return to the socket with a snap. Should it not do so, the superintendent should take the thigh high up towards the trunk in his hands, and raise it, and use a round towel, passed under the limb and over his neck, and raise it, *twisting it outwards* at the same time.

Dislocation Backwards. *Symptoms.*—The symptoms are nearly the same, except that the shortening is less, and the turning in of the

foot less also ; but both the symptoms exist in a degree. The head of
the bone lies lower down, and is less apparent to the hand when pressing on it.

Treatment.—The reduction is effected by the application of nearly the
same means. The limb should, however, be drawn rather more over
the opposite limb. When fully extended, it should be turned outwards,
when the head will slip into the socket.

Dislocation Downwards and Inwards. *Symptoms.*—The leg is
a little lengthened, and is drawn forwards on the trunk ; or, if placed
straight downwards on the ground, the trunk will be bent forwards as in
a stooping posture ; the toe points a little outwards. The line of the
thigh, when compared with its fellow, is directed too much inwards towards the middle of the body, and also too far backwards. The thigh
should be moved in all directions slightly, to ascertain that it is fixed in
this position.

Treatment.—Apparatus applied as before, patient lying on his back ;
extension to be made *downward* and *outward*, and when brought down,
after some minutes' extension, the thigh should be forced in its upper
end, outwards, by the hand, or the towel being placed between the thighs
and drawn in the direction opposite to that of the dislocation, namely,
upwards and outwards.

Dislocation upwards and Inwards. This is the most formidable of all these dislocations.

Symptoms.—The leg is shortened, and, like the last dislocation, drawn
forwards on the body, as though in the act of stepping to walk. Both
these last symptoms are more strongly marked than in the former. A
swelling, caused by the head of the bone, is apparent at the groin, and
the bone is firmly fixed.

Treatment.—The same means as before, and nearly the same direction
as the last accident, except that the limb should be drawn outwards and
more *back-*   *to* These two last dislocations may be reduced in the
sitting n°   f the patient, and in that position drawn around a bed-
post   ah's rest is required, or even more.

L   _ation of the Knee-Pan. The knee-pan (*patella*) may be
forced off the end of the thigh-bone either outwards or inwards ; but the
latter is very rare. Displacement outward is generally caused by sudden
and violent action of the muscles of the thigh.

Symptoms.—The appearance of the bone on the outside of the knee-
joint, instead of in front, attended with pain, stiffness of the knee, and
inability to walk without much pain.

Treatment.—The leg must be bent forwards on trunk, and the knee
straightened as much as possible ; the bone is then to be forced back
by the pressure of the hand. When it is returned, the knee should be
very slightly bent, and placed over a pillow. From three weeks to a
month will be required.

Ruptures or Hernia. This is a common weakness. It is caused by a portion of the bowels or their covering slipping out of their natural position in the cavity of the abdomen. If happening after birth, a rupture shows itself as a swelling suddenly appearing in the groin after violent exertion ; remaining distinct while the person stands upright, disappearing when he lies down, and returning again when he gets up. It also usually fills out when he coughs. If let alone it continues increasing in size, so that instead of the bowels being contained, as they should be, in the belly, the greater part drop into the swelling, which may become of an enormous size.

The proper treatment for a rupture ·is the wearing of a well-fitting truss. It should never be laid aside for a day. The bowels should never be allowed to become costive.

When through neglect of precaution the bowel "comes down" and will not return, the rupture is said to be "strangulated." This is a dangerous condition, and one should be able to recognize it. When a person has been costive two or three days, and he becomes violently and frequently sick, at first throwing up stuff like coffee-grounds, and after some hours like stools, and very offensive ; if there be a feeling of a cord tied round the midriff, constant feeling of sickness, much uneasiness and anxiety, there is great reason for supposing that this is something to do with a rupture. The inquiry should be made, and if there is a rupture, and it has fallen down, immediate treatment is required.

The patient may be put into a warm bath up to his neck, and kept there till he feel very faint ; he may then attempt, according to his own usual method, to put the rupture up, by pressing it gently, if it be in the groin, or by lifting it up if in the purse, and gently squeezing it toward the belly, but no violence must be used, or the gut will burst.

If this do not succeed, cold may be applied over the swelling, by filling a bladder with pounded ice and a handful of salt, or with a freezing mixture consisting of Glauber salt and salammoniac, to which some water must be added. Either of these, after being kept on some hours, will occasionally cause the return of a rupture, but they require to be used with some caution, as if the skin become frosted it may mortify. If neither ice nor the materials for the freezing mixture can be obtained, a wet rag may be put on the part, and evaporation encouraged by a continued stream of air from a pair of bellows, repeatedly wetting the cloth as it dries ; by these means almost as great a degree of cold can be produced as by ice.

Some surgeons have of late strongly recommended attempting reduction of a rupture by reversing the position of the body ; in other words by holding the patient head downwards, or nearly so, and they state that in many instances this method has succeeded.

Sprains. A sprain is a wrenching of a joint, whereby some of the ligaments—the bands which unite the bones—are torn or severely stretched. In many cases there occurs also an injury to the bones.

The severity of the injury varies extremely. A severe sprain, while containing no element of danger to the life of the patient, is nevertheless a serious injury, which may result in the permanent impairment of the functions of the joint.

Treatment.—The first item in the treatment of a sprain is *perect rest* of the limb. In many cases it is advisable to apply a splint in order to prevent any unconscious movement of the part. The splints are essentially the same as those which have been described in treating of fractures.

A most valuable feature for reducing the swelling and pain consists in wrapping the joints with cloths saturated with water as hot as can be comfortably endured. These fomentations should be continued for three or four hours. So soon as the pain and swelling have somewhat subsided an elastic bandage or cap should be placed around the joint.

Care should be taken in avoiding any violent movements of the affected joint for some weeks after the injury. Some stiffness may occur, which can be overcome by having the joint moved regularly every day by an assistant. The restoration of the motions can also be furthered by the use of some stimulating liniment, such as the ammonia liniment. In some cases serious diseases of the bone follow a sprain. This is the result of the original accident and cannot be averted by treatment.

Toothache. The most frequent causes of toothache are:

First.—Decay of the tooth extending to the nerve contained in the pulp of the tooth.

Second.—Inflammation of the membrane surrounding the root of the tooth. This usually causes a swelling of the gum, and the formation of an abscess or a "gum boil."

Third.—The general condition of the body, which predisposes to neuralgia.

Treatment.—The treatment depends upon the source of the pain. There is no one remedy which can be relied upon to cure toothache in general.

If the tooth be decayed, a dentist should be consulted and his opinion sought as to the advisability of removing the tooth.

In many cases it is possible for him to devitalize the pulp of the tooth, fill the cavity, and retain a useful member.

Until the services of a dentist can be procured the pain must be alleviated by the application of oil of cloves, creosote, chloroform, laudanum, or Jamaica ginger. The cavity of the tooth should be cleaned with a little cotton ; a few drops of one of the substances named is then placed upon a small wad of cotton, which is then gently inserted into the cavity of the tooth.

If the toothache be caused by inflammation at the root of the tooth, it can sometimes be quieted by painting the gum with a mixture of tincture of aconite and tincture of iodine in equal parts. Sooner or later, however, matter will form at the root of the tooth; the abscess should be opened at once. Such teeth are usually of no service, and may be extracted.

Toothache which originates not in any local difficulty around the tooth, but in a constitutional condition, must be treated by internal remedies. In many cases advantage will be derived from the internal use of quinine, two grains of which may be taken three times a day.

Toothache is sometimes of malarial origin; it may come on every second day, just like the chills and fever of ague. In such cases it must be treated by quinine.

Drowning. This frequent accident every one should learn how to treat. The body should be recovered as soon as possible from the water; the face turned downward for a moment, and the fore finger of a bystander thrust back into the mouth to depress the tongue, to favor the escape of any mucous or water that may be obstructing the throat.

The practice of rolling a person over a barrel, or hanging him head downward, to permit the escape of water from the lungs, is of no use. The body should be conveyed to the nearest house, a messenger having been previously dispatched to make the arrangements involved in the following:—As soon as the body arrives it should be stripped of the clothing, rapidly dried, placed on a bed previously warmed, the head, neck, and shoulders raised a very little, if any; frictions with the dry hands used to the extremities, and heated flannels kept applied to the rest of the body.

If artificial breathing can now be carried out for some time, it may be that the natural respiration can take place. Two methods are usually employed for the purpose, the first known as "Sylvester's Ready Method."

This consists, after the above suggestions have been carried out, in pulling the tongue forward, which better favors the passage of air along the base of the tongue into the trachea (windpipe), and then drawing the arms away from the sides of the body and upward, so as to meet over the head, by means of which the ribs are raised (expansion of the chest) by the muscles (pectoral) running from them to the arms near the shoulder. A vacuum is thus created in the lungs, the air rushes in, and the blood then is purified by the passage of the impure gases in the blood vessels to the air, and by the giving up by the air of a portion of oxygen to the blood. The arms are now brought down to the sides, and the elbows made to almost meet over what is called the "pit of the stomach." This produces contraction of the walls of the chest, and expulsion of the impure air from the lungs.

These two movements constitute an act of respiration, and should be persisted in, without interruption, at the rate of about sixteen to the minute. In other words, each complete movement should occupy about four seconds, which is about the natural rate of respiration in health.

The second " Ready Method," as it is called, is that of Marshall Hall.

The person whose breathing is to be restored is placed flat on the face, gentle pressure is then made on the back, the pressure removed, the body turned on its side, or a little beyond that. The body is then turned again on its face, gentle pressure again used to the back, then turned on the side. This should be done about sixteen times in a minute.

Both of these methods have the same object in view ; either may be exclusively used, or one may be alternated with the other. Most physicians express preference for the first described (" Ready Method of Silvester)". Both of the procedures might be practised in advance by the reader, because such practice might be more easily remembered than a concise rule.

In speaking of the restoration of persons drowned, it is often said that he was a good swimmer, and must have been attacked with "cramp." This is a spasmodic contraction of the muscles beyond the control of the individual, and occurs after exhaustion of the muscles from over-exertion. Persons suffering from debility, especially the debility peculiarly affecting the nervous system, should never be induced to go beyond depth in the water, or out of reach of immediate assistance. There is no warning in advance of the seizure, and the person sinks at once. Many lives are lost each season, in shallow as well as in deep water, from these seizures, which could have been avoided had the bather, perhaps just recovering from an attack of sickness, or even of indisposition, not neglected the precautions stated.

Recovery from drowning can scarcely be expected to take place after an immersion of five or six minutes, although there are well authenticated cases where restoration has taken place after an immersion of as much as twenty minutes. The efforts ought to be made, and persisted in, for at least a couple of hours. As soon as returning vitality permits, a few drops of brandy, in a little water, may be given ; and as the strength of the person is usually completely exhausted, from muscular efforts of the most violent and continued character, to save himself from drowning, some beef tea, or other easily digested nourishment, should be given.

The accident of drowning is so frequent that we go still further into the treatment of it, and copy into our pages the detailed rules published by the Life Saving Society, of New York. At the risk of repetition, we show their method. *Three* steps are to be taken :

1. Remove all obstructions to breathing. Instantly loosen or cut apart all neck and waist bands ; turn the patient on his face, with his head down hill ; stand astride the hips with your face toward his head, and locking your fingers together under his belly, raise the body as high as you can without lifting the forehead off the ground, and give the body a smart jerk to remove mucus from the throat, and water from the wind-

pipe ; hold the body suspended long enough to count one, two, three, four, five, repeating the jerk more gently two or three times.

2. Place the patient on the ground, face downward, and maintaining all the while your position astride the body, grasp the points of the shoulders by the clothing, or, if the body is naked, thrust your fingers into the armpits, clasping your thumbs over the points of the shoulders, and raise the chest as high as you can, without lifting the head quite off the ground, and hold it long enough to slowly count one, two, three. Replace him on the ground, and his forehead on his flexed arm, the neck straightened out, and the mouth and nose free.   Place your elbows against your knees, and your hands upon the sides of his chest, over the lower ribs, and press downward and inward with increasing force, long enough to slowly count one, two.   Then suddenly let go, grasp the shoulders as before, and raise the chest ; and press upon the ribs, etc. These alternate movements should be repeated ten or fifteen times a minute for an hour at least, unless breathing is restored sooner.   Use the same regularity as in natural breathing.

3. After breathing has commenced, restore the animal heat.   Wrap him in warm blankets, apply bottles of hot water, hot bricks, or anything to restore heat.   Warm the head nearly as fast as the body, lest convulsions come on.   Rubbing the body with warm cloths or the hand, and slapping the fleshy parts, may assist to restore warmth, and the breathing also.   If the patient can surely swallow, give hot coffee, tea, milk, or a little hot sling.   Give spirits sparingly, lest they produce depression. Place the patient in a warm bed, and give him plenty of fresh air ; keep him quiet.

Things in the Eye.   A great deal of pain is often suffered from various substances getting in the eye   The best way to remove them is by holding a knitting needle over the upper lid, close to and just under the edge of the orbit, then, holding it firmly, seize the lashes of that lid by the fingers of the disengaged hand, and gently turn the lid upward and backward over the needle, or substitute used.   Movement of the eyeball by the sufferer, in a strong light, usually reveals the presence of the intruding body, so that by means of a corner of a silk or cambric handkerchief, it can be detached and removed.

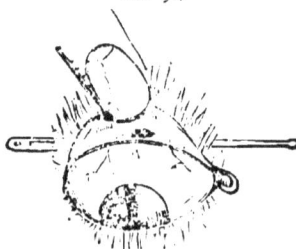

FIG. 92.

Removing an Object from the Eye.

Should the foreign body be imbedded in the membrane covering the eyeball or the eyelid, a steady hand and a sharp-pointed instrument will usually lift it out.

The foreign body often cannot be seen, but the person assures us he feels it. Often he does not really feel the presence of the body, as much as the roughness left by it. In such a case, or even if the body has been seen and removed, a soothing application to the injury is as useful as the same thing applied to a wound of the hand. Take a spoon or cup, heat it, and pour in a few drops of laudanum. It will soon become dense and jelly-like. A few drops of water added will dissolve this gummy material, and the liquid thus formed may be applied by the fingers to the "inside of the eye," as they say. The laudanum is opium dissolved in alcohol. The alcohol is somewhat irritating, but is easily evaporated by the gentle heat, leaving an extract of opium, which is dissolved by the water afterwards added.

After an injury of this kind to the eye, or any other which causes pain and inflammation, this precious organ should be bandaged, and excluded from the light.

Things in the Ear. Insects sometimes get into the ear, and cause much inconvenience, even if they do not sting and produce further mischief. The best mode of proceeding, in such case, is to fill the ear with sweet oil, which will kill the animal by stopping up its breathing pores, and generally floats it out. But if it be not thus dislodged, it must be washed out with a syringe and warm water.

When peas, beans, pebbles, or such like are found in the ears of children, attempts should never be made to get them out with a knitting-needle, or a stick. The rounded end of a hair-pin may sometimes be used. But the safest is to employ a syringe with a narrow nozzle, which will throw a good stream of water *behind* the object, and thus force it out.

The *earwig* is a small winged insect, which has its name from the frequency with which it has been the cause of trouble by entering the ear. Pouring in oil will soon destroy it. In cases where this means is not at hand, as in hunting, blowing tobacco smoke in the ear will kill or stupify it and similar intruders.

Things in the Nose. Children sometimes amuse themselves with poking things with which they are at play into their noses. If peas, beans, or any other seed or substance be thrust in, which swell as they moisten, no time should be lost in getting them out, otherwise as they enlarge, they become more firmly fixed, and more difficult to be removed, are attended with great pain and suffering, and may even cause dangerous consequences. Hard substances, as shells, which remain unchanged in bulk by moisture, are of less consequence, and may remain some days without causing much inconvenience, and often drop out of themselves.

If the pea or shell be in the nostril, the child should be made to draw his breath in deeply, and then stopping the other nostril with the finger, and closing the mouth firmly, to snort forcibly through that side of the nose in which the substance is lodged. If this be done soon after the accident, two or three efforts usually shoot the unwelcome lodger out.   But if this does not succeed, the nose must be lightly nipped with the finger and thumb above the pea or shell, so as to prevent it getting further in, and then the eyed end of a bodkin or probe, having been a little bent, must be gently insinuated between the bottom of the nose and the substance, and when introduced sufficiently far, must be gently used as a hook to bring it down.

CHAPTER IX.

The Urinary Bladder,
showing its muscular
fibres.

Section of the Kidney.

8, Left Ureter. 9, Left portion of Seminal Vesicles. 11,
11, Lateral Lobes of the Prostate Gland. 14, Urethra, tied
with a cord.

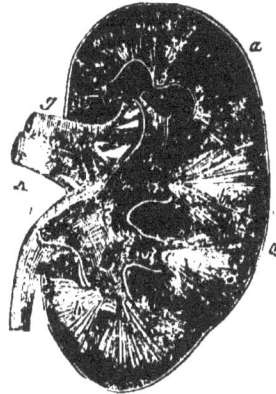

Acute Inflammation of the Kidneys. The existence of this
disease may be known by a sense of heat and sharp pains about the
loins, and a dull, benumbed feeling down the thigh.

Treatment.—Get the skin and bowels to act as soon as possible. For
this purpose give the patient a hot bath, and after drying, wrap him in
hot blankets and give him the following powder :

Jalap,   -   -   -   - 5 grains.
Cream of Tartar,   -   - 1 teaspoonful.

Repeat this morning and evening.

146

In order to promote the excretion of the urea and other poisonous materials, the patient should be permitted the free use of water, lemonade and other bland liquids for which he may express a desire.

It is quite important that the diet be judiciously selected during the disease, especially since the stomach is apt to be irritable. By employing chiefly milk and eggs we can, to a certain extent at 1st, diminish the work required by the kidneys.

In addition to these measures, the celebrated Dr. Grainger Stewart recommends early in the disease the following :—

Infusion of Digitates,	- -	1½ ounces.
Spirits of Nitrous Ether,	- -	6 drachms.
Simple Syrup,	- -	½ ounce.
Water to	- - -	6 ounces.

Take a tablespoonful three times daily.
Later on in the disease he gives :—

Tinct. of the Perchloride of iron,	-	2 drachms.
Spirits of Nitrous Ether,	-	4 drachms.
Infusion of Quassia to -	-	6 ounces.

Take a tablespoonful three times daily.

Diabetes, or an immoderate flow of urine, may be sugary diabetes, or simple diabetes.

Cause.—This disease is caused by injuries to the head ; the abuse of alcoholic liquors ; exposure to cold and dampness ; sudden checking of the perspiration ; emotion ; fevers ; diseases of the brain and spine ; and in the sugary form, the inordinate use of food containing sugar.

Symptoms.—The simple form, or *diabetes insipidus*, generally comes on suddenly, and the patient finds himself constantly passing his water, and on each occasion a large quantity, amounting to many quarts each day. The thirst is intense, and the patient drinks immoderately, both day and night, being compelled to rise frequently to urinate, and also to quench his thirst. He becomes weak and thin, with a harsh, dry skin.

The sugary form, or *diabetes mellitus*, usually comes on more slowly, and is attended with general languor, uneasiness, emaciation ; gradually the desire to pass the urine becomes very frequent, and the amount becomes alarming, accompanied with great thirst, and often with a voracious appetite The symptoms are similar to the other form, the general powers fail, hectic sets in, dropsy of the limbs, diarrhœa, the lungs become involved, and death closes the scene.

Treatment.—The treatment of this affection is yet, to a great degree, a matter of uncertainty. It is believed that the absolute prohibition of sugar in the diet, or of articles prone to form sugar, will aid greatly in checking this disease. Hence, the diet must be composed of meats eggs, butter, bran-bred, cabbage, onions, celery, lettuce, spinach ; and these must be varied by turns, lest distaste occur, and thus an additional cause of debility be induced. The forbidden fruits would be, all fruits, wheat bread, potatoes, beets, milk, liver, sweet breads, etc. Liquors must only be allowed in case of great debility, and then only in small quantities, preferring whiskey, or sherry or claret wine.

A great variety of medicines have been proposed, but few seem especially useful. Cod-liver oil, both by the mouth and inunction, may prevent the rapid progress of the debility. Perhaps the best results obtained have followed the " skim-milk treatment."

The patient is restricted to the milk carefully skimmed for a month at least, and then allowed in addition two to four pints of curd made by the use of rennets, gradually. As improvement occurs, lean meat, and green vegetables are given ; then eggs, fish, fowl, etc. There is always a loss of weight for the first few weeks after which improvement commences.

Gravel. A disease depending on the formation of stony matter in the kidney.

Treatment.—The general treatment should consist in a hot bath and warm fomentations ; a dose of castor oil should be administered, and when the bowels have acted, if there be much pain, the following may be given : Solution of acetate of morphine, one drachm ; spirit of hydrochloric ether, two drachms ; syrup of roses, half an ounce ; camphor mixture, four ounces. One-fourth part to be taken at bed-time. Linseed tea or barley water should be drunk freely. The following may be also used with good results : Infusion of buchu, seven ounces ; tincture of musk seeds, one ounce ; sal volatile, two drachms. Mix ; dose, two tablespoonsful once or twice a day. Or this : Essential oil of spruce, one scruple ; spirit of nitric ether, one ounce ; mix ; dose, a teaspoonful two or three times a day, in a teaspoonful of the decoction of marshmallow root. Or the following may be used : Rectified oil of turpentine, sweet spirit of nitre, oil of juniper, balsam of sulphur, of each half an ounce ; mix ; dose, fifteen or sixteen drops in a wineglassful of water three times a day. Or this : Alicant soap, eight ounces ; fresh lime, finely powdered, one ounce ; oil of tartar, one drachm ; mix with a sufficient quantity of water for a mass, and divide into five grain pills, from three to four of which should be taken daily. The following remedy has been highly recommended for this complaint : Parsley Blakestone (of the herbalist), ten cents worth, stewed down in a pint of water to half a pint ; when cool, add a wineglassful of gin. Take a wineglassful of the mixture every morning, until relief is afforded.

Stone, or calculus in the bladder, would be indicated by sudden stoppage of the stream in urinating ; pain in the bladder, more or less acute ; but could only be positively known by the use of a sound passed into the bladder and striking the foreign body. A prominent symptom, especially in children, is itching or pain at the end of the penis, causing the child to be constantly pulling at or rubbing the organ. This sometimes causes an elongation of the foreskin, and great annoyance by the collection of the secretions in this part.

Treatment.—The removal of the stone by an operation, either cutting or crushing, is almost all that can be done. A vegetable diet should be preferred, and the bowels and kidneys kept in prompt action.

Incontinency of Urine. This is rather a troublesome than a dangerous complaint: young children and aged persons are most liable thereto.

Causes.—Most generally from a relaxation of the governing sphincter muscle of the bladder, from weakness or paralytic affection, but sometimes it is caused by some irritating substance in the bladder; in children, some say, from sleeping on their backs.

Symptoms.—The water comes away in drops, sometimes involuntarily.

Treatment.—Dash cold water on the loins and genitals; a blister on the spine is useful, and the following are useful:

Stimulant tonic drops: Tincture of steel, six drachms; tincture of cantharides, two drachms; tincture of henbane, one drachm; mix, take thirty drops, three times a day, in water.

Or the following may be used with good effect: Sulphate of zinc, one drachm; powdered rhubarb; one drachm; Venice turpentine, two drachms; mix, divide into sixty pills, take one three times a day, and therewith a wineglassful of the decoction of leaves of bear's whortle, or bilberry.

Suppression of Urine. If there is a frequent desire of making water, attended with much difficulty of voiding it, it is called strangury. If none is made, suppression of urine.

Causes.—Inflammation of the urethra or passage, or sores, or severe inflammation about those parts; a lodgment of hard matter in the last gut or rectum, spasm at the neck of the bladder, exposure to cold, taking to excess cantharides, or blistering back, excess in drinking, stone in the kidneys or bladder, and enlargement of prostate glands.

Symptoms.—A constant desire, or feeling of necessity to make water and cannot, or if parted with, much pain and difficulty in passing it; much enlargement of the bladder. If stone in the kidney be the cause, often nausea, vomiting and acute pain in the loins; if from stone in the bladder, the stream of water will be divided into two, or suddenly checked.

Treatment.—If much inflammation and irritation exist, all straining to expel the urine should be avoided, and it let off by a cathetar every six hours, or, as it is commonly called, drawn. The following will be found very useful remedies.

Anodyne diuretic draught: Mucilage of gum acacia, six ounces; olive oil, one and a half ounces; mix well in a marble mortar, then add six drachms of spirits of sweet nitre; laudanum, one and a half drachms; fennel water, three ounces; mix, and take three tablespoonsful every three hours; or this:

Demulcent diuretic draught: Acetate of potash, two drachms; laudanum, one and a half drachms; syrup of marsh mallows, one and a half ounces; fennel water, eight ounces; mix, take three tablespoonsful every three hours.

The bowels must in all cases be kept free by using the following often:

Emollient clyster: Balsam copivi, two drachms; yolk of an egg; rub this and the balsam together; then add castor oil, half ounce; laudanum, one drachm; compound decoction of marsh mallows (that is well boiled), eleven ounces; mix, inject up the rectum; this foments and soothes the parts.

Inflammation of the Bladder. *Causes.*—It is seldom a primary disease, but is in consequence of inflammation in the neighboring parts; it is, however, sometimes caused by retention of the urine, and consequently over-distension of the bladder, or by a large stone in the bladder.

Symptoms.—Acute pain and tension of the part, frequent desire to make water, but difficulty in passing it, or a complete retention of it; and tenesmus, and frequent desire to go to stool to no purpose.

Treatment.—The diet must be light and thin; the drinks in all bladder diseases must be linseed tea, barley water, solution of gum arabic, marsh mallow tea, and the like; bleeding by leeches, if very bad, and this anodyne clyster: Linseed tea and new milk, each half a pint; laudanum, forty drops; mix and inject; this foments the internal parts. The bowels may be kept open by this mild aperient draught: Tartrate of potash, three drachms; tincture of senna, one drachm; manna, half an ounce; warm water, one and a half ounces; mix and take at once.

GENERATIVE ORGANS.—In the human race, as throughout the greater part of the animal kingdom, generation is accomplished by fecundation, or the effect of the vivifying fluid provided by one class of organs upon the germ contained in the seed or ovum formed by another class, in the opposite sex. This germ, when fecundated, is termed the embryo. The process consists of impregnation in the male—conception in the female.

The organs of generation in the male are -1. The testes and their envelopes, namely, the scrotum, or cutaneous envelope; the dartos, which corrugates or ridges the scrotum; and the fibrous or vaginal tunics; we must also here include the epidermas, above the testes; the vas deferens, or excretory duct, and the spermatic chord. 2. Vesiculæ seminales, forming a canal situated beneath the bladder. 3. The prostate gland surrounding the neck of the bladder and the commencement of the urethra. 4. Cowper's glands, a pair situated below the prostate. 5. The ejaculatory ducts. 6. The penis, which consists of the corpus cavernosum, the urethra, the corpus spongiosum, which terminates in the glans penis ; then there are the vessels, nerves, and a cutaneous investment, which by its prolongation forms the prepuce.

The female organs are : 1. The vulva or pudendum, the external parts, comprehending the labia ,pudendi (lips), clitoris, situated at the middle and superior part of the pudendum ; the nymphæ or alæ minores ; the urethra, which terminates in the meatus urinarius, opening into the vagina, which is occupied by the hymen, a semilunar fold, or the carunculæ myrtiformis, its lacerated remains after the first act of copulation. 2. The uterus, whose appendages are —the ligamenta lata (the broad ligaments), and the round ligaments, commencing immediately before and below the Fallopian tubes, which extend to the ovaria.

THE PERINÆUM—The space between the anus and the external parts of the generative organs. The operation of cutting for stone in males is usually performed here, and here it is that serious injury sometimes occurs, when persons fall with their legs astride of any object, or get a bruise in that position, as on horseback ; bloody urine, or complete stoppage may be the consequence, arising from inflammation of the bladder or urethra.   Rest and warm fomentations, with leeches, and the use of the catheter, if necessary, must in this case be resorted to ; with low diet, aperients, and cooling medicines, to keep down any tendency to fever there may be.

We abstain from giving cuts of these parts and organs for sufficiently obvious reasons ; in a book intended for family use they would be altogether objectionable.   With regard to the diseases which more immediately affect them, a few simple remarks will be made under their several heads ; but we would here impress upon our readers the necessity of at once seeking medical advice for all affections of the genital organs.   It is in the treatment of this peculiar class of diseases that advertising empirics reap their richest harvest, entailing the greatest present sufferings, and most fearful after-consequences upon their too credulous dupes.

Bleeding from the Bladder—Bloody Urine.   *Causes.*—Falls, blows, bruises, or some violent exertion, such as jumping or the like ; sometimes from small stones in the kidneys, ureter, or bladder, which wound those parts.

Symptoms.—The blood parted with is somewhat coagulated, and deposits a dark brown sediment resembling coffee grounds.   When the blood is from the kidney or ureter it is commonly attended by acute pain, and sense of weight in the back, and some difficulty in parting with it.   When the blood is from the bladder immediately, it is usually accompanied by a sense of heat, and pain at the lower part of the body, and the blood is not so much coagulated.

Treatment.—Empty the bowels with cooling purges, and take the following astringent tonic mixture :—

| Tincture of steel, | - | - | - Three drachms. |
| Infusion of roses, | - | - | - Six ounces. |

Mix, take two tablespoonsful every three hours ; and physicians generally recommend that the drink should be thick barley water, solution of gum arabic, or decoction of mallows sweetened with honey.

Mild aperient draught :—

Tartrate of Potash,	-	-	- Three drachms.
Tincture of senna,	-		- One drachm.
Manna,	-	-	- Half ounce.
Warm water,	-	-	- 1½ ounces.

Mix, and take at once.

Clouded, Thick or Dark-Colored Urine.
Take the following antacid diuretic mixture :—

Liquor potash,	-	-	- Two drachms.
Tincture of cubebs,	-	-	- Two ounces.
Infusion of buchu leaves,	-	-	Thirteen ounces.

Mix, take two tablespoonsful four times a day.

The following will usually effect a cure :—

Dilute nitric acid,	-	-	- Two drachms.
Syrup of lemon,	-		- Four drachms.
Water,	-	-	Eight ounces.

Mix, take one tablespoonful three times a day ; or take half a teaspoonful of citric acid in water four times a day.

Gonorrhœa or Clap.
Treatment.—In the mild form of this disease, and in the first stage when the discharge is fully developed, and the inflammation confined to the first inch and a half of the urethra, the first thing to be done is to open the bowels briskly. This may be effectually accomplished by administering the following powder :

| Powdered Jalap, | - | - | - | - 20 grains. |
| Calomel | - | - | - | - 4 grains. |

Mix. To be given in something thick at bed-time. Animal food, all stimulating drinks, such as ale, spirits, and wine should be carefully abstained from. Great cleanliness should be observed, the penis should be bathed several times a day in hot water, allowing it to soak for a few minutes each time, and taking care to wash off all discharge which might be collected between the foreskin and glans of the penis. The patient should rest as much as possible, and he should wear a suspensory bandage to keep the penis out of the way of all friction. His diet should consist of light farinaceous food, such as arrow-root, sago, or bread puddings ; and for his ordinary drink, barley water or toast and water. Broths of an unstimulating character, such as mutton and chicken might be both allowed occasionally. He should then take the following powder three times a day :

Cubebs Powder,	-	- - -	1 drachm.
Powdered Gum Arabic,		- - -	1 scruple.
Carbonate of Soda,	-	- - -	10 grains.

Make a mixture. To be taken in a little milk or water. This treatment should be continued for a few days, after which the doses of cubebs might be increased to two drachms three times a day. Should the discharge still continue after persevering in this plan for eight or ten days, and when the active stage of the inflammation has subsided, the following mixture may be administered with advantage :

Balsam Copiaba	-	- -	1 drachm.
Sweet Spirit of Nitre,		- -	4 drachms.
Tincture of French Flies,		- -	2 drachms.
Water,	-	- -	3 drachms.

Mix, shake up, and take one teaspoonful three or four times a day, in a wineglassful of water.

Later on in the disease the following injection should be used :

Sulphate of zinc,	
Subacetate of lead (of each),	- 16 grains.
Water,	- 4 ounces.

Inject one ounce once or twice daily. If this injection causes pain, dilute it with an equal quantity of water.

Dr. Sturgis.

One of the most painful and sometimes most troublesome consequences of gonorrhea is Inflammation of the Testicle. This affection, usually termed "swelled testicle," may occur at any period of the disease ; and although its occurrence may be favored by improper treatment or mode of living, is in most cases independent of such causes. It arises from extension of the inflammation from the urethra down the spermatic canals to one or both testicles, but usually attacks only one at a time. It is best to be avoided by careful attention to regular living and quiet during the inflammatory stage of the gonorrhea. It commences sometimes with pain in the testicle itself, and sometimes the pain is felt first in the groin, in the situation of the spermatic cord. If its approach is thus perceived, the application of numerous leeches in the groin, or of cupping to the loins, with rest in the recumbent posture, and suspension of the scrotum in a proper bandage, will frequently prevent the extension of the inflammation to the testicle itself. Should the inflammation, however, have reached that organ or commenced in it, the most immediate relief will be obtained by carefully surrounding the swelled testicle with narrow strips of adhesive plaster, together with per-feet rest, the testicle being further supported in a bandage; and should the pain extend to the groin, the application of leeches in that situation will usually put a stop to the disease in a few days. As many, however, will be unable to apply the strapping in a proper manner, and it is only applicable in the early stage of the affection, it may be as well to say that usually the inflammation will subside spontaneously in a few days, if the patient will keep quietly lying on his back with the testicles supported in a proper bandage, and fomented either with hot water, or with cold water, as his feelings may dictate. The bowels should be kept open by saline purgatives, such as Epsom salts, etc., and the diet should be low. If there is much pain in the groin, flank and back, leeches should be applied in the former situation, or cupping in the latter, and a full dose of Dover's powder should be taken at bed-time. In extremely painful cases, great relief will be experienced by the application of a tobacco poultice to the scrotum. This may be made by mixing equal parts of tobacco and meal together, and moistened with hot water.

Mercury is never requisite in this affection, and leeches should never be applied to the scrotum itself. The swelling of the testicle in most cases leaves hard swelling on the back of the gland, which is gradually removed in process of time; but during its existence, care should be taken to keep the testicles well supported in a suspensory bandage, as relapses under neglect of this precaution are not unfrequent.

In Phymosis, the glans of the penis frequently become excoriated from the irritation of the matter from the urethra, and warty excrescences grow between the glans and the foreskin. In order to prevent such effects, great cleanliness should be observed, the foreskin should be drawn back as far as possible and the matter washed off, and warm water should be thrown under the foreskin several times a day by means of a syringe. If excoriation or warts exist, black wash will be of the greatest service—it should be used in a similar manner to the warm water. Black wash is made by mixing thirty grains of calomel with two ounces of lime water—to be well shaken when used. The bowels should be kept gently open by means of the common black draught.

Treatment of Chordee.—We have observed before that chordee consists in a painful erection of the penis, produced by the non-extension of the spongy cellular body surrounding the urethra, while all the other parts of the penis are distended with blood. This want of harmony between the parts occasions the penis to be bent downward, and also the pain which is experienced by the patient during an erection. In order to obviate this the penis should be rubbed with strong solutions of opium, such as the tincture; or pledgets of linen, wet with the tincture of opium, should be constantly applied, taking care to exchange them as often as they become warm; or it may be rubbed with the following application, which is found of great service in this affection:

Extract of Belladonna,	-	-	-	- 2 drachms.
Camphor,	-	-	-	- 10 grains.

Rub up the camphor into a fine powder, having previously dropped on it a few drops of spirits of wine, then add the belladonna ; about the size of a pea of this, rubbed along under the surface of the penis, and upon the frænum and bridle, quickly brings down and relieves pain. All lascivious ideas should be dismissed from the mind. The bowels should be kept open by a mild aperient. As the erection generally comes on more frequently when the patient becomes hot in bed, the best means of temporarily relieving it will be to bend the penis downward with the hand, and to apply cold ; but the most certain means of preventing it will be to administer at bed-time the following draught :

> Tincture of Opium,   -   -   -   -   20 drops.
> Camphor Mixture,   -   -   -   -   1½ ounces.

Mix. This draught to be taken at bed-time, and to be repeated in three or four hours, if not asleep or if in pain.

In the treatment of sympathetic Buboes accompanying gonorrhea, little will be required to be done, as they depend on the amount of inflammation in the urethra, and will increase and diminish in size according as the original disease becomes worse or better ; however, as they sometimes enlarge very much and become very painful, it may be found necessary to apply leeches once or twice a week. The patient should rest as much as possible, and pledgets of linen wetted in Goulard water should be constantly applied. The bowels hould be kept freely open. If they should not yield to this treatment, but should proceed to suppuration, poultices should be constantly applied until matter is formed, when it may be evacuated by the lancet.

In cases of retention of urine following gonorrhea the patient should be placed in a warm bath, and a large dose of laudanum administered. If this treatment does not succeed in relieving the bladder, the catheter should be introduced.

Syphilis, or Fox, is usually accompanied by three distinct characters of sores or ulcers ; first, the common primary venereal sore ; secondly, the phagedenic or sloughing sore ; and thirdly, the true syphilitic or Hunterian chancre. The common venereal sore usually appears in three or four days after connection ; the patient feels an itching about the tip of the penis, finds either a postule or an ulcer, situated either upon the prepuce externally or internally, at its junction with the glans or on the glans itself, or at the orifice of the urethra at its union with the bridle or frænum.

The form of this ulcer is generally round or circular, and is hollowed out, presenting a dirt brown, hard, lardaceous surface, which secretes a puriform matter. When this ulcer is situated on the prepuce, it becomes raised, particularly at its edges ; when in the fossa, or at the root of the glans of the penis, it is ragged ; and when on the glans, it is excavated. Its progress is first destructive, and then healing ; and, if not in-

terfered with in favorable cases, usually runs its course in about twenty
days—the destructive or ulcerative stage lasting about ten days, and the
granulating or healing stage lasting the remaining ten. This sore is un-
accompanied by any thickening or hardened base in the first stage, unless
interfered with by mal-treatment, dissipation, or the abuse of caustic.
This sore is frequently productive of swelling and inflammation in the
groin, and is followed by warts and growths of an unhealthy character
situated between the buttocks.

Treatment.—In the first stage—that is, before the crust falls off, or
where the ulcer is very small—the sore should be touched with lunar
caustic ; this frequently stops the ulcerative stage, and causes it to take
on a new action by which it heals ; the same application, but weaker,
will be necessary if the sore becomes indolent. During the ulcerative
stage, or that stage in which the ulcer increases instead of diminishes,
great attention must be paid to cleanliness ; the sore should be washed
three or four times a day with warm water ; a piece of lint or fine linen,
covered with spermaceti ointment, or wetted with black wash, should be
applied to it after washing. The bowels should be kept open, and five
grains of blue pill, or five grains of Plummer's pill, administered night
and morning, taking care not to produce salivation. When the sore assumes an indolent character, great benefit will be derived from the application of the following wash :

Lunar caustic,   -   -   -   -   -   -   5 grains.
Distilled water, -   -   -   -   -   -   1 ½ ounces.

Mix. A piece of lint or linen, wetted in this lotion, to be applied to the
sore three or four times a day.

Black wash is the best application for those warts and growths which
spring up about the anus and buttocks. The swelling in the groin, arising from the common venereal sore, seldom requires any treatment ; but
if it should prove troublesome and painful, leeches may be applied,
followed by fomentations and poultices. The patient should rest as
much as possible, and make use of a plain, unstimulating diet.

In the treatment of phagedenic or sloughing ulcer, no specific rules
can be laid down, the sores at one time requiring a stimulating and at
another time a soothing method of treatment. This sore usually commences from an excoriation, or a pustule, as in the case of common venereal sore, or it may follow that form of the disease. It is known
by that process of extension by which its edges appear to melt
away ; "the action is chiefly confined to the margin, which the destructive process having undermined, overlaps with an irregular and ragged
edge." In this form of ulcer, the reparatory action commences as soon
as the destructive is exhausted, so that the two processes advance
together at opposite edges, the sore ulcerating at one part and healing
at another at the same time. In the commencement, the sore may be

touched with nitric acid, or diluted nitric acid, upon two or three occasions, and if found not to agree, the stimulating treatment should be laid aside, and the soothing substituted. It may now be washed with warm water, and various applications tried, as it is impossible to say what form of wash will answer best. Those in most repute are the black wash, yellow wash (yellow wash is made by adding six grains of corrosive sublimate to four ounces of lime water), diluted nitric acid, Peruvian balsam, and solutions of the nitrate of silver.

For the treatment of the true syphilitic or Hunterian chancre, mercury is the sheet anchor, and must be employed either internally or externally, or, where circumstances require it, by both means. This sore, unlike the preceding, seldom appears before a week or ten days, and is sometimes not detected for four or five weeks after connection. It appears in the form of a red, raw, superficial ulceration, placed on a circumscribed elevated, hardened base. This base is firm, incompressible, and inelastic, and is as hard as cartilage ; it is destitute of pain and very slow in its progress. This form of the disease is generally accompanied by true bubo—that is inflammation of one or two glands in the groin, distinct and circumscribed in their outline, and totally dissimilar to those swellings in the groin arising from gonnorhea or the common venereal sore.

As soon as the sore is detected, the patient should commence taking five grains of blue pill, and a quarter of a grain of opium, made into a pill, night and morning ; and he may, at the same time, in order to bring the constitution as soon as possible under the influence of mercury, rub in, twice a day, along the inside of the thigh, about the size of a nut, an ointment composed of blu or mercurial ointment and camphor. The following is the formula : Rub down twenty grains of camphor on a slate with a spatula, having previously saturated it with spirits of wine, and then mix it up with mercurial ointment. This treatment should be continued only until the mouth and gums become slightly affected, when it should be left off for a short time. The patient should be kept under the influence of the medicine for three or four weeks, and then the decoction of sarsaparilla and the hydriodate of potash administered. Five grains of the latter, in a common-sized tumblerful of the former, may be taken three times a day, and continued for a month, according to circumstances. The sore, in the meantime, should be kept clean, and such applications employed as may happen to agree with it best ; these consist of washes of nitrate of silver, black wash, and spermaceti ointment.

When the ulcer cicatrizes, or heals, and any hardness remains, mercury should be given to promote its absorption, and the skin destroyed by the direct application of the nitrate of silver.

When the disease has been neglected, or a sufficient quantity of mercury has not been given, the constitution becomes affected in a time

varying from six weeks to three months, which manifests itself by producing sore throat, disease of the skin, and inflammation of the eyes.
These diseases must be severally treated by the remedies already recommended.

When a bubo becomes troublesome and painful, it should be well
leeched, fomented and poulticed ; and should it proceed to suppuration,
the matter must be let out by a free incision with a lancet, as soon as
fluctuation is felt.

During this disease, the patient should be warmly clad, he should rest
as much as possible, and live on plain, unstimulating food.   In the commencement, he should refrain entirely from spirits, wine, or fermented liquors ;, he should not expose himself to wet, damp, or night-air, and he
should pay strict attention to his bowels.

The syphilitic poison, when it has once entered into the system, is
with great difficulty eliminated, and sometimes shows itself in children
several generations removed from the person originally infected.   It may
be communicated by a pregnant woman to the child in her womb through
the medium of her blood, by which the fœtus is nourished ; and thus,
as in numerous other cases of disease, the children suffer for the sins of
the parent.

Gleet.   By this name is designated a chronic discharge from the
urethra, which occurs as the sequel to gonorrhœa   In many cases the
discharge does not cease after the gonorrhœa has lasted four or five
weeks ; it becomes gradually less and finally amounts only to a few drops
in the course of the day.   These drops are thin and watery, and the discharge occasions the patient no other annoyance than the mental anxiety and uneasiness.

In many cases the discharge will be noticed only when the patient
rises in the morning.   There is no pain when the bladder is evacuated,
no frequent desire to pass the water, in fact nothing wrong except the
slight watery discharge.   This is, however, most obstinate and difficult
to get rid of ; it frequently lasts for months, and in many instances even
years may elapse before the patient may become entirely free from this
last vestige of his indiscretion.

One of the uncomfortable features about gleet is the fact that excesses
of various kinds are apt to increase the discharge so that its quantity becomes almost as great as during the original gonorrhœa.   This is especially often the case after excessive sexual indulgence, but it may also
follow immoderate use of liquors, especially of beer, or even physical
or mental exhaustion from over-work.   After the discharge has broken
out a few times it becomes extremely difficult to check it completely.

Treatment.—The treatment of gleet should never be undertaken by
the patient himself.   It is, in fact, even in the hands of the ordinary physician, a most obstinate and puzzling affection.

Gleet often depends on the existence of a stricture, and can be relieved only when the stricture has been cured. In other cases, again, it is due to ulceration of the urethra, or to inflammation at the neck of the bladder. For these conditions the physician alone must be consulted.

Where the patient is weak and debilitated he should use the following :—

Tincture of the chloride of Iron,	6 drachms.
Tincture of Nux Vomica,	6 drachms.
Compound Tincture of Iron	4 ounces.

Take a teaspoonful in half a wineglass of water before meals.

Impotence. An inability to perform the sexual act is one of the commoner derangements of the genital organs in the male.

Treatment.—Temperance in all things is the first great principle to be recognized in treating this difficulty. Avoid over-exercise and stimulation, bathe the parts each evening in cold water, and use the following :

Tincture of Sanguinaria,	3 drachms.
Flues Extract of Stillingia,	5 drachms.

Take 15 or 20 drops in water 3 times daily.

Dr. Bartholow.

Or,

Tincture of Cantharides,	6 drops.
Tincture of the Chloride of Iron,	15 drops.

Take this three times daily in water.

Dr. H. C. Wood.

This prescription has been found to cure the above condition so speedily as to commend itself to the use of all medical men in the treatment of their cases.

Nocturnal Emissions. These, to which young men are sometimes especially liable, often cause more alarm than there really is any occasion for ; they are involuntary discharges of the seminal fluid, and are likely to occur when the organs are excited by dreams, or imaginations of a certain character. Unless they become frequent and profuse, there is no reason for regarding them with the morbid feeling of anxiety which they commonly occasion ; still such discharges should be attended to and checked as much as possible. They generally indicate a debilitated system, and are in most cases, perhaps, the result of criminal self-indulgence and venereal excesses, from which those thus affected should rigorously abstain. A course of tonic medicines should be taken ; nothing is so good as the muriated tincture of iron with quinine, about one grain of the latter with ten drops of the former, in a little water, three times a day. Sea bathing or the shower bath, regular but **not excessive**

exercise, a sufficiently nourishing but not a stimulating diet, with gentle aperient medicines if required (avoiding aloes), are the proper remedial measures.

Dr. Bumstead gives the following prescription for its special tonic effect upon the genital organs :

Tincture of Iron,   -   -   -   - 3 ounces.
Extract of Ergot (Squibb's Flued),   -   - 3 ounces.
Take a teaspoonful in water after each meal.

As a direct means of diminishing the frequency of the emissions he recommends :

Bromide of Potash,   -   -   -   - 1 ounce.
Tincture of Iron,   -   -   -   - 1 ounce.
Water,   -   -   - 3 ounces.

Take one or two teaspoonsful in water after each meal and at bedtime.

The avoidance of tobacco in all its forms, cleanliness of mind and body, laxatives for the bowels when needed, and, in a word, attention to the rules of hygiene are to be strictly enjoined.

Masturbation (*Self-Pollution, Onanism*).—This destructive vice is indulged in to a frightful extent by the youth of both sexes. Often the habit is indulged in without its victim having the slightest knowledge of its destructiveness, and only when nature is so outraged that the system refuses to perform its offices, does the victim become conscious of the evil. A grave responsibility rests upon parents towards their children in these matters. Every child, male or female, should be carefully watched, until it is old enough to understand the subject, and then it should be carefully explained to it. The earlier this is done, and the stronger the impression made upon the mind of the child of the wickedness of this abuse, the better. It is truly a matter of life and death, and squeamishness is as much out of place as if the child were really dying.

Treatment.—The habit must be abandoned at once ; unless this be done no treatment will be of any avail. The moral character must be strengthened. All things of a sensational character must be avoided, the company of the good and virtuous cultivated, and the mind kept engaged in some elevating study or useful employment. Avoid all stimulants—wine, coffee, liquors, novels, love-pictures, balls, theatres, and sleeping on the back. Use a hard bed, and light not too nutritious food. Take whey, acidulous drinks, fruits, and a vegetable diet. Take a bath morning and evening, and exercise till quite fatigued. Avoid all aromatic articles fish, eggs, jelly, game, salad, mushrooms, cantharides, aloes, and all stimulants except camphor. If there is irritation in the cerebellum, by heaviness or heat, cut the hair very short, wear **no** cap, use a hard pillow, ice applications on the nape, with hot foot bath, dry or narcotic friction on each side of the vertebral column, also cold liquid applications.

In extreme cases, where the habit has overcome the reason of the patient, he should not be left alone day or night. Let him go to bed only when much fatigued, and rise the moment he awakes. Let the bed be hard and cool, with light covering. Attend to the evacuating of the bowels and bladder. Dashing cold water on the genitals, with the free use of the vagina syringe for females, will assist much in restoring the tone of the organs.

CHAPTER X.

POISONS, BITES AND STINGS.

General Directions For Treatment of Poisoning—Acids—Alcohol—Aconite—Ammonia –
Antimony—Arsenic—Baryta— Belladonna— Bismuth—Bitter-Sweet—Camphor—Copper
—Corrosive Sublimate— Digitalis—Henbane –Iodine--Iron Sulphate—Lead--Mushrooms
—Salt Petre—Lunar Caustic—Opium—Oxalic Acid—Poison Vine—Prussic Acid—
Phosphorus—Oil of Savine—Stramonium—Strychnine—Tobacco –Laying out the dead
—Bites and stings of Insects.

Hardly any accident is more common than poisoning, either by intention or by mistake. Often, there are symptoms of poisoning when the patient cannot or will not say what it is he has taken. Therefore the importance of some

General Directions for the Treatment of Poisoning.

1. Make the patient *vomit* at once. To do this give him a teaspoonful of ground mustard, in a teacupful of warm water, every minute, until he throws up. Or a tablespoonful of common table salt in the same quantity of warm water. Or tickle the inside of the throat with a feather or the finger.

2. After he has well vomited, let him take the antidote for the poison, when any one is given in the following pages.

3. Rest and quiet, a low diet, and the reclining position, should be kept for several days. Barley water, linseed tea, chicken broth and such articles should be the main staples of food for a few days.

Proceeding now to the particular poisons which one is liable to be called upon to treat, we speak of them in alphabetical order :

Acids. *Mineral.*—These are nitric acid, or aqua fortis ; sulphuric acid, or oil of vitriol ; and muriatic acid, or spirits of salt. Commence with a vomit. Then, give a tablespoonful of *lime-water* in a wineglassful of water, every minute, until the burning pain is relieved. Common soap may be made into a strong suds, and a wineglassful of this given frequently.

Alcohol. This, in the form of brandy, rum, gin, whisky, or other intoxicating liquors, is a dangerous poison. Persons who become "dead-drunk " are liable to be dead in earnest, unless restored. Give an emetic, or tickle the throat, to make them vomit. Then pour cold water, from a height, on their heads. When awakened, give five grains of carbonate of ammonia, in a wineglass of water, every quarter of an hour.

162

Aconite. Called also markshood and wolfbane. Give emetic at once, and if the patient is stupid, keep up the breathing by artificial respiration. This dangerous poison is much used in liniments, which are sometimes taken by mistake.

Ammonia, and other alkalies. By the latter name chemists call lime, soda (washing soda), potash, lye, and similar materials. Spirits of hartshorn, or aqua ammonia, is a well-known strong irritant. When taken internally, give, at once, table vinegar, by the dessertspoonful, till the pain lessens. Lemon-juice will also answer. Olive oil will afterwards be beneficial.

Antimony. This is contained in tartar emetic, and antimonial wine ; also in "hive syrup," sometimes used for colds in children. It causes violent vomiting. The antidote is tannin or tannic acid, nutgalls, or powdered oak bark. A teaspoonful of tannin, in water, may be given. A cup of strong green tea is also useful as an antidote, and is readily prepared.

Arsenic. This common poison is found in ratsbane, Paris green, fly poison, Fowler's solution, and other familiar preparations. The first step is to give an emetic and vomit freely. Then the patient should drink plenty of milk, white of egg and water, or flour and water. The antidote is freshly prepared hydrated peroxide of iron, which can be had of any apothecary.

Baryta. This substance, largely used to adulterate certain paints, is sometimes accidentally swallowed in poisonous doses. The antidote is water, acidulated to about the strength of lemonade, with sulphuric acid, which converts the baryta into an insoluble compound, which must be dislodged from the stomach by an emetic.

Belladonna. *Deadly Nightshade.*—The berries are sometimes eaten by children. Empty the stomach with an emetic, pour cold water, from a height, upon the head, if there is stupor, and give ten drops of laudanum, every quarter of an hour, for two hours (to an adult, two drops to a child).

Bismuth. Often used in toilet powders. Give an emetic, and when it has acted, copious draughts of milk.

Bitter-Sweet. *Woody Nightshade, Dulcamara.*—Proceed as for belladonna.

Camphor. Give an emetic, followed by draughts of warm water, flaxseed tea, gum-arabic water, milk and the like.

Copper. Cooking in copper vessels, or allowing acid fruits to remain in them, may poison the food. Blue vitriol is a common salt of copper. After free vomiting, give milk, or white of eggs, in water. Ordinary baking soda, or iron filings, a half teaspoonful every five minutes, should be given, to the extent of four or five doses, if the symptoms are severe.

Corrosive Sublimate. The bichloride of mercury (corrosive sublimate), often used as a solution, in houses, for destroying vermin about beds, is one of the most active poisons, when taken internally. The red oxide of mercury (red precipitate) is another dangerous salt of the same metal. When swallowed, the white of eggs should at once be given, and often repeated. In the absence of this form of albumen, common milk can be used, or wheat flour beaten up with water.

Digitalis. *Foxglove.*—Treat as for belladonna. Twenty or thirty drops of aromatic spirits of ammonia, in water, will aid in restoring the strength of the heart.

Henbane. *Hyoscyamus.*—Tr ' as given above for belladonna.

Iodine. The common tinc. iodine, used for external application, is the usual form of this poison. Starch, in water, may be freely given until vomiting is secured by an emetic.

Iron, Sulphate of. *Copperas or Green Vitriol.*—This is an irritant poison. After vomiting, let the patient take carbonate of soda (baking soda). as recommended for copper poisoning.

Lead. The form from which poisoning by this substance usually takes place is the acetate of lead (sugar of lead). The carbonate of lead, the " white lead " of the painters, and the red oxide (red lead), are also sometimes swallowed in poisonous doses. They all act as irritant poisons.

The treatment of such cases consists in giving, as an antidote, water acidulated to about the strength of lemonade, with sulphuric acid (oil of vitriol).

Sulphate of magnesia (epsom-salts), or the sulphate of soda (Glauber's salt), in water, are also reputed antidotes. After the antidote has been given, in poisoning by lead, an emetic should be given.

Lead poisoning, in the forms of painters' colic, and lead palsy, follow from much exposure to the metal. Cosmetics containing white lead, and hair color restorers, containing sugar of lead, water which is contaminated by lead piping, and eating food preserved in leaden cans, may cause them. The free use of milk will often prevent these bad effects.

Mushrooms. When poisoning from eating mushrooms takes place, the contents of the stomach should at once be evacuated by an emetic. After vomiting has commenced, it should be promoted by draughts of warm water, barley water, but particularly by drinking copiously of warm milk and water, to which sugar has been added.

Nitrate of Potash. *Saltpetre.*—In large doses, say half an ounce or more, taken internally, is followed by poisoness symptoms. There is pain, with heat in the stomach, vomiting, and purging of blood with great prostration, and other symptoms denoting the action of an irritant poison.

No antidote is known. The treatment consists in rapidly evacuating the contents of the stomach by an emetic, and the free administration of mucilaginous drinks, with some paregoric, every little while.

Nitrate of Silver. *Lunar caustic.*—Used in hair dyes and indelible ink. The antidote for this violent poison is common salt, which acts promptly and efficiently. A strong brine should be swallowed as soon as possible.

Opium. *Laudanum, Morphia, Soothing Syrup.*—This is the most frequently used poisonous agent. The first step is to give an active emetic, like ground mustard, salt and water, or ipecac.

The narcotic effects upon the brain, at the same time, as far as possible must be attended to. If the respiration is yielding to the poison, that is, falling much below the standard of about twenty to the minute, it must be sustained by assistance. The exposed body of the patient should be dashed with cold water, not neglecting the head, face, and chest. After the cold water has been sufficiently used in this way, the body should be dried, removed to a dry spot, and hot applications made to the extremities and other parts. This is necessary, owing to the heat-producing power of the body being impaired by the suspended or diminished respiration.

If the respiration is not suspended, but is going on at a diminished rate, say six or eight to the minute, artificial respiration is not required, unless the number of respiratory movements of the chest falls below that ; but the other measures may be used. In addition to these, a strong stimulant, in the shape of twenty or thirty drops of aromatic spirits of ammonia in a tablespoonful of water, may be given three or four times, at intervals of a couple or more minutes. It is better than brandy, or anything alcoholic, because the mode of the action of brandy is much the same upon the brain as opium, and it might be rather adding to instead of taking from the poison that is at work. The aromatic spirits of ammonia will give the advantage, without the disadvantage. A few tablespoonsful of very strong freshly made coffee is a useful thing to give in such cases.

Among measures to keep in activity the circulation and respiration, as well as to promote the elimination (casting out) from the blood of the poison acting as a narcotic, there are few things more useful than muscular exercise. The patient should be walked or run up and down the room constantly.

Oxalic Acid. Often taken by mistake for epsom salts—a dangerous mistake. Give at once po. red chalk, calcined and powdered magnesia, or strong lime-water. After these have been administered for a time an emetic will empty the stomach.

Poison Vine. *Poison Oak, Poison Sumac.*—These are species of *Rhus,* and abound in many parts of the United States. The juice, or even the touch of the leaf when the dew is on it, brings about in many persons redness, itching, swelling, and blisters. The person so affected should take a dose of epsom salts or cream of tartar, to empty the bowels, and bathe the parts with lead wash. A wash of a teaspoonful of baking soda in a tumbler of water immediately after exposure will prevent the eruption. When the latter has appeared, painting the parts with tincture of iron will usually check it. A solution of blue vitriol or sulphate of copper, a teaspoonful to the pint of water, is also an efficient lotion.

Prussic Acid. This substance is so rapidly fatal that little can be done to avert death. If possible, give an emetic of mustard, and follow with stimulants.

Phosphorus. Sometimes taken in rat and roach poison, and in matches. There is no antidote known. Some calcined magnesia may be given in plenty of water, to be rapidly followed by an emetic, and then an abundance of mucilaginous drinks.

Savine, Oil of. This substance in large doses inflames the stomach and bowels. Give olive oil in tablespoonful doses, and empty the stomach with emetics.

Stramonium. Usually known as thorn apple, or jimson weed; belongs to the same natural order in botany as belladonna, and when taken internally in improper quantities, is to be treated by similar general means. Children often gather the seeds and eat them.

Strychnine. *Nux Vomica.*—This dangerous substance destroys life quickly, with severe convulsions. The patient should be made to vomit without delay. Chloroform should then be given in teaspoonful doses, in water, every quarter of an hour. Artificial respiration may be tried if apparent death has set in.

Tobacco. The oil of tobacco is a violent poison, and the leaf when swallowed causes nausea and vomiting. This should be encouraged with warm water, after which twenty-drop doses of ammonia, in a tablespoonful of water, will be of benefit.

Laying Out the Dead. When a person dies, the eyes should be closed by gentle pressure with the fingers for a few minutes, or a small weight—a penny or similar coin—may be used to keep up the pressure.

The limbs should be straightened out carefully, and a bandage applied under the lower jaw, to support it; the arms should be placed by the side, and the lower extremities kept in position by means of a bandage connecting the great toes.

The clothes should then all be removed, and after the body has been thoroughly washed, be replaced by a clean bed-gown.

Bites and Stings of Insects.—The most frequent wounds of this kind are those made by bees or wasps. These are not of course dangerous, unless many be inflicted at the same time, or unless the sufferer be a young child. Single stings are, however, quite painful, and occasion much swelling if inflicted round the eyes or in the mouth.

When a large number of bees attack an animal, they inflict injuries which are usually fatal. Men, as well as horses, have been repeatedly stung to death by an infuriated swarm of bees.

In some parts of the country there are found certain other small animals which inflict painful and severe wounds. In the southern and western parts of our country individuals frequently suffer from the bite of a large spider called the *tarantula*. In the northern part of the country there is a small black spider which is often found in the neighborhood of old logs and trunks of trees, and which inflicts a painful wound.

Treatment.—The bites of spiders and the stings of bees and wasps usually require no other treatment than measures to allay the pain. There are various popular remedies employed for this purpose. Sometimes hartshorn is applied to the skin in the vicinity of the wound; some people consider a cabbage leaf the best possible application. The fact is, that anything which serves to cool the surface diminishes the irritation and pain. Cloths wet with cold water, or a mixture of equal parts of water and hartshorn, are usually very grateful to the sufferer; or a solution of ordinary baking soda, a teaspoonful of which is stirred up in a glass of water, will make a cooling and pleasant application.

If a person be stung in the mouth or throat, the swelling which results is apt to be so great as to embarrass the breathing. In such a case the patient should even before the parts are much swollen, employ faithfully gargles of hot water containing a little borax. A popular remedy is a mixture of vinegar and water, which is heated and used as a gargle. The swelling is sometimes so great as to render surgical interference necessary in order to prevent suffocation. The tongue may be punctured with a sharp penknife in several places, and the use of the gargles should be continued.

In many of these cases the pain is so great that opium must be given to alleviate it. For this purpose twenty drops of laudanum may be taken every two hours until three or four doses have been administered.

CHAPTER XI.

Women, in addition to the diseases incidental to both sexes, are sub-ject, from their peculiar organization, to a number of distressing com-plaints ; and in many instances, through a mistaken sense of delicacy, their lives are shrouded in sadness and pain, from a want of proper in-formation relating to their peculiar ailments. To woman is entrusted a most sacred charge—the germ of a new being, whose position and use-fulness in life will be greatly influenced by her prudence or indiscretion.

We shall treat under their various heads the principal forms of disease and suffering that commonly affect the women of civilized life.

Menstruation : its Physiology and Functions. The func-tions of the uterus, by which the menstrual, catamenial, or monthly dis-charges take place, generally commence between the fourteenth and six-teenth years of age, although we have known them to begin as early as eleven or twelve. A considerable period may elapse between the ap-pearance of the first and second menstrual discharge ; but, when they are properly established, their recurrence at regular periods may be cal-culated on with great certainty, unless some functional or other derange-ment of the system interferes with them. Ordinarily, a lunar month of twenty-eight days is the intervening period ; but with some women the discharge occurs every third week. The fluid discharged resembles blood in color, but it does not congulate. The quantity is from three to five ounces, and the process occupies from three to seven days.

The cause of this monthly flow is the ripening and expulsion of the egg from the ovaries. We quote from Professor C. D. Meigs, of Phila-delphia, a reliable and competent authority on these matters :—

" ' *Omne vivum ex ovo*' (every living thing comes from an egg, or germ), is the universal law of reproduction. This can be shown as well in the

168

vegetable as in the animal kingdom. The sturdy oak from the acorn, the ear of corn from the grain planted by the farmer, the robin and the elephant, all springing from germs, go to prove the truthfulness of this law. Every seed, every egg, contains a germ, which, when brought under proper influences, will produce of its own kind. Thus far all is plain enough, but where do these germs originate? It has been ascertained that each animal, as well as each plant, is provided with an organ for the production and throwing off of these cells or germs. In the female, this organ is the ovary. The ovaries are two in number—small oval bodies, about one inch in length, a little more than half an inch in breadth, and a third of an inch in thickness. This measurement will differ in some cases, but will be found generally correct. Each ovary is attached to an angle of the womb, about one inch from its upper portion, by a ligament. The whole physiological function or duty of the ovary, is to mature and deposit its ova or eggs every twenty-eighth day, from the age of fifteen to that of forty-five, or for about thirty years. This function is suspended only during pregnancy and nursing, but sometimes not even then. There are numerous cases on record where the woman has had her courses regularly during the time she was pregnant, and there are many with whom lactation does not at all interfere. During the maturation or ripening, and discharging of the ovum into the canal or tube which conveys it into the womb, the generative organs become very much congested, looking almost as if inflamed. This congestion at last reaches such a height that it overflows, as it were, and produces a discharge of bloody fluid from the genitalia, or birth-place. As soon as the flow commences, the heat and aching in the region of the ovaries, and the weight and dragging sensation diminish and gradually disappear. Thus you will see that menstruation consists merely in the ripening and discharge of an ovum or egg, which, when not impregnated, is washed away by the menstrual fluid, or blood, poured out from the vessels on the inner surface of the womb. It will also be seen that a woman can become pregnant only at or near the time of her menses. The marvellous regularity of menstruation has always excited great wonder, but why should it? When we look around, we see that both animal and vegetable life have stated and regular times at which germ production takes place. Fruits and vegetables ripen, and animals produce their young at certain periods. It is a law of nature, and why should not woman obey it in her monthly term? Now since we have shown that menstruation consists in the ripening and regular deposit of an egg—the flow being but the outward visible sign of such an act—it is possible that a woman may menstruate regularly without having any show. To prove this, there are many cases on record where a woman has married, and become pregnant without having had the least show, which would be impossible if she did not menstruate. Again, a woman who has always been regular may have several children, without in the meantime having had any sign. This may be explained by her becoming pregnant during the time she was nursing her first child, carrying it to the full term, again becoming pregnant, and so on, until being no longer impregnated, her courses return, and are regular thereafter.

" Menstruation commences at about the age of fourteen or fifteen in this country. In warmer climates it appears earlier, and in colder ones later. Menstruation, menses, courses, catamenia, monthly periods, and "being unwell," are some of the terms by which this function is designated. Those who are brought up and live luxuriously, and whose moral and physical training has been such as to make their nervous systems more susceptible, have their courses at a much earlier period than those who have been accustomed to coarse food and laborious employment. The appearance of the menses before the fourteenth year is regarded as unfortunate, indicating a premature development of the organs ; while their postponement until after the sixteenth year is generally an evidence of weakness, or of some disorder of the generative apparatus. If, however, the person has good health, and all her other functions are regular, if her spirits are not clouded, nor her mind dull and weak, it should not be considered necessary to interfere to bring them on, for irreparable injury may be done. The first appearance of the menses is generally preceded by the following symptoms : Headache, heaviness, languor, pains in the back, loins, and down the thighs, and an indisposition to exertion. There is a peculiar dark tint of the countenance, particularly under the eyes, and occasionally uneasiness and a sense of constriction in the throat. The perspiration has often a faint and sickly odor, and the smell of the breath is peculiar. The breasts are enlarged and tender. The appetite is fastidious and capricious, and digestion is impaired. These symptoms continue one, two, or three days, and subside as the menses appear. The menses continue three, five, or seven days, according to the peculiar constitution of the woman. The quantity discharged varies in different individuals. Some are obliged to make but one change during the period, but they generally average from ten to fifteen.

" It is during the menstrual period that the system, especially of young persons, is more susceptible to both mental and physical influences. Very much depends upon the regular and healthy action of the discharge, for to it woman owes her beauty and perfection. Great care should therefore be used to guard against the influences that may tend to derange the menses. A sudden suppression is always dangerous ; and among the causes which may produce it may be mentioned sudden frights, fits of anger, great anxiety, and powerful mental emotions. Excessive exertions of every kind, long walks or rides, especially over rough roads, dancing, frequent running up and down stairs, have a tendency not only to increase the discharge, but also to produce falling of the womb."

The quantity and duration of the emission varies greatly in different women, and unless the former is either very scanty or excessive, these do not appear important particulars ; but the regular recurrence of the issue is important to health. This should be borne in mind, and due care taken not to suppress the discharge by exposure to cold or wet, or by violent exertions of any kind about the time when it may be expected. It is desirable that young females should be properly informed by their mothers, or those under whose care they are placed, of what may be expected at a certain age, or they may be alarmed at the first appearance of the menses, taking it to be some indication of a dangerous disease or injury, and, perhaps, by mental agitation, or a resort to strong medicines, do mischief to themselves.

DISEASES OF THE MENSTRUAL FUNCTION.

Delayed or Obstructed Menstruation. If the menses do not appear at the usual age, or for some years after, no alarm need be felt, provided there is no constitutional derangements which can be attributed to this cause. If the girl has not developed about the hips and breast, and feels not the changes peculiar to this period, it would be very injurious to attempt to force nature. If, however, she is fully developed, and her general health suffers, a course of treatment will be necessary.

Causes.—An undeveloped state of the germ-producing organs ; an impoverished condition of the blood ; habitual costiveness ; or the womb may be closed, or hymen be imperforate.

Symptoms.—Discharges of blood will sometimes occur from the nose, mouth and gums, or from the stomach and bowels. Nearly always there will be unnatural heats and flushings, headache, tendency to faint, and hysterical symptoms,

Treatment.—The patient must be very attentive to her diet and regimen. Much exercise should be taken in the open air. Avoid late hours, rich food, and exciting pursuits. If the retention proceeded from costiveness or bad condition of the system, use the means as directed under the several heads. If from a mechanical cause a physician must be consulted. When it results from defective action of the ovaries, give the following :

Carbonate of Iron, -	·	-	- 1 drachm.
Extract of Gentian, -	·	·	- 1 drachm.

Mix, and make into thirty pills. Dose, one pill two or three times a day.

Suppressed Menstruation (*Amenorrhœa*).—Suppression is the stoppage of the menses after they have been once established. It **may** be either acute or chronic.

Causes.—Sudden cold, wet feet during the flow, fear, strong emotions, anxiety, or any cause that affects the general health. Chronic suppression may result from the acute, or from defective nutrition of the organs; from the early determination of menstrual functions, or from the weakness occasioned by a profuse discharge of whites from the uterus.

Symptoms—The symptoms usually present in a well developed body are all those mentioned in delayed menstruation, in a more aggravated form. In chronic suppression, failure of the general health, loss of appetite, pains in the head, back, and side, and constipation are the usual symptoms. At the regular periods when the menses ought to appear, there will be great excitability, and an aggravation of the above symptoms. With those of full habit, there will be a strong, bounding pulse, with acute pain in the head, back, and limbs ; with the feeble and sickly, extreme langour, tremblings, shiverings, and pale visages.

Treatment.—Care must be taken that pregnancy is not the cause of the stoppage, or the health may be seriously injured by treatment for their restoration. Where the flow has stopped suddenly from exposure, the patient must take warm diluted drinks, saline aperients, till the bowels are freely opened ; have hot bran poultices applied to the lower part of the abdomen ; immerse the feet and legs in hot water, rendered stimulant by the addition of mustard. If the pain is extreme, take an opiate draught every four hours, and have an injection, with one drachm of turpentine and half a drachm of tincture of opium thrown up. The patient must be kept as quiet as possible. If it can not be brought on, wait till the next period, and use the hip-bath a few days before the period. Every other night the bath should be made more stimulant by the addition of a little mustard ; and on every occasion, active friction with dry coarse towels should be used. A lavement containing two drachms of spirits of turpentine may also be useful ; and a leech or two applied to each thigh, on the upper part, as near to the situation of the uterus as may be. Also give the following, which seldom fails if persisted in :

Barbadoes Aloes,	-	-	-	- 1 drachm.
Sulphate of Iron,	-	-	-	- 1 drachm.
Powdered Cayenne,		-	-	- ½ drachm.
Extract of Gentian,		-	-	- ½ drachm.
Simple Syrup, sufficient quantity.				

Mix, and make into sixty pills. Dose, one pill night and morning.

The warm hip-bath should be used about the proper period of menstruation ; and it would be well to give some uterine stimulant, such as a mixture composed of spirits of turpentine, made into an emulsion with a yolk of an egg, sugar, and essence of juniper, about six drachms of the first and the last, in a six-ounce mixture. One ounce to be taken three times a day. Attempts to promote the discharge in any case must not be prolonged much beyond the menstrual periods, between which all possible means must be taken to strengthen the system—good diet, plenty of active exercise, the use of a shower-bath, or cold or tepid sponging ; steel mixture with aloes and iodine, in one or the other of its forms, are the proper remedies.

If the amenorrhœa proceeds from a want of energy in the uterine organs to secrete the red discharge, as is often the case after frequent miscarriages, child-bearing, or inflammation of the womb, as well as after leucorrhœa, or "whites," there will probably be the usual signs of menstruation, followed by a white discharge only and accompanied by acute pain at the bottom of the back, vertigo, hysteria. Weakly young women, before accession of the menses, and elderly ones, at the time of their cessation, or "change of life," as it is commonly called, are often so affected. In such a case we should prescribe hot baths and tepid injections, pills of sulphate of iron and aloes, with balsam of copaiba, ten or twenty drops in milk, three times a day ; or powdered cubebs, from a scruple to half a drachm ; good diet and a recumbent position as much as possible during the periods. If the patient is of full habit, apply leeches, ten or twelve over the sacrum, to be followed by a blister, with restricted diet, and, for a time, avoidance of sexual intercourse.

Painful Menstruation (*Dysmenorrhœa*).—This is the rule with some females, but the exception with most. It does not seem to be in any way connected with the quantity of the discharge, and it may attend both the secretion and the emission ; or but one or other of the processes, and but partially coming on in paroxysms, or continually, during the whole process. The matter discharged is often thick and membraneous, and sometimes has in it clots and streaks of blood.

Causes.—The cause of this is not very clear. It has been observed to occur after strong mental emotions, a cold caught during the menstrual period, a fright or other shock to the system, and would seem to indicate an irritable state of the womb.

Treatment.—Use the warm hip-bath and friction : fomentation of the parts with warm water ; diluent drinks, saline aperients, and a spare diet, must be followed ; also injection of warm water high up into the vagina ; and take the following mixture :

Tincture of Aconite leaves,	-	-	-	2 drachms.
Best Spirits of nitre	-	-	-	-	1 ounce.
Morphia	-	-	-	-	-	3 grains.
Simple syrup	-	-	-	-	4 ounces.

Mix. Dose, one teaspoonful every half hour till relieved.

Profuse Menstruation. This consists either in the too frequent return, or too long continuance of the periods ; or in an excess of quantity during the natural periods ; or in the character of the discharge being other than it should be, such as thick, fibrous, or bloody.

Causes.—This is in consequence of irritability of the uterine system, probably produced by over-exertion, luxurious living with insufficient exercise, or excesses of any kind ; too rapid child-bearing, frequent miscarriages, or protracted lactation. The habitual use of tea and coffee will also produce it.

Symptoms.—It is generally accompanied by pain across the loins, great languor and debility, throbbing of the temples, headache, and vertigo. When there is much hemorrhage, there is an aggravation of these symptoms, sometimes followed by dropsy of the cellular tissue.

Treatment.—In persons of full habit, where the menses are not bloody, the following may be taken :

Sulphate of iron, - - - - - -	12 grains.
Dilute Sulphuric Acid, - - -	1 drachm.
Sulphate of Magnesia, - - - -	6 drachms.
Cinnamon Water, - - - - -	12 ounces.

Mix. Take two tablespoonsful three times a day. If there is much pain, add tincture of henbane, two drachms ; or compounded infusion of roses may be taken, with sulphate of magnesia ; or ten or fifteen drops of the muriated tincture of iron in water, with or without the salts, as the bowels may require, two or three times a day. Sponge the loins and pudenda with vinegar and water, use the hip-bath, but let it be cold water, with a little salt in it, to strengthen the system as much as possible, and avoid all enervating influences. If there is blood in the discharge, use cold vaginal injections, with alum and opium in them, or the latter with gallic acid, about a drachm of each to a quart of water. Apply hot bran-poultices to the breasts : keep the feet warm, but let the loins be lightly covered ; take gentle exercise, bitter ale, and tonics, especially iron.

Cession of Menstruation.

As the accession of the menses shows when the womb is in a fit state for conception, so then, cessation gives notice that the period of child-bearing is past. With females of our age and country they commonly, continue up to the age of from forty to fifty; sometimes they cease at about thirty-five, and in a few instances have been known to continue up to the age of sixty. This cessation marks what is commonly termed the turn or change of life in women, and with those of average health it occasions little or no disturbance of the general system. There may be flushings of the face, and a sense of fulness in the head, with occasional giddiness ; but with those who are weakly and nervous, or suffering under any organic disease, we generally see a marked change at this period—it may be for the better or worse, according to circumstances. With most persons the stoppage of the menses is a gradual process,—the quantity decreases, or the intervals become protracted, and it is probably superseded by a white discharge, which also will by and by disappear ; with some the cessation is sudden and complete.

Women generally consider this an eventful period of their lives, and attribute all sorts of wonderful effects to it ; but we can not learn that a sickly constitution was ever renovated at this time, or a strong one ever broke down in consequence of the change ; indeed, fewer women than men die at the age when it usually takes place. Diseases of the genital organs, and of the breasts, which are sympathetically associated with them, require special attention at this time, as they are likely to be stimulated into activity. When there are no complications of disease connected with the change, little or no medical treatment is required. It is best to observe an abstemious diet, and to keep the bowels moderately open with rhubarb or colocynth pills ; powdered aloes, with canella, commonly called hiera picra, is a popular opening medicine, and as good as any for such an occasion, except the patient be of a very full habit, in which case it should be a saline aperient like the following : Dissolve two ounces of epsom salts in a pint of warm water, add one drachm of essence of peppermint, and take a wineglassful every morning, or twice a day if required. If there is flatulency or hysteria, add to each dose twenty drops of the fœtid spirits of ammonia, or the same of ether.

The Womb and its Diseases. This most important organ in woman is situated in the cavity of the pelvis,—from whence, when distended in pregnancy, it rises into the abdomen, with the general lining membrane of which the pelvis, called the peritoneum, it is covered. It is of a flattened pear-shape, and is held in its place by elastic ligaments. In its unimpregnated state it is about three inches in length by two in breadth across the broadest part, and one in thickness. At the period of puberty it weighs about one and a half ounces ; after parturition, from two to three ounces : and in the ninth month of utero-gestation, from two to four pounds. It is supplied with glands, vessels, and nerves, the latter of which consti tute an extensive network over its entire surface.

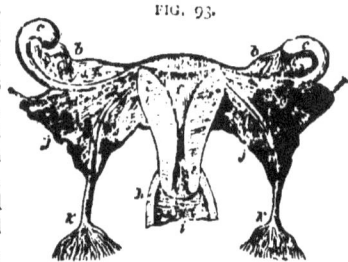

FIG. 93.

The Womb and its Appendages.

a, Right Ovary. *b, b*, the Fimbriæ. *c, c*, the Fallopian Tubes. *d*, an Ovum being grasped by the Fimbriæ. *e*, an Ovum descending the Fallopian Tube to the Womb. *f*, Cavity of the Womb. *g, g*, Walls of the Womb. *h*, Wall of the Vagina. *k, k*, Ligaments of the Organs.

The ovaries are two in number—one on each side of the uterus, in the groin. They are nearly as large as the male testicles, and perform a corresponding function. When the germ (or ovum) has been perfected in the ovary, it is cast out, and seized by the extremity of one of the fallopian tubes, through which it is conducted to the uterus.

Falling of the Womb. (*Prolapsus Uteri*)—Is the falling of the womb from the weakening of its membraneous supports and the pressure of the viscera above, generally increased by tight lacing, the pressure of the clothing, sustained by the abdomen and adding to its weight upon the uterus, and by the pressure of a load of fæces in the constipated rectum, and the daily efforts to expel them. These causes all acting together, press the uterus down the vagina until it sometimes comes out externally. As nearly all women are exposed to some of these causes of falling of the womb, nine in ten have more or less of it. Even young girls, eighteen or twenty years old have falling of the womb. Very few entirely escape it, for very few women are entirely well.

Treatment.—Avoid tight corsets and heavy skirts ; suspend the under garments from the shoulders, and not from the waist, as is usually done ; avoid fatigue, and lie down as much as possible ; use the cold hip-bath once or twice a day, . d inject cold water into the vagina with a syringe; use plain vegetable diet, and avoid tea and coffee, spirituous drinks, &c. If the womb has descended to the external orifice, it is often necessary to restore it to its natural situation by pressing it upward and backward by a finger or two passed into the vagina. If there be any pain in the operation, the vagina should be well washed by injections of thick flaxseed or slippery-elm bark tea for a day or two before the astringent washes are used.

When the womb has passed completely out of the vagina, which is always drawn down and inverted, the parts sometimes become suddenly so swelled that it would be impossible, as well as improper, to return them at once. The inflammation to be reduced by leeches, ice-water, or warm fomenting poultices of bread and milk, or hops and flaxseed, continually applied until the swelling and pain subside ; then, with the hand well oiled, and the patient's hips well elevated upon a cushion or pillow at the edge of the bed, the organ is to be placed carefully within the vagina, and restored to its natural situation. The bowels and bladder must be regularly evacuated ; but the patient should not be allowed to rise for several days, and should even then assume the upright position very gradually and cautiously, after having used injections composed of the following ingredients :—

Take one drachm of alum, and dissolve it in half a pint of clear water ; or, half an ounce of the inner bark of the black oak, with three gills of water ; boil down to a pint and strain. Two ounces of either of these preparations should be injected into the vagina by means of a vagina-syringe. This operation should be repeated twice a day, for a week or more—the syringe being always well lubricated with lard or oil, that it may be introduced without difficulty or pain.

If there is much sensibility, use from thirty to forty drops of laudan-.m in the injection, and repeat the operation till it is removed. If there re frequent relapses, a pessary must be worn.

The womb is also liable to fall either backwards (*Retroversion*) or forwards (*Anteversion*), but the treatment of these conditions must be confided to an experienced physician.

Inflammation of the Ovaries. (*Ovaritis*).—This disease is characterized by pain, heat swelling, perhaps redness, in one or both groins. It is to be treated as any other inflammation—sitz-baths, with rest, and a strict diet. The bowels must be occasionally opened by a gentle aperient, as castor-oil. Injections of tincture of belladonna and hysocyamus are very useful and soothing.

Inflammation of the Womb. The treatment is the same, with the addition of injections, both to the rectum and vagina, cold if they can be borne, or with the chill off.

The ovaries, uterus, and fallopian tubes are so closely connected in situation and function, that they are generally inflamed together. The cause may be weakness, causing a local determination of some general disturbance, such as cold or irritation of these organs. It usually follows childbirth, abortions, or excessive and violent sexual intercourse.

Ulceration of the Neck of the Womb. Ulceration of the neck of the womb is produced by corroding discharges and the irritation of continual sexual intercourse. It is readily cured by abstinence, vaginal injections, and direct application to the parts of a strong solution of nitrate of silver once in five or six days ; or the ulcers touched with solid nitrate of silver once in five days.

Flooding (*Uterine Hemorrhage*).—This commonly occurs after abortion, in the puerperal stage of labor, or it may be occasioned by disease of the womb. Immoderate flow of the menses is also called flooding, and to this some women are very subject. It is extremely weakening to the system, and should be checked as soon as possible.

Treatment.—The best treatment is perfect quiet, and astringent and tonic medicines like the following :—

> Tincture of the Sesquichloride of Iron, - - 2 drachms.
> Infusion of Quassia, - - - - - - - - 6 ounces.

Mix, and take a tablespoonful every four hours. If there is much pain and irritation, add tincture of conium, or hyoscyamus, two drachms. Should this not have the desired effect, consult a medical man, as there may be disease of the womb.

Polypus of the Womb. When a woman has been wasting away for some time, under a more or less copious discharge of blood, and the remedies recommended under the head of flooding, have been faithfully but unsuccessfully used,—when, during this time, she has remained free from burning and pain in the part, but has merely complained of a sense of weight in the womb,—there is great reason to suppose that she has a

L

polypus excrescence growing there, and the best advice should be at once procured.

Cancer of the Womb. *Symptoms.*—A sudden pain which shoots through the bottom of the abdomen, and either disappears entirely, or leaves after it a dull aching or a gnawing sensation, accompanied by more or less discharge of a fluid, which is sometimes pale and thin, but soon becomes thicker, yellower, perhaps streaked with blood, and very offensive. This pain is gradually rendered more severe and almost constant, and an exhausting hemorrhage sets in at times, perhaps continuing until checked by fainting. In other cases, a burning heat, followed by a fœtid discharge of matter mixed with streaks or spots of blood.

Treatment.—Cleanliness, fresh air, plain nutritious diet, regulation of the bowels, and tranquillity of mind, are all that can be recommended in a work like this. The woman who has the misfortune to be visited with this affection, must resolutely determine to retire early from the active duties of life, and be content to abstain from indulgences which would heat the system, excite her passions, and increase the circulation of blood. Bland, soothing nourishment, and local applications, are all that can be administered until she can have judicious and experienced medical assistance. Her bowels should be kept open by the mildest laxatives that will effect the object ; the fœtid and erosive discharges should be washed away by injections of flaxseed tea, castile-soapsuds, or a solution of chloride of lime or soda, with a little hop or camomile tea. When the hemorrhage becomes very profuse, the vagina should be plugged up with a fine sponge or a strip of soft cotton or linen rag, imbued with strong alum-water.

Formidable as the last two diseases are, they are not always beyond the reach of surgery.

Whites. *(Leucorrhœa).*—There is no disease so common among women as this complaint. Few married women, particularly if they are mothers, escape its attack. Very generally this troublesome discharge is associated with general debility, especially if it has continued profuse for any length of time. Hence it is very desirable that attention should be paid to it at the commencement : for, if neglected, it may seriously impair the constitution, and grow from a comparatively mild affection into an inveterate and dangerous disease.

Causes.—Over-exertion of the uterine organs, irritation of the rectum from loaded and constipated bowels. It may also be brought on by diarrhœa, piles, worms, irritation of the bladder or of the nervous system, excessive sexual intercourse, miscarriage, abortion, and displacement of the womb. Weakness, too, is a cause of *fluor albus*, as well as a consequence of its long continuance ; confinement in a warm atmosphere, luxurious living, and chlorosis must likewise be numbered among its exciting causes.

Symptoms.—This disease may be distinguished from gonorrhœa by the absence of local irritation and swelling of the external parts, and the glands of the groin ; also by the discharge being less regular and copious. In leucorrhœa this is commonly at first white and pellucid, or it may be opaque and thick, coming away now and then in lumps. After a while the color will perhaps change to green, yellow, or brown, and sometimes it will become very acrid, causing abrasion and smarting of passing urine. In this stage it is apt especially during pregnancy, to cause a gleety discharge from the urethra of one having sexual intercourse with the patient. Ere long, if the disease is not checked, we get great local irritation and constitutional disturbances : there will be costive bowels, pains in the loins and back, great lassitude, with nervous and hysterical affections. Menstruation, too, will be irregular, at one time being altogether suspended, and at another too abundant.

Treatment.—If the patient is of full habit, saline aperients should be taken, and a spare diet observed ; local ablutions should be practised three or four times a day, using occasionally a decoction of poppies for the purpose ; the hip-bath, and an injection of goulard water, with a scruple of powdered opium in each pint, will also be found serviceable. The recumbent position should be preserved as much as possible, and the parts kept cool. The practice of wrapping them up is objectionable, as it heats and weakens them. Local treatment will be of little avail in cases of long standing, unless the general health be attended to. To keep the bowels gently open, take five grains compound rhubarb pill, as often as required ; and to strengthen and cool the system the following mixture :—

Sulphate of Iron	-	-	-	- 12 grains.
Diluted Sulphuric Acid,	-	-	-	1 drachm.
Sulphate of Magnesia,		-	-	4 drachms.
Cinnamon-Water	-	-	-	- 12 ounces.

Mix, and take two tablespoonsful two or three times a day. In obstinate cases, there should be an injection into the vagina of a solution of alum and sulphate of zinc, three drachms of the former and one drachm of the l· to a pint of water ; three or four ounces to be thrown up while the patient lies with the hips rather elevated ; this position to be retained for some time, with the parts covered by a cloth or sponge, so that the fluid may be kept in. If there is itching and irritation of the parts, it may be allayed by an injection composed of carbonate of soda, two drachms, in a quart of bran tea. If the simple alum and zinc injection proves ineffectual, add a drachm of powdered catechu to each pint, or use decoction of oak-bark as a vehicle for the above salts. When there is much debility, with suppressed or scanty menstruation, preparations of iron (as the above mixture) with compound steel pills, or some compound of Canada balsam, three grains, and a half grain of quinine ;

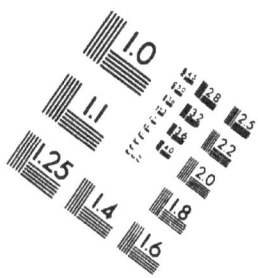

**IMAGE EVALUATION
TEST TARGET (MT-3)**

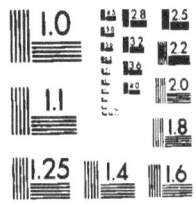

or a half drachm of quinine with one drachm dilute sulphuric acid, in six ounces of gentian or cascarilla ; a tablespoonful to be taken two or three times a day.   Should there be profuse menstruation, nothing is so likely to be effectual as the iron and acid mixture, with or without the sulphate of magnesia, according to the state of the bowels.   Mustard poultices to the lower part of the back or stimulant liniments, rubbed well in every night, for a time, will often prove useful.

Women who are likely to have leucorrhœa should avoid all predisposing causes of the disease, such as wines and other stimulants and hot tea and other slops taken in large quantities ; luxurious living, and sensual indulgence of all kinds, especially much sexual intercourse, and anything that has a tendency to enervate and enfeeble the frame.   Early rising and regular open-air exercise, warm and comfortable clothing, good food and tonic medicine, with use of the shower-bath and bathing,—these will prove the best preventives.

Herbal or Electric Treatment for Womb Diseases.   For the whites care must be taken not to arrest the discharge too soon, or bad consequences may ensue.   Use every means to improve the general health.   Avoid hot rooms, excessive exertion, and strong tea and coffee.   A decoction of the roots of comfrey-root, boiled in milk, is highly recommended.   Take a teaspoonful three or four times a day.   Injections of alum-water or decoction of oak-bark, are very good.   A preparation of one ounce of aloes and two drachms muriated tincture of iron well mixed, and forty drops taken three times a day in a little water, has been found of great advantage.

For falling of the womb, an infusion of white-oak bark, or an infusion of equal parts of peach-leaves, Solomon's seal, and hops, as an injection, will produce excellent results.   Where heat and difficulty in making water exists, give a drink of infusion of marsh-mallow and spearmint.

Pregnancy and its Disorders.   Utero-gestation, or the period of child-bearing,—that is, from the time of conception to that of delivery,—extends over a period of forty weeks, or two hundred and eighty days.   It is commonly set down as nine calender months, but this would make only two hundred and seventy-five days ; or, if February be included, two hundred and seventy-two days ; that is, thirty-nine weeks only, instead of forty, or nine calendar months and a week.   In making the necessary provision for the coming on of labor, it is best to calculate from midway between the last occurrence of menstruation and the one which would have followed if conception had not taken place, and allow nine calendar months from that time.   Thus, if menstruation had taken place on the first of January, labor might be expected some time about the middle of October.

The Signs of Pregnancy.   The chief signs of pregnancy are as follows :—

1. The cessation of the menses,—although this is by no means an un-failing sign, for sometimes this discharge will cease from other causes, and sometimes it will continue after conception has taken place.

2. Morning sickness, which generally commences about the fourth or fifth week, and lasts to about the fourth month. With some this is but slight, and causes but little inconvenience; but with others it is more continuous and serious, sometimes causing the rejection of nearly all the food for a very considerable period. This symptom, again, can not be taken as a proof of pregnancy; it is merely a suspicious circumstance, to be watched in connection with others.

3. Enlargement of the breasts, which generally increase in size about two months after conception. They also become tender and sore; they throb and burn, and, when pressed by the hand, have a hard knotty feel, in consequence of the swelling of the glands by which the lacteal fluid is secreted. The nipple also becomes more prominent, and increases in diameter, while the areola around it assumes a purplish tinge, and has on it several little raised pimples of a yellowish-white color.

4. Enlargement of the womb and abdomen, which, in the fourth month, becomes very perceptible,—the womb, which may now be felt in a firm rounded body, having ascended above the bone of the pubes, and pushed the bowels up into the abdomen.

5. A tendency to flatulent distension of the stomach, towards evening especially, rendering insupportable a pressure of stays, etc., which in the morning could be easily borne.

6. "Quickening," which is the mother's first perception of the second life within her. There is at first, probably, a very slight tremulous motion, like a mere pulsation. This day by day grows stronger, until it becomes quite distinct, often painfully so. It is as though the child, to use a scripture phrase, "leaped in the womb." These movements can be distinctly felt by the hand placed upon the abdomen.

There are other and less obvious signs, which only the professional man would be likely to detect. All may notice, however, the change which generally takes place in the countenance. The mouth and eyes seem to enlarge, and the nose becomes what is generally termed more or less "pinched up." There is an alteration, too, in the color of the eyes, which becomes somewhat paler,—especially is this perceptible if they are blue eyes. Then the patient is generally fidgety, peevish, and restless, exhibiting a high degree of nervous irritation; she has odd fancies, and longings after out-of-the-way things and articles of diet, which should be procured for her if possible. At such a time she requires soothing and humoring; harsh and unkind treatment will be likely to have a most injurious effect, both upon her and her offspring.

Management and Conduct of Mothers during Pregnancy.

A pregnant woman should be made aware that the advantages obtained by well-regulated habits are by no means exclusively conferred upon her, but that others equally important are likewise conferred on the child, for whom a larger supply of nutrition, and of a better quality, will thus be provided ; and so, being nourished by sound and healthy fluids, will commence its career of life strong, vigorous, and less liable to those morbid debilities and derangements which affect the children of the indolent, the pampered, or the debauched.

From the moment, therefore, that conception has taken place, a new and most sacred duty involves upon the female. She is bound by all ties of maternal sentiment, of humanity, and of moral and religious obligations, to protect the nascent being in her womb against every circumstance, under her control, which might have an unfavorable influence on its delicate organization.

The state of pregnancy is one peculiarly liable to disease and injury ; and we daily witness much suffering and danger incurred both to the mother and the child, from the influence of causes which, with proper care, might have been avoided altogether, or, at least, rendered inoffensive.

Diet during Pregnancy. The peculiar tendency to febrile irritation and general plethora, in pregnancy, renders it especially proper to avoid undue excitement and nourishment during this period. Not that the pregnant female is to be dieted like a valetudinarian ; but that moderation and simplicity of food is especially proper in her case. As the appetite is frequently very craving during this period, an inordinate indulgence in rich and high seasoned food is among the most common errors ; and this mistake is the more readily fallen into from the erroneous idea which many entertain, that, as the fœtus draws its nourishment from the maternal system, a greater quantity of aliment is required in pregnancy. The exercise of caution in the selection of proper food, appears to be particularly important towards the termination of gestation. When the stomach is in a weak and irritable state, rice, barley, arrow-root, oatmeal, the lean part of mutton, tender beef, soft-boiled eggs, and stewed apples constitute appropriate articles of nourishment. But it is always to be recollected, that the temperate use of food is of more consequence than any very cautious selection as to its kind. Coffee and tea may be moderately used ; but all vinous or alcoholic drinks should be studiously avoided. The temptation to indulging in small portions of cordial, or brandy, in the early months of gestation, is often very strong ; but it ought to be firmly checked, or the deplorable habit of solitary dram-drinking may be the result of indulgence in these potent stimuli.

The appetitive sensibilities of the stomach, in some instances, undergo extraordinary derangement, especially in weak and delicate females. Articles of food which, previous to pregnancy, were very grateful and congenial, become highly disagreeable, and an almost irresistible craving for singular and even disgusting substances, is experienced. This remarkable irregularity of the appetite is usually called *a longing*. If the substances longed for be not evidently of an injurious character, they should not be withheld ; in some instances these longings may be regarded as instinctive calls of the stomach for articles favorable to the health of the individual. Thus, when a strong desire for eating chalk, charcoal, or clay, is manifested, we are admonished that the digestive powers are feeble and that there is a tendency to acidity in the stomach. In such cases the means of relief are alkalines, mild laxatives, and tonic vegetable bitters, with a suitable regimen.

Dress and Exercise. The custom of wearing tightly laced corsets during gestation can not be too severely censured. It gives rise to functional disorder of the stomach and liver, as well as to uterine hemorrhage and abortion in the mother ; it likewise impedes the regular nourishment of the fœtus in the womb. The clothing should always be sufficient to protect the body against the injurious influence of cold and atmospheric vicissitudes ; the abdomen and feet especially should be guarded against injury from these causes. In winter, or cold and damp seasons, the use of a flannel bandage or roller around the abdomen will be found very beneficial and comfortable. All kinds of agitating exercise, such as riding in carriages rapidly over rough roads, dancing, lifting or carrying heavy loads,—in short, all masculine and fatiguing employments—whatever ought to be avoided by pregnant women ; and the more so as gestation approaches the term of its regular completion. During the eighth and ninth months of pregnancy unusual exertion or fatigue is particularly apt to excite premature labor. It is to be observed, however, that if injury is apt to result from too much exercise, injurious consequences may also arise from too much indolence and inactivity. Riding in an easy carriage on even roads, or moderate walking, may be enjoyed with great propriety, and usually with obvious advantage during gestation. Sexual intercourse should be avoided after conception : it is useless to reproduction, and is interdicted by moralists and physicians, as prejudicial to the parents and their offspring.

Moral Influences. Tranquillity and cheerfulness of mind are of prime importance during pregnancy. Convulsions, severe hysteria, spasms, syncope, hemorrhage, and abortion, may be produced by violent anger, terror, or jealousy, during pregnancy. Intense grief will occasion debility, indigestion, jaundice, and various other functional disorders. A strong excitement of the imagination is supposed by some to be capable of producing impressions on the fœtus in the womb.

The Breasts. The breasts and nipples should be particularly attended to during the latter months of gestation, in order to prepare them for suckling the infant. For some weeks before the expected termination of gestation, the nipples should be daily washed in lukewarm water, then dried by exposing them to the free air, and afterwards gently rub bed for five or six minutes with a soft piece of flannel, or with the extremities of the fingers. When the skin of the nipples is very delicate and sensitive, they may be washed with brandy and water, or a wash composed of two drachms of the tincture of myrrh, one drachm of laudanum, and two ounces of water. In using this, the nipples should first be bathed with lukewarm water, and dried and rubbed as before directed, and then washed with the lotion. Compression of the breasts by corsets, or any other artificial means, is carefully to be avoided. When the nipples are very small, or sunk in the breasts, they should be drawn out by means of a suction pump with a common clay tobacco-pipe. This process should be repeated several t mes daily, until they have acquired a sufficient degree of prominency.

DISEASES OF PREGNANCY.

Morning Sickness. This complaint is sometimes very troublesome and obstinate.

Treatment.—The patient should have breakfast in bed, and remain in a recumbent position for some time after. Small lumps of ice put into the mouth and allowed to dissolve will sometimes give relief. Give. if the sickness is troublesome, three times a day, a mixture composed of one scruple of bicarbonate of potash or soda, dissolved in a wineglass of water. Take while effervescing, with a tablespoonful of lemon juice.

Constipation. During the latter months of pregnancy, constipation is nearly always present, the pressure upon the lower bowel being the cause. Neither aloes nor any violent cathartic should be taken. A moderate dose of castor oil may be administered about every other day, or as often as necessary; but if the stomach nauseates at repeated doses of this, try the following mixture:

Sulphate of Magnesia,	-	-	-	-	- 1 ounce.
Infusion of Roses, -	-	-	-	-	- 6 ounces.
Cinnamon Water, -	-	-	-	-	- 2 ounces.

Dose, a wineglass every morning early. If, as is sometimes the case, diarrhœa supervenes, give the following :—

Chalk Mixture,	-	-	- 6 drachms.
Aromatic Confection, -	-		- 2 drachms.
Tincture of Opium,	-	-	- ½ drachm.

Dose, a tablespoonful every three or four hours.

Heartburn. This may be removed by moving the bowels with a little magnesia, and taking a wineglassful of lime-water in milk two or three times a day ; or carbonate of potash and magnesia, of each ten grains, in cinnamon water, with one drachm of tincture of gentian.

Incontinence of Urine. The frequent desire to make water, arising from irritation of the bladder, should be attended to, as long retention of urine may cause retroversion of the womb and abortion. An abdominal belt will be found of great service in the renal affections of pregnancy. Effervescing draughts, with ten grains of nitrate of potash and the same of magnesia, will also be found serviceable ; and if there is much pain, add five grains of laudanum, and apply hot fomentations or use the hip-bath.

Cough. If there is cough, which frequently attends pregnancy, give any soothing pectoral mixture. If the cough is attended with pains in the chest, or headache, apply in the former case mustard-poultices over the sternum.

Varicose Veins. For cramps and pains in the legs, with swelling and varicose veins, sponge the legs with cold vinegar and water, and put on roller bandages or elastic stockings, and rest in a recumbent position.

Itching of the Private Parts. Itching about the vagina, with gleety discharges, call for the use of the hip-bath, and a slightly astringent injection, such as Goulard water, a weak solution of alum, or an infusion of green tea.

Restlessness at Nights. For dreams and restless nights, extract of hemlock, or henbane, five grains at bedtime, with strict attention to the condition of the bowels.

Melancholy, Despondency, &c. Despondency frequently seizes upon those who are about to become mothers ; but generally, if the health be pretty good, it is shaken off as the great trial approaches. There are some women who are never so well and cheerful as during the time of pregnancy, but many there are to whom it is indeed a period of trial and suffering ; and especially is this the case with those who are about to become mothers for the first time.

False Pregnancy. It is necessary to the completeness of our subject that we say a few words here about false or spurious pregnancy, a condition of the female system of a remarkable kind, most frequently observed about the turn of life, when the catamenia becoming irregular, previous to their final cessation, are suppressed for a few periods ; and, at the same time, the stomach being out of order, nausea or vomiting is experienced, the breasts enlarge, become sensitive or even slightly painful, and sometimes a serous or acrolactescent fluid exudes from the nipples and orifices of the areolar tubercles ; the abdomen grows fuller and more prominent, especially in women of full habit and constitutionally disposed to *embonpoint*, and the abdominal enlargement progressively increases, partly from deposition of fat in the integuments and in the omentum, but still more from distension of the intestines by flatus, which, passing from one part to another, communicates a sensation like that produced by the motion of a fœtus ; the nervous system is generally much disturbed, and the woman feels convinced that she is pregnant, an idea which, at the time of life alluded to, is cherished by the sex with an extraordinary devotion, and relinquished with proportionate reluctance ; and not unfrequently at the end of the supposed gestation, the delusion is rendered complete, and almost assumes the character of a reality, by the occurence of periodical pains strongly resembling labor.

The Breast and its Diseases. We use this term in its restricted sense, as applied to the fleshy protuberance common to women, in which is situated the mammary glands, for the secretion of the milk by which the infant is nourished. Its full development depends greatly upon habit and constitution, being in some much more early in advancing womanhood, of which it is one of the most remarkable signs, and prominent in full maturity, than in others. In the earlier stages of pregnancy, its fullest development commences : the breast swells, and the nipple enlarges, and by, or near the time of delivery it is filled with lacteal fluid, which passes readily, on suction, into the mouth of the child. Too frequently the proper enlargement of the breast, and increase of the nipple, is retarded by tight lacing.

FIG. 94.

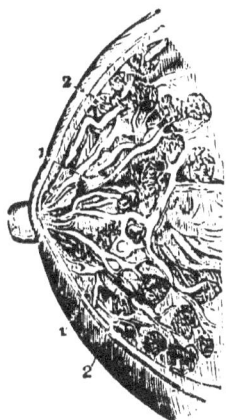

Section of Mammary
Gland.

1, 1. Galactophorous Ducts.
2, 2, Lobuli.

The consequences, sometimes, are hardened and congested states of the tissues, an insufficient supply of milk, or a failure of it altogether; or a nipple so flattened and pressed into the chest that it cannot be taken hold of by the mouth of the infant. Abscess, cancer, and other evil consequences may also ensue from undue pressure upon such tender parts.

Inflammation of the Breast. This affection is common, and frequently results in abscess. Various causes may produce it, such as a blow, exposure to cold or wet, great mental excitement, unnatural distention by an accumulation of milk, or too much pressure by corsets. It may occur at any period between early and advanced womanhood, but most commonly it does occur within a week or two of childbirth, and is the result of some obstruction in the flow of the milk, or change in its normal character. Such a change will be sure to occur if the milk is suffered to remain long in the breast ; therefore, should the infant be unable to relieve it at all, or insufficiently, artificial means must be taken to do so.

FIG. 95.

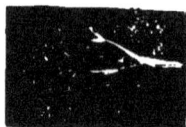

Origin of the Milk Ducts.

FIG. 96.

Ultimate Follicles of the Mammary Glands.

a, a, the secreting Cells. *b,* the Nuclei.

A simple and cheap form of breast-pump is made with a stout elastic bag with a glass mouthpiece, a wide-mouthed bottle sufficiently capacious to hold two quarts. This is dipped in hot water, and the mouth immediately applied to the breast ; the heat will have rarefied the air within, which, as it cools, contracts and leaves a vacuum, causing suction, which draws the milk into the bottle. Some nurses have the art of drawing the breast with the mouth ; and it is well to let them do so, as no instrument can effect the object so thoroughly.

Abscess in the Breast. When there is an inflamed state of the tissues of the breast, there are shooting pains, and often febrile symptoms The part will become hard and exhibit knotty protuberances, indicating the formation of an abscess. These symptoms will be succeeded by throbbing and a sensation of weight,—the skin gradually assuming a thin and red appearance, and becoming thinner until it finally breaks, and allows the escape of the matter.

Treatment.—When the premonitory symptoms of mammary abscess are observed, recourse should at once be had to remedial measures. Let the breast be well yet gently rubbed with a soft hand, into the palm of which is poured fresh olive or almond oil ; the friction should be continued for about ten minutes, and repeated every four hours or so. Goose-grease and other fatty substances are recommended, but simple oil is best, the friction being the principal agent for good. Between the intervals of this the breast should be kept covered with a tepid-water dressing —having over it oiled silk to prevent evaporation. Care must be taken during this treatment to keep the bowels gently open, and to keep under the febrile symptoms.. A mammary abscess will frequently continue discharging for a considerable period, and, during this time, the patient should be supported by a nourishing, although light diet.

A warm bread-poultice is best for the abscess ; it should be changed about every four hours, and covered with oiled silk. When the discharge has nearly ceased, simple tepid-water dressings may be substituted. The breast, during all this time, should be supported by a soft handkerchief tied round the neck An application of collodion all over the part has sometimes been used ; it forms a thin coat which, contracting as it dries, affords the necessary support, if the breast is not very large and heavy. If some amount of pressure is required, strips of strapping crossing each other will effect this object. After all danger of inflammation is over, a more generous diet may be allowed. Should the breast remain hard, friction with soap liniment should be resorted to. A drachm of compound tincture of iodine to each ounce will render it more effectual.

Sore Nipples. Very painful and distressing cases of sore nipples frequently occur after childbirth. Sometimes they cannot be avoided, but frequently they arise from too great an anxiety on the part of the mother, who is constantly meddling with them, applying the mouth of the child, and resorting to all sorts of expedients to draw them out. Nipple-shields, with Indian-rubber teats, may be readily procured, and should be used when the nipples are too sore and tender to bear the application of the infant's mouth. In this case the milk must be drawn from the breast by one of the contrivances above mentioned, and given to the child in a feeding-bottle.

Glycerine has been found a good application for chapped or otherwise sore nipples. It must be applied with a camel-hair brush, first wiping this dry with a soft piece of linen. If obtained pure, there will be little or no smell in it to annoy either mother or child. Collodion is also useful, but it causes considerable smarting. If, as is sometimes the case, there be suppuration, warm bread-poultices must be applied, and after them tepid-water dressing. Infants, a few days after birth, sometimes have the breasts distended with a thick, milky-looking fluid ; youths just arriving at the age of puberty have hard and painful swellings about the nipples. In both cases warm fomentations only are required ; the parts should not be pressed or rubbed ; for the child, a little cold cream or simple ointment, after fomenting, is desirable.

When the nipples are cracked or abraded the following is one of the best applications, and may be obtained at the drug store :

Tannic Acid,	-	-	-	-	- 3 grains.
Glycerine,	-	-	-	-	- 1 drachm.
Simple Cirate,	-	-	-	-	- 1 ounce.

This may be applied to the nipple and into the fissures three or four times a day, care being taken to remove it and cleanse the nipple before the child is applied to the breast.

Milk Fever. An aggravated form of the excitement which takes place at the onset of lactation.

Causes.—The cause may be a cold, or over-heating the apartment, too stimulating a diet, or any obstruction to the flow of milk from the breast.

Symptoms.—Its first symptoms are increased heat of the system, preceded by shivering, and sometimes accompanied with vertigo and slight delirium. These are followed by severe headache, thirst, dry tongue, quick pulse, throbbing of the temples, and intolerance of light.

Treatment.—Spare diet, perfect tranquillity, subdued light, cooling drinks, and saline aperient medicines ; the head should be kept somewhat elevated, and bathed with cold water or evaporating lotions. If the symptoms should become worse in spite of this, apply half-a dozen or more leeches to the head, and put the feet in a warm mustard bath. Most lying-in women have more or less of this fever, which is no doubt an effort of nature to rouse the hitherto dormant mammary organs to secrete a proper quantity of milk. If, however, it is not checked, the arterial action runs too high, and no milk at all is secreted.

Confinement. (*Parturition*).—Every prudent woman, who has the power of doing so, will make all necessary preparations for an approaching *accouchment*, as the French term childbirth, or delivery.

Few women, who are near their confinement, are sufficiently cautious of exposing themselves to unnecessary fatigue and atmospheric changes. They will "keep about until the last," and it is well for them to do so, provided they take only gentle exercise, and avoid getting wet or chilled, or heated in crowded assemblies, and the like. Miscarriages, difficult labors, and frequent lasting injury to mother and child, if not death of one or both, is not unfrequently the result of imprudence at this critical period. Therefore we would impress upon all our readers who are likely to become mothers, the duty which they owe to themselves, their friends, and their future offspring, of refraining, when *enceinte*, as much as possible from the more exciting pleasures and laborious occupations of life, and of preparing for the pains and cares which will shortly come upon them.

Let all the preparations for the little stranger be made in good time, and the services of an experienced nurse engaged. Let the mother or some female, very near and dear, be at hand to aid and counsel, and, above all, to cheer and encourage the often sinking heart, not only at the actual period of labor, but for some time previously. And let the mother in expectancy be treated with all possible love and gentleness. She may be fidgetty and whimsical,—what of that!—provided they do not run into outrageous extremes, let her very whims be indulged. She is frequently in a state of great nervous excitement,—her body may be racked with pain, and her mind unhinged. Let her be soothed and tenderly dealt with. She has that to go through, at which the strongest man might well tremble, and shrink aghast.

We will suppose that the inevitable hour has come, and that the labor-pains are regular, and that the work of delivery proceeds properly, although perhaps slowly. In due time—it may be in two hours, or four, or six, or even, in the case of a first child, twenty-four hours—the infant is born, and treated according to the directions given for the management of infants. But we are getting on too fast, and must go back to explain what has been, or should have been done to bring about the desired consummation of a safe delivery, and what is yet of more consequence, the safety of the mother and child, and the gradual recovery of the former from the shock which, under the most favorable circumstances,

her system will receive. If she be a strong, healthy woman, and no unusual complications arise to disturb the natural process, but little aid or
interference may be required. There will be the usual warning symptoms; intermitting pains in the back, slight at first, but increasing in intensity. There will probably be a slight discharge of mucus, stained
with blood, and perhaps also a considerable discharge of a clear fluid,
popularly called " the waters." This is an albuminous liquid filling up
the membrane in which the fœtus floats, and so preventing pressure. It
sometimes does not escape until labor has actually commenced by the
falling down of the child into the pelvis. When this takes place, the recumbent positions should be assumed. Previous to this, it is best for
the patient to sit upright or walk gently about, and to assist the action
of the uterus.

When the labor-pains become very great, the patient should be placed
on the bed, previously guarded by some waterproof material on her left
side, and not far from the edge, so that needful assistance can be easily
rendered. She should have a tightly rolled pillow placed between her
knees. If there is no unnatural obstruction to the delivery, it is best left
to nature Should the patient in the struggle become very faint and
weak, a little brandy and water may be administered at short intervals,
but this must be stopped as soon as the labor is over, or inflammatory
action may ensue.

As soon as the child is born, and the umbilical cord,—or, as it is
commonly called the navel string,—by which it is attached to the womb,
has been tied and cut, a broad bandage or towel should be passed
round the body of the mother, so as to cover the hips, drawn tightly,
and pinned or tied, so as to sustain a pressure upon the womb, and stimulate the vessels to return to their normal condition. Before this is done,
however, it will be best to pull that part of the above-named cord which
remains attached to the uterus very gently, and by this means to accom-
plis.t if possible the removal of the placenta, commonly called after-birth,
which sometimes comes away with the child, or immediately after, and
is sometimes only removed with great difficulty. If, at the expiration of
a couple of hours or so, this still remains in the womb, where it will
cause irritation, the hand of the nurse or medical man, previously well
oiled, must be carefully passed in, so as to grasp, and without breaking
it, detach it gently from its adhesion, and bring it away, waiting to complete the process until an after-pain comes on. Generally the natural expulsion, or the artificial removal, of the placenta is attended with hemorrhage, sometimes to a frightful extent. For directions how to proceed in
this case, see article on *Flooding.*

For at least six hours after labor, the patient should be disturbed as
little as may be. We have seen fussy nurses very desirous of making
"missus" comfortable, and begin to put things to rights about her, when
she, poor soul ! only wants perfect rest and quiet. Let her have it. And

if the pulse is thin and feeble, and the cheeks are colorless, and the breathing scarcely noticeable, so that life seems almost ebbing away, put a little a very little, brandy and water, warm and sweet, between her lips now and then; but stop instantly if it produces flushing or restlessness; and do not give it at all unless there seems urgent necessity for a stimulant. At the expiration of the above time, if a revival has taken place, soiled bed-clothes and body-linen may be changed; but all this should be done very carefully and gently, or the fatigue may occasion a relapse. If the after-pains continue severe at the expiration of the above time, an anodyne draught may be given. It may be composed of from twenty to thirty grains of tincture of opium, or a quarter of a grain of morphine, in an ounce of plain or spearmint water.

For eight or more days after labor, the recumbent position should be strictly maintained; and the same rule holds good after a miscarriage. Some women feel so well and strong in a day or two, that they will sit up, and sometimes even get out of bed, and make themselves useful in the house. We have seen a woman at the wash-tub three days after she had been confined; and we have heard of females undergoing the pains of labor under a hedge by the roadside, and in a few hours proceeding on their journey with their babies at their breasts. But these women were semi or entire barbarians; they had not been delicately nurtured. With the immense advantages, we must also take some of the disadvantages of civilization, and those who give birth to children surrounded by all its comforts and luxuries, must not attempt to emulate the Indian squaw. If they do, they will inevitably suffer for their temerity. Getting about too early after childbirth is, perhaps, the most fruitful of all sources of uterine disease. The consequences may or may not show themselves at once, but whether or no, bad consequences there most likely will be; therefore we warn all mothers to keep their beds long enough; but little exertion should be made until the end of the first fortnight. If there is a necessity for getting about earlier, of course it must be done, for necessity has no law; but unless there is, the risk should not be run. Delicate women especially do wrong to attempt it, and the strong will be likely to render themselves weak by the practice.

Abortion, or Miscarriage, the premature expulsion of the foetus from the womb,—that is, before the seventh month. After that period, if delivery occurs before the ninth month, it is called premature labor.

Causes.—A sudden shock to the system by a fall or a fright; straining, or over-reaching; the administration of strong purgatives or emetics; excessive indulgence in venery, or ought which may tend to debilitate the system; malformation of the generative organs; fevers and severe inflammation; syphilis or constitutional disease of any kind; the growth of polypi or tumors in the cavity of the uterus, or adhesion to the surrounding viscera; too great contractibility of the uterine fibres and blood-vessels. Most frequently, perhaps, it is a diseased condition of the foetus itself, which, wanting the elements of growth and vitality, is rejected as a useless and troublesome incumbrance. Two classes of females, very different in constitution and appearance are more than commonly liable to abortion, namely, those of a voluptuous and plethoric habit, and those of a weak and irritable frame. Those who continue to suckle after conception has again taken place render themselves liable to it, because a certain amount of nutriment required by the fœtus goes to the formation of the lacteal fluid.

Miscarriage is generally attended with much pain. It weakens the system, and often severely tries the constitution of the sufferer, whose liability to the accident increases with each occurrence. The periods at which it is most likely to take place are said to be about a month after conception, again in twelve weeks, and again in the seventh month —the liability increasing in those stages which correspond with the periods of menstruation. Some people invariably miscarry at a certain stage; and thus, although often in the way to become mothers, are never blessed with offspring.

By this it will be sufficiently plain that pregnant women ought to avoid all violent exercises of the body, strong mental excitement, over-indulgence of sensual appetites, exposure to wet, or any extremes of weather, or aught which may tend to constitutional derangement of whatever kind; and those who have once aborted should be doubly careful, on account of their greater liability.

Symptoms.—These vary considerably, according to the more or less advanced stage of pregnancy, and state and condition of the patient; but usually she feels at first slight pains in the loins, and parts about the womb. There is a sense of bearing down, a frequent desire to make water, or to evacuate the bowels, and a feverish state of the system generally. A discharge of blood commonly follows, sometimes in clots, at others in gushes, at longer or shorter intervals; and this will continue until the fœtus is expelled. As the patient can not be considered out of danger until relieved of the ovum, the discharge ought to be carefully watched, and preserved for the examination of the medical man, should he not be present during its progress, which is much to be preferred.

Treatment.—The first object, when the premonitory symptoms above mentioned set in, is, if possible, to *prevent* abortion. To this end, the patient should at once assume a recumbent position, and on no account be suffered to move more than may be absolutely necessary. For a few days, use only cold drinks, and at bedtime take a pill composed of one grain of opium and two grains of sugar of lead.

If there is much heat in the abdomen, cloths wet with vinegar and water, in equal portions, should be applied thereto, and removed as often as they get warm. When the hemorrhage becomes at all profuse, all hopes of prevention are at an end, and the efforts should be directed to relieve pain, prevent utter exhaustion of strength, and finally to remove, as quickly as may be, the ovum from the womb. To effect the latter object, mechanical means are sometimes resorted to, but only one thoroughly acquainted with the anatomy of the parts should attempt this. As the flooding proceeds, the patient should be kept as cool as possible; she should be exposed to, and suffered to breathe, cold air : acidulated drinks should be administered ; if ice can be obtained, let it be used to lower their temperature. Should fainting ensue from loss of blood, cordials may be given, but not hastily, or frequently, a teaspoonful of brandy, or fifteen drops of aromatic spirits of ammonia, in half a wineglassful of cold water, is the best stimulant for the purpose. When the discharge is very profuse, lint, wadding, or a piece of sponge, dipped in a solution of alum, and then in olive-oil, may be introduced into the vagina, or an injection of the same gently thrown up by means of a syringe ; or a decoction of oak-bark may be used for the same purpose.

Should these means fail to check the hemorrhage, make up eighteen grains of sugar of lead into twelve pills, with crumb of bread, and give one every two hours, with a draught of vinegar and water, or dilute sulphuric acid, fifteen drops in half a wineglassful of water being a sufficient dose. Opiates may be given with advantage when the pain is very severe, especially before the flooding comes on, or after it has continued too long. Suppositories, consisting of about a grain of powdered opium, made up into a softish mass, with a few grains of powdered gum, or extract of henbane, are also useful. These latter may be introduced when miscarriage is likely to ensue. With rest and proper care, they will sometimes prevent it.

The best preventives of miscarriage are the frequent use of the cold hip-bath, and sponging the lower part of the belly with cold vinegar and water; strict attention to diet, and avoiding all violent purging medicines; moderate gentle exercise, and entire abstinence from sexual intercourse during the first month of pregnancy.

We can say nothing here about abortions voluntarily produced except to warn women of the folly and danger of resorting to unprincipled empirics, or the use of powerful drugs, to hide the consequences of an unlawful gratification of the passion... Death has frequently resulted from the employment of such means as are necessary to produce abortion, and far better is it to bear the shame and disgrace of being the mother of illegitimate offspring than to incur the risk and sin of being possibly the destroyer of self, as well as of the embryro of a human being over which the parental instinct alone ought to stimulate a tender care and watchfulness.

Anæmia. This is a condition of the constitution in which there is a deficiency of the red globules, or coloring matter, in the blood. It is marked by extreme pallor in those parts, such as the lips, which are generally suffused ; and is not uncommon in young females of a weak or scrofulous habit. It appears to arise from a deficiency of vital energy

M

in the system, either constitutional or brought on by want of nourishment, breathing impure air, or great loss of blood. In any case a cure may be effected by good generous diet, pure air, moderate exercise, and strengthening medicines.

Treatment.—Any of the various preparations of iron may be taken in combination, if the appetite be bad, with some bitter tonic, such as infusion of gentian, with a little quinine Should there be emaciation, cod liver oil, taken in orange wine, will be of service. The pores of the skin should be kept open by tepid sponging, and the bowels moderately so by a rhubarb or colocynth pill now and then. Strong purgatives should be avoided, and especially salines. In young females the absence of the monthly discharge need cause no uneasiness ; with returning strength, that will most likely return. Should it not do so, however, when this treatment has been persisted in for a time, and should the pallor, languor, sleeplessness, headache, confined bowels, swelling of the feet, &c., which generally distinguish anæmia, continue, a medical man ought to be consulted, as it is likely there may be consumption, or other organic disease, at the root of the mischief.

Barrenness is the defect of power in the female to produce offspring.

Causes.—It is caused sometimes by want of tone or strength in the system ; nervous debility ; sometimes the result of malformation of structure in some part of the generative organs ; and sometimes of functional disorders from local or constitutional causes.

Symptoms.—Want of issue ; and, in married women, frequently continued ill-health.

Treatment.—Cold bathing, or dashing cold water on the loins daily ; general tonics, or strengtheners to the system ; electricity or galvanism applied locally. A milk and vegetable diet is recommended, and abstinence from sexual intercourse for a time. Take plenty of exercise early in the morning in the open air, and take the following :—

Compound Aloetic Pill,		2 scruples.
Compound Rhubarb Pill,		2 "
Sulphate of Iron,		2 "
Extract of Henbane,		2 "

Mix, and divide into thirty-two pills. Take one every night, and the following in the daytime :—

Compound Tincture of Valerian,		½ ounce.
Compound Tincture of Lavender,		1 "
Aromatic Spirits of Ammonia,		½ "

Mix, and take a teaspoonful twice a day in two tablespoonsful of infusion of cascarilla.

Green Sickness. This disease has obtained its name from the pale and greenish cast of the skin of the patient. It is one of the forms of anæmia, and chiefly affects young girls, although adult and even married women, and young delicate males, are subject to it.

Causes.—The disease appears to rise from a defect in the blood of red particles, and other solid constituents, and this is caused by defective assimilation. Those young persons of sedentary habits, or who work in crowded factories or shops, or who live in underground kitchens, and like places, are particularly subject to it.

Symptoms.—In addition to the pallor of the skin, which is common to all the forms of anæmia, this has some peculiar symptoms, such as hysterical paroxysms, and extreme nervousness, pain in the side, swelling of the ankles, headache recurring at certain periods ; there is also frequent. ly depraved appetite and a disinclination for wholesome food altogether- If the case is long neglected, the symptoms become greatly exaggerated, the secretions are unhealthy in character, and deficient in quantity ; the limbs swell, the pains in the head and face are more severe, and so weak is the patient that every exertion, even the slightest, is laborious ; the depraved appetite becomes more remarkable—cinders, chalk, slate-pencil, and articles equally unfit for eating, are eagerly sought for, and masticated with avidity.

Treatment.—Change of air, tonics, and the course of treatment prescribed under the head of *Anæmia*, is the best in such cases. Exercise, fresh air, and nourishing diet, are the grand restoratives. Iron is the best tonic, alone, or in combination with quinine. It should be given in the least nauseous form, and at least one hour before meals.

Puerperal Fever. (*Childbed Fever*).—This is one of the most fatal diseases which attack lying-in-women. It is a fever of a very high character, arising from inflammation of the serous membrane, and often of the womb itself, and of its veins and absorbents. It runs a very rapid course, and is commonly fatal. It assumes the character of an epidemic, and frequently causes great mortality in lying-in hospitals. Whether it is really contagious or not is yet an open question. The mere probability that it may be should render persons extremely cautious in their intercourse with those who are suffering under it.

Symptoms.—There is usually an anxious countenance, sickness, hurried respiration, a furred tongue, and a stoppage of the secretions, especially of the milk. When these symptoms occur soon after childbirth, no attempt should be made at domestic treatment. Let the medical man be summoned immediately, if he be not already in attendance.

Puerperal Mania, or Nervousness. This disease frequently attacks women either a little before, during, or shortly after childbirth, and sometimes during nursing.

Symptoms.—Great nervous irritation ; the face is commonly pallid, the eye troubled, the tongue white, and skin hot ; the mind wanders, and conduct very irregular.

Treatment.—Give a purge of senna and salts, and keep the bowels regular by the compound rhubarb pill. Keep the room darkened, and let the patient be kept quiet, and free from the interruption of friends. If she is restless at night, give her an anodyne, such as twenty drops of hartshorn, or one grain of opium in a solid pill.

White Leg, or Milk Leg. This troublesome disorder is apt to follow childbirth in some constitutions, and is of long duration.

Symptoms.—It may commence two or three days after delivery, or it may not for some weeks. There is a little fever, and the parts about the thigh and groin feel hot, stiff and painful ; swelling commences, which extends over the whole limb, which does not, however, change color, except it be paler or whiter than natural. At this time the pain is usually very severe, after a time the symptoms abate a little, but the limb remains for a long time swollen and comparatively useless.

Treatment.—Cooling purgatives, such as magnesia, and salts and senna, and warm fomentations and poultices. Judicious bandaging will be of great service.

Itching of the External Genital Organs. The delicate internal lining of the external organs of generation sometimes becomes the seat of a most distressing itching, to relieve which, the parts may be so irritated by friction as to become violently inflamed. Leeches have been used sometimes with benefit : so has the application of cold, such as ice-water, or even lumps of ice introduced into the vagina. When there is an eruption like that in the sore mouth of children, injections of a strong solution of borax have been very useful ; thick starch water, with a solution of lead, injected into the vagina, and retained for an hour or two, have been also of great utility in a few cases under our care. This irritation sometimes arises from disease of the womb, pregnancy, the presence of a stone in the bladder, or worms in the bowels. The original affection must first be attended to in these cases, as elsewhere directed.

In every instance, except where there is considerable abrasion of the skin from scratching or other cause, one of the following lotions may be applied on cloths to the skin :

Sugar of Lead,	·	·	·	· 1 drachm.
Carbolic Acid,	·	·	·	· 10 drops.
Laudanum,	·	·	·	· 1 ounce.
Water,	·	·	·	· 1 pint.

Mix, and apply as directed.

Borax,	·	·	·	· 1 ounce.
Morphine,	·	·	·	· 5 grains.
Water,	·	·	·	· ½ pint.

Duration of Pregnancy. For ordinary purposes it is sufficient to accept the popular idea as to the duration of pregnancy, namely : that it occupies a period of nine calendar months. To be accurate, however, it must be remarked that the average of a large number of cases observed with considerable accuracy is 280 days or ten lunar months—a period, therefore, equivalent on the average to nine calendar months plus one week. The important thing to be borne in mind is, however, that this process, pregnancy, is not limited by iron regulations ; that a certain amount of variations from the average period, whether greater or less, is the rule. In fact the term 280 days is given as the average, not because it is the actual time in the majority of cases, but because the average time of those periods, greater and less, is about 280 days. A convenient rule for reckoning the probable time of confinement is as follows :—Count three months back from the time of the last menstruation and to this add seven days and the result will approach very nearly the date of confinement (of course in the following year). As an example, suppose a woman to have menstruated last on the 1st of May, 1888 ; counting three months back from this date and we get February 1st ; to this add seven days and this gives us February 8th, which will be the probable date of confinement in 1889.

The Relation Between Lactation and the Sexual Functions. Since the nourishment of the child by the secretion of the breast is a part of the reproductive process—and one, too, which abundantly taxes the physical powers of the woman—we would expect to find what we know to be the case, namely, that the other reproductive organs, the womb and ovaries, usually remain in a quiescent state until lactation is completed ; hence it is that conception rarely occurs until the child is weaned. But it should be remembered that this intimate sympathy between the breasts and the sexual organs is one which may react upon the former as well as upon the latter ; and that undue sexual excitement is apt to be followed by injurious influence upon the secretion of milk. For this, among other reasons, it is a matter of the utmost importance, that during the early period of lactation there may be a total abstinence from intercourse, which should be allowed not earlier than the third month after delivery.

Another most important reason for such abstinence is the fact that a failure to observe it often interferes seriously with the involution of the uterus, which is not always completed within two months after confinement. If the period of nursing be not unduly prolonged, if the child be weaned between the tenth and twelfth month, menstruation is usually deferred so long as lactation continues. Yet exceptions to this rule do occur ; either because of undue sexual excitement, or for some undetermined reason, menstruation occasionally begins five or six months after delivery. In such cases the woman is often urged to discontinue nursing for the sake of her child. In regard to this we would simply say that the appearance of the menstrual discharge does not necessarily demand the discontinuance of lactation ; the decision must rest entirely upon the health of the child. If this be evidently impaired from the time when menstruation began, it is always advisable that the infant be weaned, otherwise not. It is observed that in the majority of cases, when menstruation begins so late as the fifth or sixth month, lactation may be prolonged without interference during the usual period.

Quite otherwise is it, however, if conception occur while the mother is still nursing. For in this case her energies, unequal to the increased demand made by the attempt to perform both functions, will be devoted to the child in the womb to such an extent as to interfere seriously with her nursing. This does not necessarily imply that the milk will be dried up, since the contrary is often observed, but the effect upon the child is always evident. The milk plainly suffers such a change in quality as to render it injurious to the infant, and in every case the child should be weaned just so soon as the fact of conception is apparent. In some cases, indeed, it has been observed that a previously healthy child thriving upon the milk of a healthy mother, becomes suddenly and unaccountably ill. Obstinate derangements of digestion, resisting all remedies, either in the shape of drugs or in attention to the mother's diet, transform the previously rosy babe into a deathly caricature of its former self. The mystery has been only explained a few months later by the evident pregnancy of the mother.

Marriage. In selecting a partner for life many factors, religious, social, mental and moral, perhaps I should say also pecuniary, enter into consideration, which it is not the province of the medical adviser to discuss ; yet there are certain facts bearing upon the physical basis of marriage which it is the physician's duty to impart, and which may therefore be properly presented here.

First.—It should be remembered that marriage implies as its natural result the production of offspring ; and that a due regard for the welfare of such possible and probable offspring should be taken into consideration as a by no means unimportant element. It is, therefore, evident that marriage can be complete only when the parties to the contract are physically competent to fulfil the sexual relation, and, more than that, when the woman is capable of maternity. Now, while the girl is frequently capable, even in the earlier years of puberty, of becoming a mother, yet it is a fact patent even to the unprofessional mind, and well established by medical observation, that the girl is physically unfit for maternity, and that the disastrous results of premature motherhood are often visited, not only on the youthful mother as physical injuries; but are also apparent in the puny bodies and limited intellect of her offspring. The girl, in other words, is not made a woman by her first menstruation, for in the years to follow there must occur not only the development of her sexual organs, but also the increase in size and change of form of her whole frame, particularly the part included between her hips—the *pelvis* —whereby the germ of a new life may be fitly and fully developed within her body, and at the proper time permitted to pass through the pelvis to the outer world. For the too youthful wife marriage often proves a pain, not a pleasure; a grief, and not a joy. The imperfectly developed womb and ovaries, which might well have attained perfection if permitted to remain unmolested, unable to meet the demands of matrimony, are goaded into a state of irritation and disease. Her nervous system is often thereby enfeebled and she is prone to general prostration, as well as to those diseases peculiar to women. If she become a mother, she experiences more risk of injury during and subsequent to her confinement; and when called upon to nourish her infant as well as her own still growing body, it is not surprising that she often breaks down entirely.

It has been found that in our latitude and climate women usually continue to grow and develop up to the age of 20 years; though there are, of course, numerous exceptions in which maternity occurs earlier as well as later than this period. On the other hand, there are certain physical disadvantages accompanying over-maturity in the bride; for it is a well-established fact that women who experience the first confinement at an age exceeding 28 or 30 years furnish a larger mortality from child-birth than those who become mothers between 20 and 30 years of age. From the physical standpoint alone, therefore, matrimony seems most advisable as a rule between 20 and 25 years—an age, too, previous to which the mental development is not usually such as to demand marriage.

In the choice of a husband no adviser can influence the dictates of a woman's heart; and it is not our purpose either to usurp the duty of the parents in suggesting ordinary discretions and previous acquaintance with the mental and moral, as well as the physical, characteristics of the suitor; nor to pad our pages with romantic, sentimental, and utterly absurd advice, so interesting to imaginative young ladies, as to just how tall and heavy and graceful and manly he should be; as to what should be the color of his eyes, etc. It should be, however, remarked that certain physical characteristics ought, in the interest of the girl herself, to constitute insuperable obstacles to matrimony: It cannot be too emphatically insisted upon, that a man and a woman presenting the same hereditary taints, suffering from the same constitutional disease, or tendency to disease, should not, as they value their own happiness and that of their possible children, marry. In our land this is particularly true in regard to consumption and insanity. Were our laws made with the same rigid regard for physical health as prevailed in ancient times, we would doubtless forbid marriage to all suffering or likely to suffer from consumption; and while we are in these days more humane; while we take into consideration, in the estimation of conjugal happiness, the mental and moral as well as the physical welfare of the participants, yet we must remember that consumption is an eminently hereditary disease, and that the child's chances of becoming a victim to it are greater if both parents be born of tainted stock than if one at least be healthy. The same remark may be applied to insanity, epilepsy and other diseases of the nervous system ; for we may be sure that while children may escape if the tainted be mixed with healthy blood, yet the most aggravated and numerous cases of obstinate nervous diseases are found in families where both parents exhibit a tendency to the disease.   In this general fact, too, we have a solution of that much discussed question, whether relatives, particularly cousins, should be allowed to marry.   With reference to this, we may say that the simple fact of relationship—when not nearer than that indicated—constitutes no physical impediment to marriage, yet there usually exists in these cases a physical objection ; for the physical imperfection, if any exist—hereditary taints and tendencies to disease—will probably be found in both members of the family, and these defects and taints would in all probability be condensed and aggravated in their children ; and while we may say there is no physical objection to the inter-marriage of cousins as such, provided both be healthy, yet there will usually be found upon closer scrutiny a family tendency, the aggravation of which by inter-marriage, would be disastrous to happiness.

It need scarcely be remarked that closer and repeated inter-marriage among relatives is, from the physical point of view, undesirable.   It is a law, true of man as of other animals, that the most vigorous qualities of a given stock are best maintained by an admixture of foreign blood ; and it is a fact of observation, that marriages between Americans—those whose ancestors have lived in this country for several generations—are less productive in at least the number of children than marriages between a native American and a European : though it must be admitted that since the size of the family is influenced by many other circumstances than simple fertility of the parents, we are not justified in drawing the same conclusions from the fact just stated as might follow such observation upon animals.   It is specially interesting in this connection to note the peculiarity of the Jews ; they, as is well known, marry, as a rule, only members of the same race, and yet are remarkable for both physical and mental vigor of their numerous progeny.   The peculiar traits, mental and physical, we may indeed say moral, are retained and perpetuated by inter-marriage, and yet a sufficient latitude of choice is allowed to secure a proper admixture of stock.   It must, however, be remembered that the religious tenets of Israel provide not only for the health of the soul, but contain also admirable regulations for the health of the body ; to which perhaps their fertility and general health are to be in part attributed.

In selecting the time for marriage, certain physical facts should not be lost sight of amid social considerations.   The health of the wife and of her possible offspring is furthered by consummation of the marriage rite in the spring or in the fall ; for entrance upon this new life is beset with physical and mental trials, which are certainly all the more trying amid the heat of summer or the cool of winter.   One important consideration gives spring an advantage over autumn ; that if a child be born within a year its chances will be far better for surviving the trying period of teething, since the most critical part of the process will then occur in cool weather, and not in the heat of summer. The wedding should occur about the middle of the interval between two menstrual periods.

Although custom ordains that the newly-married pair shall start at once upon a wedding tour, yet it is generally understood that this tour need not be extended a greater distance than suffices to remove them from the immediate and critical observation of their friends ; indeed, it is a hopeful sign to observe that the wedding tour is no longer so imperatively required by society as formerly. From a physical point of view, certainly nothing could be more objectionable than a long journey immediately subsequent to the marriage ceremony. When, in addition to the annoyances inseparable from travelling, the bride is subjected to the trials incident to initiation into her new life, it becomes apparent that the girl is, during the ordinary bridal trip, subjected to a severe and in large part unnecessary physical strain, and that, too, at a time most critical and important for the security of her future happiness, as well at that of her husband. They are, it is true, withdrawn to a certain extent from the rude realities of life into an atmosphere of affection and sentiment ; yet it must be remembered that this affection and sentiment, however sincere and hearty, has a physical basis—a foundation which would be much better and more securely laid if both, especially the bride, were relieved from all unnecessary fatigue and annoyance, for at this time she has supreme need of physical perfection and at the same time of the greatest tact and discretion ; sometimes, too, she must be prepared for disappointment, for probably every man, however sensible and rational in other matters, is positively silly during the courtship and engagement : invests his fiancé with perfections of body and mind which are actually never clothed in mortal shape ; in fact he marries an ideal creature of his own imagination, and during the first week of married life must learn to substitute the actual for the idea. Hence it often happens that a certain revulsion of feeling is felt by many men, who nevertheless have sincere affection for their wives—a revulsion of feeling for which the bride is not responsible, and yet which she must anticipate and be prepared to meet. There can be little doubt, though it is a matter of course which scarcely permits of actual demonstration, that the seeds of much unnecessary discord and unhappiness are sown during the honeymoon by ignorance and lack of tact. It is therefore extremely desirable that all useless troubles and fatigues, such as those attendant upon travelling, be postponed until the wedded life be fairly begun. And it is hardly necessary to add that it is desirable to avoid the inquisitive eyes of friends and acquaintances, while on the other hand it is just as undesirable to forget and forsake the world entirely during this time ; the boy who eats jam without bread will surely have dyspepsia.

CHAPTER XII.

DISEASES OF CHILDREN.

Care of the Infant. Before birth the child is but a portion of the mother's body, enjoying the advantages of the protection and nourishment which she provides for herself. At birth the infant is deprived of these favoring influences and compelled to conduct an existence independent, to a large extent, of assistance from others. It is not surprising that the experiment is in many cases a failure ; that the tender little creature, deprived of the warmth and shelter of the maternal body, and thrown upon its own resources, compelled to eat, digest and breathe for itself, instead of having food and air furnished it as before, should succumb to the unaccustomed influences, notwithstanding the most assiduous attention. The mortality among infants is accordingly large—one of every ten dies during the first month, and fifteen of every 100 during the first year. Many of those which have survived the first difficulties of life with the assistance of the mother's breast, die when they are compelled to surrender this aid and masticate their own food. Hence it happens that at the end of four years there remain alive but three out of every four infants born. Yet even without these figures it would be self-evident that the new-born infant demands especial and intelligent care and attention.

In the course of a few days after birth the navel-string, which has been at delivery enclosed under a flannel bandage, withers and falls off. If, before this happens, there be a decided odor of putrefaction, it is necessary to cleanse and dress the string somewhat frequently. It sometimes occurs that after the stump has dropped off the navel remains unhealed, raw, perhaps even ulcerated. If this occur, the matter should be brought at once to the attention of a physician, since the result may be serious injury to the child. For some weeks after birth this spot in the abdominal wall remains weak, and may easily be made to protrude when the child strains, especially if the infant be addicted to violent crying and screaming. In such cases there may result a rupture, or hernia—a source of serious annoyance and even danger in subsequent years. To guard

202

DISEASES OF CHILDREN. 203

against this, it is advisable in every case to place a soft pad over the navel and keep it bound on by means of a flannel bandage or adhesive plaster.

While the clothing of the child is to be regulated to a certain extent by the climate and seasor., yet it will be found advantageous to place flannel next to its skin all the year round. The advantages comprise not only the maintenance of an equitable temperature, but also the absorption of the secretions of the skin, which might otherwise be a source of irritation. The regulation acquires double importance in the case of those infants prematurely born, and, therefore, less capable of maintaining an independent temperature.

A most important element in securing the well-being of the infant is strict cleanliness ; and in nothing else is the difference between an attentive and careless nurse more evident than in the management of the napkins and in the protection of the child's skin from its own discharges. It may be, in general, stated that soreness and rawness about the child's thighs indicate neglect of cleanliness either of the infant itself or of the napkins. So far as the child is concerned, there should be, and is usually, no difficulty, since the warm bath is everywhere procurable. It is, however, to be borne in mind, that the infant can be bathed to excess ; during the first weeks of its existence the child should not be put into the bath more than once a day, nor remain there more than two or three minutes ; feeble children must be bathed with still more caution, since the warm bath is quite exhausting. Yet many infants which are regularly and carefully cleansed suffer, nevertheless, from soreness of the skin ; and many a mother applies industriously, but unsuccessfully, a variety of baby powders, and is puzzled to know why the child's skin remains sore. There is just one slovenly habit which is apparently responsible for much of the trouble of this sort : many mothers, namely, seem to consider that the napkins require washing only after being soiled ; and that if only wet they may be reapplied after drying by the stove. This idea, is, of course, a mistake, since the urine contains a number of substances which are extremely irritating when applied to the skin ; the napkins should be washed in one case as well as in the other.

After the first ten or twelve days the child may be carried out thoroughly wrapped up if the weather be properly warm (70° F.) ; and unless the inclemency of the season absolutely forbid it, the infant should receive a daily airing and sunning as regularly as plants.

Food.—It need hardly be repeated that the most appropriate and desirable nourishment for the new-born child is derived from the mother's breast ; yet it may with propriety be remarked that this nourishment is of itself amply sufficient for *all* the needs of the infant ; and that sugar and water, and a dozen other mixtures which are poured into the helpless child during the first few days of its life, may do harm, but can do no good. If the secretion of milk be delayed beyond the usual time nothing should be given the child except a little water simply *stained* with cream. So soon as the secretion is established there should be no further administration of artificial food. The best assurance of health to the child during the first seven months of its existence, is an ample supply of mother's milk.

In a considerable minority of cases, however, it becomes necessary, on account of some of the causes already indicated, that artificial food should be substituted entirely or in part for the breast. These are the cases which furnish so much sickness or mortality, especially in our larger cities. It must not be understood that artificial feeding implies in itself anything injurious to the child, but as ordinarily implied it is such a poor substitute for the natural food that the results are, as a rule, very unsatisfactory.

With care and attention a bottle-fed infant may and often does enjoy the most robust health ; and there are certain combinations of artificial food which give, as a rule, excellent results as a substitute for mother's milk. Yet while physicians are accustomed to rely upon these substitutes, they are equally well aware that there must be an adaptation of means to meet the requirements of individual cases ; that the effects upon the child, as indicated by the condition of his digestive organs, must be carefully observed, as an index to possible modifications of diet, A food which may meet the requirement in four successive cases may require modification in the fifth ; and it is extremely important for mothers to realize the fact that there is no such thing as a universal infants' food ; that the nourishment of infancy must vary, not only according to age and season, but also with individual peculiarities ; and that her vigilance may not be relaxed, so soon as she has obtained from a friend or physician a formula for preparing infants' food. The best plan in every case in which difficulty is experienced is to place the matter in the hands of a physician. Yet a few directions for the composition of food, which will in many cases answer all requirements, are appended :

The simplest substitute for mother's milk is obtained by diluting cow's milk. The milk of the cow differs from that of woman in two essential details : It is considerably richer in solid constituents but contains less sugar. To approximate it to human milk it is merely necessary then to add a certain amount of water and sugar, preferably " milk sugar." The amount of dilution varies with the age of the infant ; since mother's milk is less rich in the early months of nursing than it subsequently becomes, a correspondingly greater amount of water must be added to the cow's milk to secure the proper consistence. The milk, as obtained from the cow should be mixed with its own bulk of water for a child one or two months old ; if the milk be obtained from a dealer it will rarely be necessary to add more than half its bulk of water ; the mixture may be sweetened with table sugar, or better with sugar‚of milk. This preparation often answers admirably as a substitute for the mother's breast, especi‚

ally in the country ; in large cities the plan is less often successful, since the milk cannot be obtained so fresh, and in warm weather at least, has always undergone fermentation whereby irritating compounds are formed. Yet it is oftentimes possible to obviate, to a certain extent, these disadvantages by previously boiling the milk. Yet if it be impossible to secure fresh and pure cow's milk, it will be best to resort to one of several other modes of preparing artificial food. A very popular mixture, and one which has rendered valuable service, is an attempted imitation of the natural composition of milk—a popular and ready form of Liebig's food. The ingredients required are :

Malt,	-	-	-	-	½ ounce.
Flour,	-	-	-	-	½ ounce.
Skimmed milk,	-	-	-	6 ounces.
Bicarbonate of potassium,	-	-	-	7 grains.
Water,	-	-	-	-	1 ounce.

Malt should be crushed or ground in a coffee mill. All the ingredients may be mixed, put into a clean pan, boiled for eight or ten minutes and constantly stirred ; then strained through an ordinary piece of muslin ; if the child must be fed from the very first, it will be desirable to increase the quantity of water in the above mixture to six ounces. After the first two or three weeks the quantity named will be insufficient, as the child requires two or three teacupsful. The mixture tastes quite sweet ; no sugar should be employed. The disadvantage of this food is the necessity for this somewhat complicated preparation every twelve or twenty-four hours.

A most satisfactory and generally applicable food can be obtained from arrow root and cream. Dr. Meigs, of Philadelphia, gives the following directions : " A scruple of gelatine (or a piece two inches square of the flat cake in which it is sold) is soaked for a short time in cold water and then boiled in half a pint of water until it dissolves—about ten or fifteen minutes. To this is added, with constant stirring, and just at the termination of the boiling, the milk and arrow root, the latter being previously mixed into a paste with a little cold water. After the addition of the milk and arrow root, and just before the removal from the fire, the cream is poured in and a moderate quantity of loaf sugar added. The proportions of milk cream and arrow root must depend on the age and digestive powers of the child. For a healthy infant less than a month old, I usually direct from three to four ounces of milk, half an ounce to an ounce of cream, and a teaspoonful of arrowroot to half a pint of water. For older children, the quantity of milk and cream should be gradually increased to half or two-thirds milk and from one to two ounces of cream. I seldom increase the quantity of gelantine or arrow root."

A still simpler food may prepared simply from arrow root and cream. Two teaspoonsful of arrow root are added to half a pint of water, stirred over the fire until pasty, and then strained ; a tablespoonful or cream is then added and given warm.

The most important feature in the success of artificial feeding is perfect cleanliness of the bottles and tubes employed ; and only such bottles and tubes should be used which can be readily taken to pieces and thoroughly cleansed, from the point of the rubber nipple to the bottom of the bottle. A failure to observe this simple precaution will certainly vitiate any and all attempts at artificial feeding, whatever material be employed ; for there occurs fermentation in the milk and other matters collected at the joints and in the crevices of the feeding bottle ; and these fermented matters passing into the child's stomach with the next instalment of food, must derange its digestion. For cleansing the bottles and tubes, warm water containing a little borax may be used. The success of the attempt to substitute artificial food will be ultimately measured by the welfare of the infant ; yet careful observation may early indicate the imperfections of the method in use before the child's health has been seriously impaired. Perhaps the most important indications of failure in the character of food are to be observed in the appearance of the infant's discharges, which should be always carefully inspected by the mother herself and not left entirely to the chance observation of the nurse. Indeed, it may be said as a general truth, that serious disorders of infancy—which are, in a majority of cases, derangements of digestion and their consequences—might be often avoided if the appearance of unnatural stools were always regarded as a demand for a careful supervision to the diet ; and this applies to children at the breast, as well as those artificially nourished. In this way it would be often possible to avoid that scourge of infancy—summer complaint ; the first indications of this affection should be met not so much by medicines as by a regulation of the child's food and general management.

Whether sustained by the breast or by artificial means, the infant will usually, in six or seven months, be able to digest other food also ; yet a mistake is often made in permitting the child to have such food at a too early date ; not infrequently a four months' babe is supplied with crackers and similar articles, while the mother wonders why the child is not well. As a rule, nothing should be given aside from the regular diet until the completion of the sixth or seventh month, and then it may not feed promiscuously upon whatever chances to be in its way, but must be gradually accustomed to the digestion of solid food. As a preparation, it may be well to give the child, even as early as the fourth or fifth month, some of the artificial food already mentioned, without, however, discontinuing the breast ; later, soups and broths—containing but little fat— may be administered in small quantities. The general principle should not be forgotten, that until the child has some teeth it cannot properly dispose of anything solid ; the attempt will almost certainly result in disaster to its alimentary organs.

Teething. This term is applied to the period at which the growth of the teeth causes their penetration through the covering of the gums. It is understood, of course, that their development has begun at a period previous to birth, but their growth appears to have no particular influence on the general condition of the child until the commencement of the irritation caused by the protrusion of the teeth through the membrane covering the gums. This irritation is doubtless often the cause of troubles manifested in other parts of the body ; yet there is a too prevalent disposition to ascribe all the ills which afflict babyhood during this period, to the process of teething. Any indisposition of the child, whatever its nature or wherever manifested, is often regarded merely as a manifestation of the teething process, the evil result of which is that affections dependent upon other causes which might be detected and removed, are regarded as inevitable because the child is teething. The fact is, that many a child acquires its ihilk teeth without suffering any appreciable disturbance of its general health ; and that the troubles ascribed to teething are oftentimes the result of errors of diet and improper management, which originate quite independently of the teeth, and are mere ly aggravated by the effects of the irritation in the gum.

The first teeth ordinarily appear during the sixth and seventh month, though there may be variations of several months either way. In fact, instances are recorded in which some teeth have been cut before birth. If there be any irritation, it is often manifested some weeks before the tooth becomes visible—in which fact lies sometimes the explanation of an unusual fretfulness. It is desirable to know and note the periods at which the teeth appear, since in order to avoid the complication of teething, the child should be weaned at one of the longer intervals. The teeth ordinarily appear in a certain order, and at regular periods, which may be grouped as follows : *First.* The first to appear are usually the two middle teeth of the lower jaw, technically called incisors ; this ordinarily happens in the course of the seventh month. *Second.* After a pause of one or two months the corresponding teeth of the upper jaw appear, usually followed, after a short interval, by two more, one on either side of the two central teeth. *Third.* There now occurs a pause of six or ten weeks. It is during this pause, while the child is quite free from any irritation of the gums, that weaning is ordinarily advisable. This pause occurs, it will be noticed, during the tenth and eleventh month. *Fourth.* At the completion of the first year there usually appear the first grinding teeth. *Fifth.* At about the eighteenth month, the eye-teeth appear, ordinarily in the upper jaw first. *Sixth.* The full set of twenty teeth is completed during the early part of the third year, by the appearance of the remaining grinding teeth, or molars.

So long as the process of teething proceeds naturally, and causes nothing more than restlessness, or perhaps even slight fever, no interference is demanded. The advance of the tooth into and through the gum is, of necessity, slow, but cannot be hastened by recourse to the lancet. There are cases in which, undoubtedly, the gums should be lanced; but it is just as certain that the early and frequent use of the lancet is undesirable. We may say, in general, that the gum should be lanced, first, when the child is evidently in pain, and the tooth is so nearly through that a slight incision will relieve the tension of the gum ; second, when the gums are hot, tender, swollen and full of blood, in which case an incision, even if it do not remove an obstacle to the progress of the tooth, will nevertheless relieve the congestion of the part ; third, when the irritation in the gum is so great as to disturb the child's nervous system, including, perhaps, convulsions. In this case, even though the gums be not obviously swollen, an incision will often relieve the difficulty entirely.

Weaning. The separation of the child from the breast is an epoch in its existence which is often attended by more or less constitutional disturbance. The time at which this separation should occur may be fixed by some unforseen conditions which render the mother incapable of providing sufficient and proper nourishment for her offspring. Such circumstances may arise at any time, and imperatively require that weaning occur at once. Yet, under ordinary circumstances, considerable latitude is allowed as to the choice of the time at which nursing shall cease. It might be, and by some has been, assumed that the child should be more or less sustained by the mother until it has acquired a complete set of teeth ; and it does sometimes happen that nursing is continued for two years. Another inducement for prolonged nursing is the protection thereby afforded to a greater or less extent against conception. This hope has induced many a mother to prolong lactation beyond the usual limit. While it is impossible to continue nursing for an indefinite time, yet a variety of considerations indicate that the best period for weaning is usually between the tenth and thirteenth months ; though, as will be presently explained, circumstances may require a certain departure from this rule. The first consideration must be the health of the child and of the mother. Now, as a rule, the infant begins to take other food than the breast as early as the seventh or eighth month, and usually becomes largely independent of the mother by the tenth or twelfth month. By this time, too, the quality of the milk has usually deteriorated, so that even though nursing be continued, the mother's milk furnishes but a part—usually a small part—of the infant's nourishment; at the same time the mother's health is often unequal to the task of furnishing so much nutriment in addition to that required for her own body. By the end of the first year, therefore, the child is usually abundantly able to digest its own food, while the mother should be relieved from the additional and no longer essential burden ; and since there occurs during the last two months of this time, an interval of complete freedom from the annoyance of teething, it will be, as a rule, found advisable to wean the child during the eleventh or twelfth month of its life. Yet this is by no means an inviolable rule ; indeed, there are circumstances under which a prolongation of nursing is advisable in the interest of the child. It may be stated, as a general principle, that the child should not be weaned while some of its teeth are cutting through, nor just before the hot weather of summer. If, therefore, an infant be an exception, in that these months of the first year are employed in teething ; or if the child have been born in the early summer it will be advisable, as a rule, to postpone weaning until the objectionable circumstances be removed. So, too, it is necessary to postpone weaning until any ailment which may happen to affect the child—even if only an ordinary catarrh —shall subside. If, in consequence of inability on the part of the mother, it become necessary to wean the child during any such circumstances, it may be desirable to procure a wet-nurse.

As to the process of weaning, but little need be said, except that it should be gradual ; that an interference with the child's health is far less probable by this plan than if nursing be suddenly discontinued. By withholding the breast altogether at night, and by substituting artificial food on certain occasions during the day, the infant may be accustomed to the new regime without appreciable inconvenience or bodily disturbance. If weaning be postponed until the end of the first year the mother will rarely have any difficulty with the breasts. The gradual decrease in the demand upon them will usually be accompanied by a corresponding decrease in their activity. The breast-pump should not be used ; it will rarely be necessary to adopt any other measures than simple friction with camphorated or sweet oil.

Worms. Worms are parasitical animals which infest the intestinal canal of man. They are of five different kinds—the Ascarides, or small thread-worms, varying from an eighth of an inch to one and a-half inches in length, and having usually their seat in the rectum, or last gut ; the Lumbricie or long round worms, from two or three to ten or more inches in length, and usually occupying the small intestines, and sometimes the stomach ; the Trichuris, or long, hair-tailed thread-worm, occupying the cæcum ; and the Tænia, or tape worm, of which there are two kinds, occupying the whole tract of the intestines, and sometimes thirty or forty feet in length.

Although adults are subject to this complaint, it is most common in children.

Symptoms.— Fetid breath, grinding of the teeth during sleep, picking the nose, paleness of the face, acid eructations, swelling of a portion of the belly, which is there hard and tender ; gripings, variable appetite, great irritability and itching of the lower parts of the body ; short, dry cough ; emaciation, slow fever, increasing towards night ; irregular pulse, and liability to convulsions.

N

Treatment.—A dose of castor oil, exercise in the open air, wholesome diet, and a strict prohibition of uncooked fruit, and raw and green vegetables ; salt to be taken with all the food eaten.

Any of the following remedies may be used, according to circumstances :

Grey powder,	· · · ·	24 grains.
Santouine,	· · · ·	12 grains.

Mix and divide into 12 powders. Take one each night at bedtime until the worms are passed.

Dr. Wylie.

Oil of Turpentine,	· ·	½ ounce.
Castor oil,	· · ·	1½ ounces.

Take 2 teaspoonsful at bedtime.

Wormseed Tea. Take of

Fresh wormseed leaves,	· · ·	1 ounce.
Milk,	· · · ·	1 pint.

Boil with a little orange peel. Dose, a wineglassful, morning and evening, for the expulsion of worms from the bowels.

Croup. This is an inflammation of the larynx and trachea, causing a difficulty of breathing, and a rough, hoarse cough, with a sonorous inspiration of a very peculiar character, sounding as if the air was passing through a metallic tube. It most usually attacks children of from one to three years of age, to whom it sometimes proves fatal ; very rarely are adults affected by it.

Symptoms.—The symptoms are merely those of a common cold, or catarrh ; then comes on a dry cough, with hoarseness and wheezing ; at night there is restlessness and rattling in the throat, afterwhich the croupy crow and sound above spoken of gives unmistakable warning of the disease, which goes on increasing in intensity for a day or two, or perhaps several days, before there is a really alarming paroxysm, which mostly occurs about midnight. The child, after tossing restlessly about, endeavoring in vain to sleep, will start up with a flushed face, protruding eyeballs, and a distressing look of terror and anxiety ; there is a quick vibrating pulse, and agitation of the whole frame, which presently becomes covered with a profuse perspiration. As the struggle for breath proceeds, there is clutching of the throat, as though to force a passage ; the arms are thrown wildly about, the respiration becomes more labored, the rough cough more frequent, and the characteristic croup rings out like an alarm. There is expectoration of viscid matter, but so difficult is it to be got rid of, that the efforts appear to threaten strangulation ; gradually the symptoms become weaker, and eventually the child falls into the sleep of exhaustion. It will probably wake up refreshed, and during the day may appear pretty well; but at night again, probably there will be a recurrence of the attack with aggravated symptoms, convulsions, spasms of the glottis, causing the head to be violently thrown back, in the effort to obtain a passage for the air through the wind-pipe; there is a fluttering motion in the nostrils, the face is puffed, and of a pale leaden hue; a film comes over the sunken eyes, the pulse becomes feeble and irregular; there are more gasping convulsive efforts to continue the struggle, but in vain; the powers of life at length succumb, and the patient sinks into a drowsy stupor, which ends in death. Such is the frequent course of this painful disease, and the changes from bad to worse are so rapid that there is little time for the operation of remedies, that is, when paroxysms have begun.

Treatment.—Confinement to the house, in case of threatened croup, is always advisable, unless the weather should be very warm and open, and then exposure after sundown should be avoided; a dose of calomel (about three grains) should be administered, and followed by nauseating doses of tartarized antimony, of which one grain dissolved in an ounce of warm water, and a teaspoonful of the solution given every quarter of an hour, until the effect is produced. Should the bowels be confined after this, give senna mixture, or scammony powder. Apply mustard and bran, or flaxseed poultices, to the throat. Fill the room with the vapor of boiling water—a large kettle on the stove will effect this. Leeches, if the patient is of full habit and the breathing is very labored, and a spare diet, are the other remedial measures.

In the paroxysms, the most prompt and vigorous measures must be adopted to give any chance of success; strong emetics to cause full vomiting, which often has a most beneficial effect; warm baths and blisters applied from one ear to the other. Calomel combined with ipecacuanha powder, or tartar emetic, should be given every four hours or so; and, if the danger is extreme, counter irritation by means of mustard poultices applied to the calves of the legs, &c. In leeching for croup, one leech for each year of the child's age is the general rule to be observed, and the best part is over the breast bone, where pressure can be applied to stop the bleeding, if required; over the leech-bites apply a blister, should one appear necessary. If the above powders should cause too violent an act on the bowels, add to them a little chalk and opium. Should the child appear likely to sink from exhaustion, after vomiting has been produced, stay the emetics, and give liquor of acetate of ammonia twenty drops, with five or ten drops of sal volatile, or the same of brandy in a little water, or camphor mixture, a little white wine whey may also be administered. Of course, the first endeavor in an attack of croup should be to obtain medical assistance; but if this can not be procured there must be no temporizing: resort at once to the remedies most ready to the hand, using them to the best knowledge and discretion available. Let the contagious nature of croup be ever borne in mind, and especial care taken to keep apart those affected with it from any other children in the family or house. Let it also be r membered that the great agents in producing croup are cold and moisture ; and the greatest of all the east wind, and that those who have once been attacked by it are peculiarly liable to a recurrence of such attack.

Croup is most likely to be fatal when inflammation commences in the fauces ; and this, if discovered in time, may be stopped by the application of a solution of nitrate of silver to the whole surface within sight, and to the larynx.

The following is highly recommended even in the worst forms of true croup :

Chlorate of Potash,	-	-	-	- 2 drachms.
Syrup of Lemon,	-	-	-	- 1 ounce.
Water,	-	-	-	- 3 ounces.

Dose, according to age ; if under 2 years, a teaspoonful ; from two to ten, 2 teaspoonsful ; over ten, a tablespoonful, given every 3 hours, or every half hour in urgent cases.

Dr. T. M. Drysdale.

When the child is much exhausted, in addition to the above give

Acetic Ether,	-	-	-	- 3 drachms.
Camphor,	-	-	-	- 10 grains.

Mix, and take 10 to 15 drops every quarter of an hour.

Dr. Niemeyer.

Spasm of the Clottis or Child-Crowing. This exhibits much the same symptoms as the croup. It is not, however, of an inflammatory character, but is symptomatic of some other disease commonly coming on as a result of irritation caused by hydrocephalus, teething, worms, &c. The medical man only can judge of the probable cause, and he will use such remedies as are most applicable to the peculiarity of each case.

Treatment.—The following mode of treatment recommended by Dr. Leman, of Torzan, has, we believe, been found efficacious in many cases of croup. It is simple and easy of application. We give the details as furnished by Dr. Graves : " A sponge, about the size of a large fist, dipped in water as hot as the hand can bear, must be gently squeezed half dry, and instantly applied under the little sufferer's chin under the larynx and windpipe ; when the sponge has been thus held for a few minutes in contact with the skin, its temperature begins to sink ; a second sponge, heated in the same way, should be used alternately with the first. A perseverance in this plan during ten or twenty minutes, produces a vivid redness over the whole front of the throat, just as if a strong sinapism had been applied ; this redness must not be attended or followed by vesications. In the meantime the whole system feels the influence of the topical treatment; a warm perspiration breaks out, which should be well encouraged by warm drinks, such as whey, weak tea, &c., and a notable diminution takes place in the frequency and time of the cough, while the hoarseness almost disappears, and the rough ringing sound of the voice subsides, along with the difficulty of breathing and restlessness; in short, all danger is over, and the little patient again falls asleep, and awakes in the morning without any appearance of having suffered from so dangerous an attack. I have repeatedly treated the disease on this plan, and with the most uniform success. It is, however, only applicable to the very onset of the disease; but it has the advantage of being simple, efficient, and easily put in practice, and its effects are not productive of the least injury to the constitution."

Snuffles, or Cold in the Head. Children are very liable to this distressing complaint, caused by inflammation of the lining of the nose.

Treatment.—Rubbing the nose with goose-grease, lard or tallow, will generally give relief. Keep the bowels open with a little castor-oil; and, if the stoppage in the nose is obstinate, give warm doses of catnip, penny-royal, or balm tea.

St. Vitus' Dance. This is also known as chorea.

Cause.—Chorea generally occurs in debilitated children, especially girls from the age of eight to fifteen, or when puberty commences, the occurrence of which often appears to act as a cure of the trouble. It results frequently from rheumatism in the acute form, from general debility, fright, excitement, fatigue, great mental effort.

Symptoms.—It is characterized by incessant movements of the hands, the feet, the face, the tongue, in fact, of the whole body. These are irregular, and appear beyond the control of the patient. They interfere with walking, working, and speech, but cease entirely during sleep.

Treatment.—It is well, if possible, to ascertain the cause and remove it. If the disease is due to rheumatism the patient should be put on anti-rheumatic treatment (see Rheumatism). The bowels must be kept freely open, and the child kept as much in the open air as possible, and the general health improved by exercise, good food and, if necessary, tonics. In addition to these measures the most successful drug to be used is Arsenic, in the form of the Arsenical Solution, beginning with doses of 2 drops three times daily and gradually increasing it to 10 drops. This is the treatment used by Dr. Eustace Smith. Another excellent treatment is:—

Bromide of Zinc, - - - 1 drachm.
Simple Syrup, - - - - 1 ounce.

Take 6 drops three times daily, gradually increasing it.

Dr. Hammond.

Scarlatina, Measles and Small Pox are diseases not wholly confined to children and have been already described. When they oc-

cur in children great care should be taken when the patient is recovering
or the result may be serious. The child should be kept warm in bed
until the last trace of the disease has disappeared, and only when the
child is fully recovered should it be allowed to leave its room ; and even
then a close watch should be kept that no untoward symptoms arise.
The same medicines may be used in the treatment of these diseases in
childhood as are used for adults with, of course, smaller doses. For the
proper doses for different ages see the section of this book headed
"Weights and Measures, etc.'

Chicken Pox. (*Varicella*).—Chicken pox is a very mild form of
eruptive disease, which affects a person but once in a life-time, and which
can generally be traced to specific contagion or infection ; it is mostly
confined to children.

Symptoms.—It is preceded –in most cases, but not in all—by slight
fever, which lasts for one or two days before the eruption appears, which
at first is in the form of conical pimples with a white head, mostly on
the shoulders, breast, and neck, and more sparingly over the face and
body generally. These vesicles, on the second day, appear like little
globular blisters, but with little or no surrounding inflammation ; they
now become filled with a watery fluid, which is not converted into pus,
as in small pox—to the milder kind of which this disease bears some resemblance—and, about the fifth day, the bladders shrivel up and dry
away, leaving only crusts or scales. The main distinctions between
chicken pox and small pox are the absence or extreme mildness of the
premonitory fever in the former disease, and the form and contents of
the vesicles ; those of the latter eruption being filled with dark matter,
and having invariably, a depression in the centre.

Treatment.—On the first appearance of the eruption, the patient should
be put upon spare diet ; this, and a dose or two of some cooling aperient,
as rhubarb or magnesia, is generally all that is necessary ; but should
the febrile symptoms run high, give a saline draught as the following :

Carbonate of Potash,	-	-	-	1 scruple.
Citric or Tartaric Acid,	-	-	-	15 grains.
Essence of Cinnamon,	-	-	-	½ drachm.
Syrup of Orange Peel,	-	-	-	1 drachm.
Water,	-	-	-	10 ounces.

Shake, and drink while sparkling a wineglassful as a refrigerant. To
make it effervescing, add the acid after the draught is poured out. Give
plenty of cooling drink, and, if the bowels are at all obstinate, emollient
injections. Care must be taken that the skin is not irritated by something—as if it is, painful and troublesome sores may be produced—and
also that the patient does not take a chill. If these precautions are observed, little or no danger is to be apprehended from chicken pox.

Infantile Remittent Fever. Known also as Infantile Typhoid, Wasting Fever, of children ; Worm Fever, Infantile Hectic, and Gastric Fever, of infants.

How Distinguished.—In this affection, there is always inflammation of the stomach and bowels. It may come on suddenly, the child, though apparently well on going to bed, being attacked soon afterward by fever ; the skin is hot and dry, the face flushed, the eyes red, the pulse quick, the thirst constant, with dry, coated tongue, but red at its point ; the child is very restless, and even delirious. The abdomen is tender on pressure, and hot, and there may be nausea, and vomiting of a sour greenish-yellow fluid. These symptoms continue until nearly daylight, when they begin to lessen, though they do not disappear entirely until the day is well advanced. Even then, there is languor, with great irritability, and the symptoms are again observed to be present towards nightfall, and the fever thus comes and goes, and may even become of a continued form. Often, the child is several days complaining, with thirst, hot skin, loss of appetite, fretful, etc., and gradually the disease is fully developed. Generally, there is constipation, or diarrhœa may be present, with frequent small passages. All discharges from the bowels are very offensive, dark, or clay-colored, tarry, and even bloody. In the latter case, there is more or less straining and griping. In severe cases, the child lies upon the back, with the knees drawn up, and his face shows the distress which he is suffering. In protracted cases, the symptoms assume all the more marked appearance of typhoid fever. Cough is often present, and from this and other symptoms, as the picking of the nose, etc., worms are supposed to be the cause of the affection, and hence one of its names.

Worm Fever. The child rapidly emaciates, and becomes much debilitated, falls into a stupor, with the eyes half-closed, and presents all the appearance of disease of the brain. Or the cough develops into a complete attack of bronchitis. The disease is often observed to abate its force, and then a relapse occurs, with aggravation of the bowel symptoms. Where there exists a tendency to tubercles, these develop, and a fatal result ensues.

Treatment.—It is important that the diet should be mild, nutritious, and easy of digestion ; the drinks of cold, mucilaginous form. The bowels may be operated on by a mild laxative, as rhubarb and magnesia, or by castor oil. It should be remembered that the bowels are in an irritable condition, and, therefore, they should not be unnecessarily troubled. Chalk and ipecacuanha act well in small doses, two grains of the former to a quarter grain of the latter, and generally cause all the symptoms to improve. Should the fever and pain of the abdomen be very great, leeches to the part will greatly relive these symptoms. The warm bath is of great value, or the whole surface may be sponged with warm water, much to the comfort of the little patient. Applications to the abdomen of fomentations, as hops, mush, etc., will aid in the treatment, and in protracted cases, blisters may be employed, with good results. If symptoms of brain affection occur, hot mustard foot-baths, cold to the head, leeches behind the ears, become necessary. When the discharge becomes offensive, with windy swelling of the bowels, turpentine may be given in mucilage, say five to fifteen drops, every four or five hours. If the symptoms subside, tonics may at once be employed, as quinine, eight grains, syrup of gum arabic, ammonia water, each two tablespoonsful.

Mix. As the quinine is not dissolved, but merely suspended in this solution, the bottle must be well shaken before giving a dose.

This is a useful preparation for children, the proper dose for them being a teaspoonful, containing one-half a grain of quinine.

The utmost care must be observed as to the diet, lest indulgence lead to a relapse, and every precaution should be taken to guard against exposure of the body to cold and dampness. Exercise, at first of a passive form, in the open air, should be taken as soon as it can be borne, and gradually the patient should be brought back to its accustomed habits of diet, exercise, etc. If there remain symptoms of enlargement of the glands in the bowels, what are called alteratives, as the syrup of the iodide of iron, ten drops, three times a day, will be useful ; and this may be aided by rubbing in iodine in liniment, or the tincture of iodine may be painted freely over the parts.

Scrofula, or Scrofulosis, is commonly known as the King's evil, because it was believed, in ancient times, that the disease could be cured by the touch of the king.

Cause.—This disease, which is almost solely due to hereditary tendency, may be excited to full action by want of cleanliness, foul air, deficient or improper nutrition, cold and dampness.

Symptoms.—Scrofulous children are generally of a pale, flabby appearance, about the face and hands ; with light, coarse hair ; dull in expression ; heavy and stolid ; thick lips ; teeth early decayed ; large, clammy nose, open nostrils ; a tendency to enlargement of the glands, especially of the neck. The eyes are weak, easily inflamed, with frequent attacks of inflammation of the lids, and the formation of what are called "styes." There is also a great liability to discharges from the nose, and to a diarrhœa of a mucous form. Generally, however, constipation is an accompaniment. Every sickness occurs in an aggravated form, with slow recovery, often becoming chronic. Digestion is extremely apt to be impaired by the slightest causes. Extremes of temperature are borne with difficulty. Slight injuries produce ulceration, with slow or difficult healing. The bones are easily affected, and soften or die from trifling causes. Frequently, enlargement of the joints takes place, which is with great difficulty reduced. Deformities occur, as bending or shortening of the bones, curvature of the spine, flattening of the ribs, projection of the breast bone. In short, the whole system evinces a diseased condition, with but little recuperative power. Hence we have scrofulous ophthalmia ; tabes mesenterica, or marasmus ; slow enlargement of the glands ; deafness caused by inflammation of the ear, and the destruction of its bones ; discharges from the vagina of a thin, unhealthy fluid ; diseases of the bones ; white swelling or scrofulous inflammation of the joints ; hip disease ; spine disease, etc. In all these affections there is a marked similarity in the accompanying symptoms. The child becomes listless, languid ; is disinclined to play, especially at anything requiring much motion ; swelling is observed at the point at which the disease is about to make its outbreak. The part is weak, tender, especially to the touch ; soon, a dull, heavy pain commences, an aching pain, increased by motion of the part, and also at night, when the patient is in bed. If it is an aperture, as the nose or ear, a discharge occurs, of a thin, dirty, yellowish fluid, of a very unpleasant odor, generally due to decay, and consequent discharge of portions of the bony structure. In the joints, great enlargement is seen, the ends of the bones forming knobs, and consequent deformity, and difficulty of motion. In every instance the neighboring glands are early affected, and are felt like hard peas beneath the surface. When the nose or throat is affected there is more or less coughing, with the expectoration of a foul mucus, which causes nausea, loss of appetite, even vomiting. An examination shows the tonsils greatly enlarged, filling up the throat, interfering with the aperture of the inner ear, and thus causing more or less deafness. In these cases the child is observed to suffer from deafness and difficulty of swallowing in every spell of damp weather.

Scrofulous children are also extremely liable to be affected with eruptions of all kinds, particularly during the progress of teething, and if they are caused to dry up suddenly, the disease breaks out elsewhere, in a more dangerous form.

When a discharge occurs from the vagina, in a child, it is very apt to be regarded as the result of something else, as of an attempt at rape ; for this reason great care should be observed, lest a wrong be done, and an innocent person suffer. The history of the child should be carefully learned, its predisposition, its tendencies, if any, to such discharges elsewhere. In this connection, it should not be forgotten, as intimated on a previous page, the presence of seat worms in the rectum, or their transfer to the vagina, will often cause discharges from the vagina.

When a child begins to fail, without apparent cause, the abdomen should be carefully examined, and if found swelled, tender on pressure, the limbs emaciated, the glands of the groin and those of the abdomen enlarged, as may be ascertained by careful manipulation, scrofulous disease of the mesenteric glands may be known as the cause of the trouble.

In scrofulous disease of the spine, almost the earliest symptoms **are a tendency to stumble, a clumsiness, frequent falling, a tendency to cross** the limbs involuntarily, both in walking and lying down. The power of walking is soon lost, especially if the diseased bones be low down in the spinal column. When high up, as about the neck, the child is observed to support its head on the hand, or on a table or other convenient support, and the head begins to sink between the shoulders. The bones project and are tender on pressure, the muscles of the spine are wasted, and speedily curvature is observed, the affected bones being the point of departure from the true line. This may produce pressure upon the cord and paralysis of the corresponding parts, or it may result in what is known as anchylosis, or stiffening of the joints at the seat of the disease, leaving the patient more or less crippled.

Treatment.—A patient of a scrofulous tendency requires constant care. It must be placed under the best possible influences as to pure, fresh air, cleanliness, sunlight, food, exercise, clothing. A child properly protected by warm clothing, kept in the open air as much as possible, and properly nourished, will often be enabled to recover from such a tendency, and escape the results of its constitutional taint. Fresh air and sunlight are of great importance, and are sure to prove valuable factors in improving the general health of a scrofulous patient. Nourishment, in proper quantity and form, forms an additional and valuable aid. For the young child, milk, and especially that of a good, healthy nurse, is the first and best form of food. Later, to cow's milk may be added vegetables of easy digestion, and animal food. Soft-boiled eggs are generally acceptable to the stomach, and materially aid in the nutrition. Much depends upon the cooking, as the best food is not easily digested when either too much or too little cooked. As indigestion is a frequent accompaniment, the food preparation often increases the trouble. In addition to this, care should be observed lest the food be taken in too large quantities, and this is too often the case as convalescence occurs, and thus the progress to health is retarded or prevented.

A point frequently lost sight of is the drinking-water. In the vicinity of large cities, and particularly in their crowded streets, the water used is often very impure, and frequently is itself a carrier of disease. This is at the present time attracting the attention it deserves, and it is hoped will soon lead to an improvement which has long been needed, and which will greatly aid in the prevention of disease.

It cannot be too forcibly impressed upon the attention that those who are weak and delicate require pure, fresh air as a necessity, even more than the health *.* It has so long been the rule to confine such persons to the house, as though the outer air were poisonous, that it often requires the most earnest injunctions on the part of the physician to obtain obedience to his orders.   House ventilation is avoided, from the mistaken idea of giving the child cold.   It should be thoroughly known that fresh air can do no harm, so long as it is supplied without a draught. This may be accomplished in a variety of ways, as by partially opening a window, and interposing a screen, as a chair with a shawl thrown over it, to prevent the direct current of air. Overcrowded apartments must be strictly interdicted, whether by night or day, in the dwelling or the workroom. When possible, the patient should be removed to a healthy locality, an atmosphere of a dry and elevated situation, in the country or at the sea-side. Exercise in the open air must be regular, and never omitted except in the most inclement weather. This may be on horseback, in an easy carriage, walking, etc. If so arranged as to be combined with pleasant recreation, so as to aid in the development of the muscles, greater benefit will be obtained. Fatigue should not be allowed, hence the sports, etc., must be watched and controlled, to prevent excesses.

Bathing is a valuable adjunct, keeping the skin in proper con ition, and thus enabling it to throw off disease. If the patient is easily chilled, warm bathing is preferable, and promotes the circulation. Otherwise, the tepid bath may be employed. Sponging, followed by friction with coarse towels, or towels dipped in salt water and allowed to dry, is very useful. When at the sea-shore, bathing must be practised, but its results must be watched, lest a chill follow, and thus undo all the good accomplished.

Rarely is the cold bath of service. Almost always it is followed by a want of reaction, causing more or less congestion of the more delicate organs, and hence it proves positively hurtful. With some, a cold bath daily, to an infant, is regarded as a means of invigoration, a hardening process, which too often results in the death of its victim.

It is pleasant to know that fashion is at last becoming more reasonable, and permits her votaries to dress their children according to the dictates of good sense and propriety. The child should be fully protected against chilliness, especially in a climate liable to sudden changes. In summer, the clothing should be light, avoiding either extreme, as often a child is kept in a state of perspiration in warm weather for want of a little attention. Flannel next the skin, protecting the chest and bowels, should be worn until the warm weather has fully set in, and then should be substituted by some of the finer textures until all danger of sudden changes are past, or these may be continued during the balance of the year.

As to medicine, this must be the last resort, and at first should be of the mildest nature. The bowels should be regulated by mild aperients, or, when indicated by clay-colored passages, those which arouse the liver may be employed, such as small doses of rhubarb, magnesia, calomel, blue pill, senna, or jalap.

To improve the digestion, the blood, etc., some form of iron will be of value, and here it will be important to find, by observation, which preparation is most appropriate—the syrup of iodide, ten drops, three times a day, the tincture of the chloride, five drops, three times a day, the potassio tartrate, or the pure powder of iron. Either of these may be combined with quinine, or cinchona, as in its agreeable form of elixir, which is most readily taken by children. Symptoms must be met as they arise.

If the bowels are much affected, a special point will be the avoidance of exercise and the upright position ; rest, in bed, is of the utmost value in the treatment of all forms of diarrhœa. But frictions to the entire surface, once or twice each day, should not be omitted.

The appetite may be restored by the use of mild bitter infusions, as chamomile, columba, or cinchona. Animal fats are highly useful in such cases. Hence the value of cod-liver oil ; and if to this we add the syrup of iodide of iron, or some other of the preparations now so happily prepared by the skill of the chemist, the results will be still more beneficial. Such remedies require many months to obtain their full and lasting effects.

In enlargements of the glands of the abdomen, added to the above treatment should be the application of iodine to the surface ; this may be by the use of the tincture, or of the ointment.

The same may be said of enlarged joints, with the addition of perfect rest of the part.

Recently a variety of remedies have been introduced, as iodized milk, made by dissolving one part of iodine in ten alcohol, and mixed with ninety of fresh cow's milk.

All enlargements of glands may be painted with the tincture of iodine, or coated with the iodine ointment. In inflammation of the bones, an ointment of carbonate of lead, freely applied over the seat of disease, often acts to arrest its progress. Burdock tea is an old and valuable domestic remedy in this complaint.

The following will be found of great value in this disease :

| Sulphide of Calcium, | - | • | - | - | 1 grain, |
| Water, | - | • | • | - | - | 10 ounces. |

Take a teaspoonful every hour.

Dr. S. Ringer.

Rickets, technically known as Rachitis, is a want of nutrition of the bones and muscles, causing bending or breaking of the bones from slight injuries ; muscular weakness ; crooked limbs ; curvature of the spine ; nervous irritability ; general tenderness, etc.

Cause.—Perhaps the usual cause of this disease is want of food, or food of improper kind. It would also appear to be the result of bad ventilation, and want of sunlight. A distinguished observer declares "that wherever the rachitic child is dependent upon the mother's milk, the mother will be found to have menstruated during lactation, regularly, for several months, and the degree of rachitis to be in direct ratio to the frequency, duration, and amount of the menstrual flow." In short, the disease is the result of anything which impairs nutrition.

Treatment.—Children with a tendency to rickets exhibit in their whole bony structure unmistakable evidences of the disease. The face is broad and square ; the head is large and flat ; the moulds or openings in the skull do not close until late ; the veins are prominent ; the spine is curved ; the limbs, especially the lower extremities, are bowed, shortening the child so as to give it a squat appearance ; the muscles are feeble ; the child is generally deficient in vigor, mental as well as physical ; the skin is thick, and of a dirty appearance ; the teeth are late in appearing, and the child rarely can walk or even stand, until long after the usual time. The child is spiritless, dull, languid ; the appetite is poor ; the flesh is flabby ; the passages from the bowels are loose, dirty looking, and offensive ; perspiration occurs at night, and often this is excessive ; the child gains no power over its muscles, and requires help in all its movements, though at first the muscles do not seem wasted, and preserve their contour. As the disease advances, the mental and physical powers retrograde. and a glance at the child, with its deficient mental powers, its crooked limbs, unsightly joints, and the shape of its chest, shows the whole trouble.

The glands often become enlarged, as in its congener, scrofula, but are softer, much larger, and not so easily moved as in scrofulous cases. Death occurs from exhaustion. the powers of vitality being unable to carry the system to a favorable termination ; or death may ensue from some complication, as hydrocephalus, diarrhœa, convulsions, incurable changes in the larger glands, as the liver, spleen, etc.

Treatment.—With a full knowledge of the causes inducing this affection, and of the nature of its changes, it is easily understood what would be the special line of treatment. In the early stages, when the patient is so fortunate as to come under the proper observation, much may be done to prevent its full development, and ward off the possible complications. The nourishment must be positive and of easy digestion. The greatest care must be observed, to see that food is given at proper intervals, and in sufficient quantity, as well as of a proper kind. Too much stress cannot be laid upon the value of pure cow's milk for children, after weaning, or in the event of a failure of the breast-milk. Of course, the mother's milk, or that of a healthy young person, is to be preferred in all cases where it is possible to be obtained.

Combined with milk, other articles are valuable, as eggs, in the form of custard, or soft-boiled, rice, farina, corn starch, roasted potatoes, meats, etc. The same observations apply here as in scrofula, as the great point is to remedy the difficulty by properly nourishing the bones, etc.

All exhausting discharges, as diarrhœa, etc.; all complications, as indigestion, constipation, etc., must be met as they arise, with the appropriate remedies heretofore indicated. The constitution must be toned up, strengthened, by tepid baths, sea-bathing, fresh air, always that of the country or seashore, if obtainable ; tonics, as the preparations of iron, cod-liver oil ; bone itself, as fine filings of fresh bones, in milk or rice-milk. Chalk mixture will generally relieve the diarrhœa ; and if the stools are offensive, castor oil may be given to cleanse the bowels, and then astringents such as :

Tincture of Colomba, - ` - - 15 drachms.
Deodorated tincture of Opium, - - 1 drachm.

Take half a teaspoonful in water before meals.

Dr. Bartholow.

A valuable aid in changing the character of the evacuations, is the addition of a few drops of the solution of chlorinated soda to each dose of the diarrhœa mixture. Quinine, in combination with iron, as the potassio-tartrate of iron, in cinchona tea, or the tincture of the chloride of iron, acts as well as a tonic. In many instances, the dilute acids combined with a bitter, act well in toning up the muscular system. Thus, the dilute nitro-muriatic acid, two to five drops, in gentian, columba, cascarilla, or cinchona tea, checks the excessive perspiration, strengthens the digestion, and, in fact, improves the whole system.

Occasionally, when the treatment is commenced at a late period, rest will become necessary, to prevent curvature of the limbs becoming greater, or great deformity of the spine, resulting from the softened condition of the small bones comprising the spinal column. This rest must be in a recumbent position, but must be combined with daily excursions into the open air. Or some form of apparatus may be employed to support the part until the bones have acquired sufficient hardness. Thus, when the seat of the affection is at the bones of the neck, the weight of the head may be taken from the bones by means of an appropriate sling, suspending the head from a bowed piece connected with an apparatus placed around the chest. Each deformity, or tendency thereto, will require its own apparatus, varied to meet the indications.

How Prevented.—A child manifesting a tendency to rickets should be kept from an erect position, or the early use of its feet, until the bones have assumed a proper degree of hardness. Violent exercise, in older children, should be interdicted. Cleanliness, ventilation, full nutrition, are the prerequisites for the prevention of this affection, as well as the aids in its cure when it is established. Sunlight deprivation is too often a main exciting cause of this, as well as other diseases of children. Hence, our legislators would show their wisodm by prohibiting the building of houses, as dwellings, in the rear of tall factories, as well as upon streets of such narrowness as to preclude the entrance of the sun's rays, if at all, but for a short period of each clear day.

Convulsions, Fits, Spasms. Fits are cerebral, and arise from disease within the head, or from irritation in the stomach and bowels, or from exhaustion ; or they are evidence of, and depend on, some malformation or disease of the heart.

Treatment.—The treatment must be directed, during a convulsion, to the shortening of it, and care in preventing injury to the head or limbs by the struggles. All tight clothing should be loosened, or better, entirely removed. The child should then be placed in a warm bath, to which mustard may be added ; during this time cold should be applied to the head, either by cloths saturated, or by the douche, or by the use of a bladder filled with crushed ice. Should this fail to relieve the spasm, the child should be placed upon a couch, and a mustard plaster placed the whole length of the spine, extending on either side for an inch or two. A decided impression must be made before this is removed. If there is reason to believe that foreign matters in the stomach are acting to prolong the attack, a brisk emetic, ipecacuanha, tartar emetic, sulphate of zinc, should be given, and its effect encouraged by the use of draughts of hot water with salt or mustard. Injections should be thrown into the bowel, of salt and hot water, castor oil, oil of turpentine, to stimulate the action of the bowel, and act as a revulsive to draw the blood from the brain. In obstinate cases, convulsions may be broken up by inhalation of ether or chloroform. In rare cases, with a strong tendency to a return of the convulsions, the effect of the drugs may be continued until all such symptoms disappear. Of course, it would scarcely be necessary to make the artificial sleep very complete, only sufficient to control the convulsion. Chloral, by injection into the bowel, certainly has proved of value in these cases ; a teaspoonful in a pint of water, of which a gill may be thrown up in a child often.

In the interval, or after the spasm has gone off, the indication will be to prevent its return and relieve any accompanying symptoms, or treat whatever disease may be thus ushered in. The bowels should be freely moved ; the stomach thoroughly emptied, the teeth lanced, if necessary ; restlessness subdued by narcotics, as, above all, chloral. Worms may be expelled by the appropriate medication ; fever lessened by sponging with tepid water, and cold applications to the head ; the force of the circulation diminished by small drafts of digitalis tea, a teaspoonful of the leaves to a quart of water. If there are marked evidences of a fulness of blood in the brain, leeches behind the ears, to the nape of the neck, or temples, will rarely fail to give great relief. In obstinate cases, a blister to the back of the neck, extending up and down, and allowed to draw well, will prove of great benefit.

Other complications should be met as they occur, and treated in accordance with the principles laid down.

Night Terrors, and excessive nervousness, are of frequent occurrence in childhood. Frights to children often terminate in convulsions, imbecility, or death. Those who have the care of children should be especially on their guard to use every endeavor to prevent frights, and to protect against the foolish habit of working on the fears of a child to make it behave properly.   Not only is this very injurious, but it is liable to make children deceivers and liars, and its evil results often cling to them when grown to adult age.   On the contrary, children should be kindly encouraged, and made to feel the protecting care of those around them.   They should be constantly shown the absence of all danger.

A child will often suddenly awake in the night, with a frightened cry, and evidently impressed with a vague fear of something, and fail to recognize the presence of its protectors.   This may last for several minutes, terminating in a fit of weeping, or it may again quietly fall asleep. Such attacks may occur nightly, or at irregular intervals.   Rarely it is observed to return the same night.   These are almost invariable the result of some irritation of the stomach or bowels, and are generally associated with constipation.

There is no reason to regard these symptoms as indicative of brain disease, though a continuance might eventually lead to serious results.

Treatment.—On the occurrence of such an attack, the child should be at once attended to, and in no case is it justifiable to seek to quiet it by harsh words or treatment.   Its position should always be changed by turning it from the back, which is mostly its position during such an attack, to the side, or better, by raising it in the arms, and thus endeavoring to soothe and comfort it.   Or it may be roused by washing its face with cold water, and a sup of cold water will generally be desired and refreshing.   The child should not be allowed to nurse, which is the usual panacea of all infantile troubles, until it is thoroughly roused and quieted.

To prevent a recurrence of these attacks, the bowels should be carefully regulated.   All excitement, particularly near the hour of sleep, should be carefully avoided.   Exercise of a violent nature is never otherwise than hurtful, particularly, as is too often the case, about the bed-hour.   A light in the room, and better still, the presence of an attendant, at least for some time after it retires, will greatly conduce to the prevention of these alarms.   Care should always be had not to rudely waken a child from its rest, but in every way to promote sleep.   This great restorer is needed by the child, who plays and labors with all its powers during its waking moments, and hence requires plenty of quiet sleep.

Perhaps the bromide of potassium may prove of value should an attack continue obstinate   Opiates, or the so-called soothing syrups, should be avoided, as sure to prove hurtful.   The dose of the bromide of potassium for this purpose should be five or ten grains.   Eating a raw onion just before bed-time will, with many persons, induce refreshing sleep.   A glass of hot water (not warm, but *hot*), flavored with lemon or orange peel, has the same effect.   Hop tea is another pleasant and efficient remedy for sleeplessness.   Any of those may be tried, and where one fails another will be a success.

Erysipelas. Infants are liable to a peculiar erysipelatous inflammation within a few days after birth.

Symptoms.—It generally commences on the lower parts of the body, in the form of a small red blotch, which gradually spreads over the abdomen and the thighs, presenting a swollen dark-red surface. In most cases, soon after inflammation is established, vesicles make their appearance, and the disease soon reaches a dangerous condition, the tendency to suppuration and gangrene being very great.

Treatment.—On the first appearance of inflammation, wrap up the affected parts with cloths saturated with a strong solution of the sulphite of soda, and cover with oiled silk. The mucilage of slippery-elm bark, or grated potatoes, applied, will check the spreading. If gangrene is indicated, apply a poultice of indigo-weed, or lotions of the permangranate of potash. In inflammation, give teaspoonful doses of the elixir cinchona and iron, in addition to the external application of the sulphite of soda.

Thrush. This is one of the most common diseases of infancy. It is characterized by a peculiar eruption of minute pustules, and a whitish incrustation of the tongue.

Symptoms.—There are generally much thirst, restlessness, langour, acid and flatulent eructations, loose and griping stools, drowsiness, pain, difficulty in sucking, and a copious flow of saliva from the mouth. The stomach and bowels are almost always prominently disordered, and the infant is apt to vomit after taking anything into its stomach. The abdomen is often sore to the touch, and great difficulty of swallowing is experienced. Feeble and sickly children scarcely ever escape this disease ; children, also, who are kept in crowded or ill-ventilated apartments are especially liable to it.

Treatment.—The first object is to restore the healthy condition of the stomach and bowels, if disordered. Where the ejections from the stomach are sour, and the alvine evacuations of a grass-green color, from three to four grains of magnesia with two grains of rhubarb, and one of powdered valerian, should be given every two or three hours until the bowels are freely evacuated. If there is much irritability and restlessness, after this the tepid bath, followed by a drop or two of laudanum, should be employed. The mucous membrane of the intestines is apt to become highly irritated in severe cases ; the alvine evacuations in such instances are frequent, watery, and streaked with blood. When these symptoms are present, a large emollient poultice should be applied over the abdomen, in conjunction with the eternal use of minute portions of Dover's powder with a solution of gum arabic as drink. Borax is a familiar remedy with nurses and mothers, as well as with the profession. It may be used either in the form of a powder, or in solution. If the former is employed, two or three grains of it, mixed with a small portion of pulverized loaf-sugar, must be thrown into the mouth every two or three

O

hours ; if the solution be used, a drachm of the borax should be dissolved in two ounces of water, and applied to the mouth with a soft linen rag tied to the extremity of a pliable piece of whalebone, or with a soft feather. The practice of forcibly rubbing off the eruption is extremely reprehensible ; for, when rubbed off in this way, the crust is soon renewed in an aggravated form. Where the mouth is very red, livid, or ulcerated, we must have recourse to a decoction of bark. A half-ounce of powder-bark, boiled about thirty minutes in half a pint of water, will make a suitable decoction : and of this about the third of a teaspoonful may be put into the child's mouth every hour or two.

Ulceration of the Mouth. Children are liable to ulcerative affection of the mouth, which is evidently distinct from the ordinary aphthous eruption. It consists in a number of small ash-colored and excavated ulcerations, with elevated edges situated about the frænum, and along the inferior margin of the tongue and gums and on the cheek. They usually commence in the form of small, red, slightly elevated points, attended with slight symptoms of febrile irritation.

Treatment.—Clear out the bowels with a dose of magnesia and rhubarb. A solution of ten grains of the sulphate of copper in about three teaspoonsful of water, to which four teaspoonsful of borax must be added, may be applied to the ulcers once or twice daily by means of a strong camel's-hair pencil. Solid food, especially salted meats, and fish, must be rigidly avoided during this complaint.

Summer Complaint, and Cholera Infantum, are almost wholly confined to the hottest parts of the summer months. By some, these are regarded as names of the same disease. By others, the former is considered as the ordinary diarrhœa of children, occurring during the entire summer.

Cause.—Diarrhœa, in children, is the result of indigestion, eating improper food, as unripe fruit, or fruit partly decayed, difficult teething, exposure to cold and dampness, especially wet feet. Still, at certain times, the disease prevails much more extensively, appearing to attack almost every infant in certain localities.

Cholera Infantum almost never has been met with beyond the limits of the United States and Canada, and generally prevails extensively, during the hottest months, all over the country, but particularly in large cities, though by no means being confined to the poorest and filthiest localities, but as often attacking children placed under the best provisions for light, air, cleanliness, &c. The disease rarely attacks children above two years of age, mostly during the first teething, and for this reason the second summer is always regarded with fear and anxiety. It is one of the most fatal diseases of childhood. Any cause impairing the vitality, as general constitutional disorder, deprivation of the breast, etc., is sure to induce an attack. For this reason, weaning is often postponed until after the second summer. Yet, in many instances, when the breast milk is becoming impoverished, and unsuitable for the child, it would be far better to wean the child, and commence a proper diet with it some weeks previous to the setting in of hot weather, rather than attempt to carry it through the heated term on nourishment manifestly improper. These remarks would apply where pregnancy has again commenced; where the teeth are well out, and the child is prepared for other food; where, for any cause, the breast milk is impoverished, or its supply insufficient.

Again, when the child is so situated as to rely wholly upon cow's milk, or other artificial feeding, the milk furnished should be carefully examined, to be sure that it is pure, fresh, and not weakened by water or other adulteration.

Symptoms.—Summer complaint and cholera infantum will be considered as synonymous terms, as the same principles for the treatment of the ordinary diarrhœa of childhood will apply in every instance, guided by that common sense which is eminently necessary for the treatment of all disease.

In some cases the attack is quite sudden: the child is seized with vomiting, purging, slight fever, great desire for drinks, restlessness; the stomach and bowels being emptied, the discharges become watery, whitish, ill-smelling, or even odorless. The stomach is very irritable, rejecting suddenly and forcibly everything that is taken into it. The progress is rapid; in a few days, or even a few hours, the child comes to resemble a wilted, aged person. The pulse is quick, small, like a thread under the finger; the tongue is white and slimy; the skin is dry, dirty-looking and harsh; the feet and hands are apt to be cold. Generally, the feverish symptoms increase late in the afternoon. The child may express great suffering, as if in pain, or may lie prostrate and uncomplaining; generally, there is great restlessness, with drawing up of the knees and moaning; constant tossing and changing of the position, and often sharp, shrill screams. The abdomen is more or less swollen, and tender to the touch. Frequently, the vomiting ceases early, though the diarrhœa continues, or increases in violence; everything seems to pass through the bowels without any effort at digestion, in a very short time after it is taken. Delirium or other symptoms simulating cerebral affection come on, and the attendants, even the physician, are led to believe that the brain is inflamed, though these symptoms are solely the result of the great exhaustion, and deprivation of the brain of its proper supply of blood.

Very many cases succumb within twenty-four hours, though the majority are slower in reaching a fatal termination. Here, the emaciation is extreme, the eyes sink in, the whole face is pale, the nose is sharp, the lips are thin and dry, the skin is drawn over the cheek bones and forehead so tightly as to be smooth and glistening. The child is but half conscious, lies with the eyes partially closed, and permits the dust and even insects to settle in its eye-balls without noticing them. Every symptom betokens the approach of death, which, singularly, is often long in coming, giving hope of a recovery, long after the child is really in the agonies of death.   The child generally dies in an unconscious state ; occasionally it is attacked with convulsions.

Favorable symptoms are, a decided change in the character and frequency of the passages, desire for food, quiet slumber, reduction of the pulse, moisture of the skin.

Treatment.—The first and most important measure is the removal of the child to a cooler atmosphere, of the sea-shore, best of all ; the pure country air ; or taking it out early in the morning, beyond the built up portions of the city, where it can breath a purer air.   Excursions on the water, as in the ferry-boats of our large cities, will aid greatly in inducing a favorable change.   Under any circumstances, the air of the apartment occupied by the child must be completely purified by free ventilation, the most positive cleanliness, the avoidance of many persons in the room.   The air of the room must be kept dry, and free from anything that will vitiate it, as cooking, tobacco smoke, etc.   The clothing must be light, clean, and dry, and sufficient to protect the child from atmospheric changes, but not to keep it too warm.   The bed should be cool, hence feathers should be avoided ; a mattress, or a blanket only, thick enough to make the couch comfortable.

When there is reason to suspect teething as a cause, the gums should be cut, if necessary, as will hereafter be described, and if they are hot, swollen, and tender, rubbing them with a piece of ice will greatly relieve their irritability.   The whole surface of the body should be bathed night and morning with tepid salt water, and then the skin should be excited to action by friction with a soft towel or the hand.   The diet should be restricted to the breast-milk, or, in older children, to pure cow's milk, properly prepared, not too much diluted ; " condensed milk " is highly esteemed by many as the very best food for those cases.   Overloading of the stomach must be very carefully guarded against.   Cold demulcent drinks or ice may be allowed.

The diarrhœa in its beginning may often be speedily checked by a slight astringent, as acetate of lead, one-fourth to one-half a grain, and three or four grains of prepared chalk every two, three, or four hours, according to the urgency of the symptoms.   The irritability of the stomach can often be relieved by a few grains of magnesia, rubbed up with white sugar, and placed upon the tongue ; or by three or four drops of dilute sulphuric acid in syrup and spearmint water every hour.   Other remedies for vomiting are camphor one fluid ounce to ether one fluid drachm, given in doses of two or three drops at short intervals ; or a few drops of turpentine, or creasote, in a solution of acetate of lead, say, acetate of lead five grains, dilute acetic acid three to five drops, sugar two to three teaspoonsful, and water one ounce, given in teaspoonful doses every hour until relieved. When all else fails, a blister may be placed over the stomach, followed by a poultice of bread and milk, or of flaxseed, etc. When there is a great pain in the bowels, leeches may be applied, and followed by hop or other soothing poultices.

Should brain symptoms appear, leeches may be applied behind the ears, to the temples, cold lotions or the ice-bag to the whole of the scalp, stimulants to the lower extremities.

One of the most valuable remedial agents is the application of stimulating liniments to the whole length of the spine, and the abdomen. This is done in the belief that the counter irritation will act to draw away the irritation from the stomach and bowels.

A good stimulating liniment for the purpose is,

| Camphor, | - | - | - | - | 1 ounce. |
| Olive oil, | - | - | - | - | 4 ounces. |

Rub up the camphor in the oil and apply.

Or a narrow mustard poultice may be applied along the spine.

When the symptoms begin to improve, the diarrhœa may be checked by a continuance of the acetate of lead, etc., combined with small doses of inpecacuanha.

Should the disease tend to become chronic, perhaps the best remedy will be the tincture of the chloride of iron given in syrup and water, in doses of three to five drops every two, three, or four hours, according to age. Laudanum or paregoric, in appropriate doses, may be used cautiously if required by pain. What is commonly known as "thickened milk," milk boiled and thickened with flour, or rice flour, enjoys a high reputation in these cases.

Quinine in small and frequent doses will act as an efficient tonic. The aromatic spirits of ammonia, in doses from ten to fifteen drops, in syrup and water, often proves an excellent stimulant, and quiets the irritability of the system.

The celebrated Dr. Bartholow uses the following prescription with great success in Cholera Infantum :

Sulphate of Copper,	-	-	-	1 grain.
Deodorated Tinct. of Opium,	-	-	8 drops.
Distilled Water,	-	-	-	-	4 ounces.

Give a teaspoonful every two, three or four hours

In closing this important subject we append the following list of *things to be done* when a child is suddenly attacked, in summer, with vomiting, purging, and prostration :—

1. Put the child for a few minutes in a hot bath, then carefully wipe it dry with a warm towel, and wrap it in warm blankets. If its hands and feet are cold, fill bottles with hot water, wrap them in flannel, and lay against them.

2. Place over the belly a mush poultice, or one made of flaxseed meal, to which one-quarter part of mustard flour has been added, or flannels wrung out of hot vinegar and water.

3. Give five drops of brandy in a teaspoonful of water every ten or fifteen minutes; if the vomiting persists, give this brandy in the same quantity of milk and lime-water.

4. If the diarrhœa has just begun, give a teaspoonful of castor-oil, or of the spiced syrup of rhubarb.

5. If the child has been fed partly on the breast and partly on other food, the mother's milk alone must now be used. If the child has been weaned, it should have its milk-food diluted with lime-water, or should have weak beef tea, or chicken water.

6. Give the child cold water to drink, freely.

7. Have the soiled diapers or the discharges at once removed from the room.

Falling of the Bowel, or Prolapse of the Anus, is generally observed in debilitated children, or after protracted diarrhœa. On the other hand, it may come on from constipation and the constant straining to extrude the hardened stools. In some, it occurs after every passage, or even if the child stands for a long time.

Symptoms.—The child is observed to suffer greatly, and examination reveals the presence of a red tumor at the opening, which is generally easily pushed back. If allowed to remain down long, it becomes strangulated, and gorged with blood, inflames, and may even ulcerate or become mortified.

Treatment.—The reduction should be performed at once, by anointing the parts with sweet oil, lard, etc., and then gently pressing it back, and being sure to see that it is completely returned, and no portion allowed to remain grasped by the sphincter or lower muscle controlling the opening of the bowel. If it is constantly recurring, an injection may be thrown into the bowel, consisting of bismuth and catechu. In cases where the part is much engorged and difficult to return, the free use of cold water to the part will relieve it, or a few leeches may be applied before any effort is made to return it. The parts may then be washed frequently with astringent lotions, as strong alum water; a decoction of white oak bark; or the two may be combined; or a decoction of galls may be used.

The accident must be prevented by keeping the bowels relaxed by proper food, fruit, etc., and by careful attention at the time of having an operation of the bowels. In severe cases, it becomes necessary to retain the bowel in place by a pad or compress held on by a bandage. It is proposed to reduce the size of the opening, and thus retain the bowel, by removing a portion of the projecting folds. This has proved successful in some cases, but of course requires the hand of a competent surgeon.

Constipation. Torpor of the bowels and consequent costiveness is of frequent occurrence among infants. In some instances the bowels always require to be excited by artificial means. In constitutional costiveness, a period of from two to four days may intervene between stools without the child receiving any great injury, but it is prudent to watch such symptoms, especially where there is any tendency to convulsive affections.

Treatment.—Manna dissolved in warm water to the consistency of a thick syrup is a good laxative, in teaspoonful doses. Costiveness from accidental causes is a more serious complaint. These causes may be a preternatural determination of the blood to the head, or an undue exhibition of opiates, or a rice diet, or unhealthy milk. Cold-pressed castor-oil is an excellent laxative in ordinary cases of this kind; if acidity be present, magnesia is the appropriate laxative. In moderate cases, the introduction of a soap suppository into the anus will be of service.

Diarrhœa. Diarrhœa is more common during infancy than at any other period of life; and it is also more apt to assume an unmanageable and dangerous character at this period than at a more advanced stage of childhood or adult age.

Causes.—The exciting causes of this disease are extremely various. Irritating, crude, and inappropriate articles of food or drink, are a frequent cause of diarrhœa. Children who are entirely nourished at the breast are much less liable to this complaint than such as are partly nourished by artificial food. Some infants are invariably purged when fed with cow's milk, even when considerably diluted with water; others again are purged by arrow root, although the usual effects of this substance are rather of a constipating character. Infants who are fed with solid food seldom escape suffering more or less diarrhœa. The practice of allowing them to gorge themselves with potatoes, meat, pastry, dried fruit, and other articles of this kind is particularly injurious, and often produces chronic diarrhœa. In some instances the mother's or nurse's milk gives rise to vomiting and purging. Cold, by suddenly checking perspiration, and determining the blood to the internal parts, frequently gives rise to bowel complaint in infants. Cold bathing, or washing; suffering wet diapers, stockings, etc., to remain too long on the infant; setting it down on the grass-plots, floors, steps, etc.; passing suddenly from a close and warm room into the cold external air; exposure to cold and moist weather without sufficient clothing, particularly about the abdomen;—these are the ordinary ways in which diarrhœa, from the influence of cold, is produced in infants; and cases arising from such causes are generally attended with catarrhal symptoms, more especially with cough. A high atmospheric temperature is frequently concerned in the production of this complaint; the occurrence of bowel-complaints among children is comparably more frequent during the hot months of summer than in the colder seasons of the year.

Treatment.—Immediate attention must be paid to the diet.   Do not give the child any solid food, and especially keep from it pastry, sweet-meats, and confectionary.   The most appropriate food will be plain boiled rice and milk; in many cases simple boiled milk will arrest the discharges. Crackers and milk, gruel, tapioca, etc., are also useful.   At the com-mencement of the attack give a mild purge of castor-oil or syrup of rhubarb, and use the warm bath.   If the stools are sour, dissolve a tea-spoonful of bicarbonate of soda in half a glass of water, and give a tea-spoonful every hour.   An excellent remedy for looseness of the bowels is tea made of ground bayberry.   Sweeten it well, and give a half tea-cupful once in two hours, until the child is better.

Following are two excellent preparations in the Diarrhœa of children :

Infusion of Rhubarb,	-	-	- 2 ounces.
Bicarbonate of Potash,	-	-	- 1 drachm.
Tincture of Cinnamon,	-	-	- 2 drachms.
Simple Syrup,	-	-	- 6 drachms.

Take a teaspoonful every two hours.

Dr. Farquharson.

When the Diarrhœa becomes chronic use :

Nitrate of Silver,	-	-	- $\frac{1}{8}$ to $\frac{1}{3}$ of a grain.
Distilled water,	-	-	- 1 ounce.
Simple Syrup,	-	-	- 5 drachms.

Mix, and take a teaspoonful every 3, 4 or 5 hours during the day.

Dr. Trousseau.

Colic, or Gripes, in children, is a disease of constant occurrence. It is also regarded as a neuralgia of the bowel.

Cause.—This affection is the result of indigestion, as brought on by improper food, overloaded stomach, etc., or it seems to have an inter-t tendency, recurring at a certain hour, generally late in the after-, day after day.   Or it may be brought on by cold and dampness to .c feet.   Often, carelessness, and want of proper changing of a wet diaper, will cause such an attack.

Symptoms.—Generally, there is more or less wind in the bowels, as shown by the distension of the abdomen, and the drum sound on per-cussion.   Again, very little wind may be present.   The child becomes fretful, draws its knees up to its chest, cries suddenly and becomes quiet. These actions may be repeated at intervals, often very short, sometimes of greater duration, but rarely disappear until wind has passed from the bowels, or perhaps a thin and frothy discharge occurs.   All these symp-toms may be exaggerated, the screams are piercing, the contraction of the limbs are almost spasmodic, the face exhibits great suffering, the child is bathed in perspiration, and yet, when the attack passes off, the child appears as well as usual. These attacks are often observed in the cases of children otherwise remarkably healthy.

In rare cases, the attack becomes convulsive, almost epileptic, though the intervals are marked by apparent health.

Treatment.—The paroxysm is so distressing that the earnest desire is to give the child immediate relief. This wish is met by the warm bath, hot fomentations to the abdomen, particularly flannel wrung out of hot mustard water : poultices of hops ; mustard plasters, not too long continued ; spice poultices, etc. At the same time may be given by the mouth, oil of turpentine dropped on sugar, say three to five drops every two or three hours. A favorite and highly useful mixture is chloroform, one drop, syrup of gum arabic and mint water or anise water, of each half a teaspoonful, repeated every fifteen minutes, until complete relief is obtained. The bowels should be well opened by stimulating enemas, as of castor oil with a few drops of turpentine. A few drops of laudanum, or a half a teaspoonful of paregoric, may be considered as safe and useful. Other remedies, as bromide of potassium, and chloral, may be employed, both by the mouth and by enemas, as necessity may require. In all cases, constipation must be relieved, and acidity corrected, by the appropriate remedies, as heretofore mentioned. In some cases, mustard plasters to the whole length of the spine, or stimulating frictions over it and the stomach, prove of great service.

How Prevented.—By avoiding constipation, and the consequent accumulation of wind in the bowels, much may be done to prevent attacks of colic. Thus, occasional enemas of molasses, salt and milk, or the administration of magnesia, with belladonna, or hyoscyamus and ipecacuanha, will generally keep the bowels in a relaxed condition. Then the feet should always be kept dry and warm, and the linen changed when wet, lest its coldness induce an attack.

Bed-Wetting, or Incontinence of Urine, is a most annoying
affection of children, and, singularly, is most frequent in colored children. This is often the result of habit ; want of proper teaching ; a lack of mental capacity ; or it may result from a partial paralysis of the retaining muscle of the bladder. It most commonly occurs at night, while in bed, and hence its name. Especially is this likely to occur when care is not taken to have the child pass its water before going to bed. It may be the result of the urine containing irritating salts, and thus causing the child so much distress that it half consciously allows the urine to escape. In these instances, the child is observed to use the vessel very frequently during the waking hours. Some assert that the discharge of the water only occurs when the child turns on its back, and hence propose and insist on the good results of a blister to the back, to prevent the child from assuming that position.

Treatment.—When the urine is being constantly passed, its irritating nature may be corrected by daily exercise in the air ; avoiding exposure to cold and damp; alkalies, as lime water and bicarbonate of soda, five grains thrice daily, with mild, bitter tonics, as cinchona tea, keeping the bowels free, improving the digestion, etc. Often, under this plan, the trouble will quickly disappear, and recovery be permanent.

In all cases, the child should have a light supper; be allowed to drink but little of fluids towards night, and empty his bladder the last act before retiring. As a medicine the tincture of cantharides, say three to five or ten drops, three times a day, gradually increasing the dose, will eventually relieve the difficulty. Care should be taken to stop this remedy at once if painful and difficult urination is complained of. Belladonna in repeated doses long continued, has acquired a great reputation. Two to five drops of the tincture of belladonna should be given in sweetened water every night.

Professor S. D. Gross, M.D., treats incontinence of urine in children as follows:

Strychnine,	-	-	-	-	1 grain.
Powdered Cantharides,		-	-	-	2 grains.
Sulphate of Morphine, -		-	-	-	1½ grains.
Powdered iron,		-	-	-	20 grains.

Mix, and divide in 40 pills; give one three times a day to a child ten years old.

Weaning Brash, or diarrhœa, or looseness of the bowels at the time of weaning, most frequently occurs during the summer, and is often the result of negligence as to the food. The utmost care is necessary to see that the milk given the child is pure, fresh, and not too much, if at all, diluted. Generally, the diarrhœa is checked by these attentions, and the use of prepared chalk, and calomel, followed by a mild astringent. When protracted, the same principles will apply as given under similar headings heretofore.

Milk Crust generally occurs during teething, and while the child is yet nursing, hence its name; and it is commonly thought to be incurable until all the teeth are cut or weaning takes place. It usually appears in children of scrofulous tendency, and is, by its disgusting appearance, a great source of annoyance,

Symptoms.—Milk crust is an eruption, upon the face or head, of a number of red blotches, soon covered with pustules; these, itching and breaking, exude a whitish yellow or greenish discharge, which hardens and forms a thick brown crust, beneath which the discharge continues to ooze, and constantly adds to the crusts. These crusts may appear only on the cheeks or chin, over a small space, or they may cover the whole scalp and face, forming a mask of hideous appearance. In the course of three or four weeks the discharge ceases, the crusts fall off, leaving a red, shining surface, very irritable, and liable to renew the pustules and discharge as before, This matter seems to act as a source of conveyance of the disease, as, wherever it is brought in contact with the skin, it produces more or less of a similar irritation. Fortunately, the disease leaves no scar or marks after healing.

Treatment.—In the early stages it will often speedily yield to proper regulations of the diet, mild laxatives, and soothing lotions. Small doses of magnesia and ipecacuanha, warm baths, pure, fresh air, are invaluable adjuncts in the treatment. In very greatly protracted cases, the solution of the arseniate of potassa, or "Fowler's solution," in two-drop doses three times a day, will generally effect a cure, if persisted in.

This disease is also known as Infantile Salt Rheum or Infantile Eczema. The following has been found to be a very successful treatment:

Salicylic Acid,	-	-	-	-	2 parts.
Subnitrate of Bismuth,	-	-	-	30 "
Corn Starch,	-	-	-	-	20 "
Ointment of Rose Water,	-	-	-	100 "

Spread thickly on muslin and cover carefully every portion of inflamed and moist skin. This should not be used on the scalp on account of matting the hair. For the scalp use Oil of Cade and Almond Oil, equal parts. If thick crust is present, remove by poulticing and wear an oiled silk skull cap.

Dr. G. H. Fox, New York.

Rupture or Hernia may occur in children from the time of birth, or may be developed at any age in those predisposed, after unusually violent exercise, as leaping, wrestling, or riding.

Whenever a child is observed to have a slight tumor or swelling at the navel, or in the groin, enlarging on coughing or straining, he should be warned of the risk of violent effort, and the first opportunity embraced to consult a competent surgeon. The permanent cure of a rupture, by means of a properly fitting truss, is more likely to be brought about in a child than in grown persons.

In very young children a rupture at the navel is not uncommon. It may appear a few days after birth. When a child presents this infirmity, it must be removed by compression. This is accomplished by a well-fitting bandage around the waist, containing a coin or similar body immediately over the navel, thus retaining the parts in their natural positions.

Cancer, although not specially an affection of childhood, is liable to occur at that time, and moreover, as a disease generally hereditary, may be appropriately spoken of in this connection. Cancers are of various kinds, and may occur in all parts of the body, and appear at any age. They have been described as hard, soft, open, or bleeding, black, skin and bone cancers. They usually commence with a swelling and pain, followed by a breaking of the skin, and the formation of an external running sore. The pain is a very vaiable symptom. At the outset it is rarely very severe, and may not be continuous. At times, it is described to be of an aching or rheumatic character; at others, and more commonly, it is an occasional dart of pain through the part, as if a needle had suddenly been thrust into it. Later in the disease, the suffering becomes severe and constant, even excruciating, and can only be relieved by heavy doses of opiates.

What makes cancers so dangerous, is their tendency to recur after removal, and penetrate into and poison the tissues of all the organs of the body. Hence the only hope is to treat them early, before this has come to pass. Although generally hereditary, they are, in their outbreak, entirely local, and many instances are on record where early and complete removal has saved the patient from any recurrence of the cancerous growth.

One of the most common forms of cancer is epithelial, or skin cancer. It is fre̅ ̄ently seen on the lips and face. Its surface may be dry and warty, or watery and ulcerating. The edges are thick and a little elevated, and the discharge thin. It is usually slow in progress, lasting for years before causing such serious inconvenience as to lead the sufferer to a surgeon.

The treatment of a cancer, whether in a child or in an adult, is always and only by *removal*, and the earlier and more completely this is done, the greater safety has the patient for his life. Removal may be accomplished either by the knife or by caustics. Of the first of these methods, we shall only say that in the majority of cases it is the least painful, the most prompt, and the most efficacious. Whenever it is possible, the advice of a good surgeon should be sought early, and his intervention solicited.

But many people have a great dread of the knife. They prefer to suffer indefinite agonies from strolling quacks who advertise cancer salves, and it is, of these we would speak. Whatever secrets such quacks pretend to have, they are all deceptions, and unscrupulous frauds. The caustics which are of use in cancer are all well known to the regular medical profession, and appear in the text-books of surgery. The two principal ones are the chloride of zinc and arsenic. The former, made into a paste with two parts of flour and a small quantity of morphia, is as efficient a cancer salve as any, but it is dangerous and unjustifiable for persons ignorant of surgery to employ it. Of the internal remedies given to prevent the return of cancer, the best are arsenic in small doses, long continued, and poke root. The latter vegetable preparation, given as a tea or extract, seems to have really valuable properties.

Swellings of the Glands of the Neck are often treated at home, without the aid of a physician or surgeon, until they suppurate, and are likely to leave scars. When such domestic management, however, is determined upon, the invalid should be allowed a nutritious animal diet, his bowels kept free by exercise on foot, whilst mild purgatives should be given, and the solution of iodide of iron, in doses of from ten to sixty drops, or the syrup of the iodide, in doses of a teaspoonful (a fluid drachm, which contains three grains of the iodide), should be internally administered, in a glass of water twice a day. The tumors should be treated with fomentations of salt or sea water, and friction employed twice daily, for half an hour each time. If suppuration cannot be arrested, under the improved state of health, then surgical advice must not be delayed until the abscess bursts spontaneously, for an ugly scar is likely to be the result; an event always to be regretted, especially in females.

CHAPTER XIII

Abortion or Miscarriage.

To prevent this accident, use :—

(1) Fluid Extract of Ergot, -   -   5 drachms.
     Deodorated Tincture of Opium, -   -   3 drachms.
     Syrup of Lemon,   -   -   -   1 ounce.

Mix and take a teaspoonful thrice daily.

Or :—

(2) Tannic Acid,   -   -   -   -   15 grains.
     Powdered Ipecac,   -   -   12 grains.
     Extract of Opium,   -   -   3 grains.

Mix and divide into 12 pills. Take one every six hours.

Abscess.

In addition to the local treatment (such as poultices, etc.), which has been described, use :—

     Syrup of Hypophosphites,   -   -   6 ounces.

Take a dessertspoonful thrice daily.

Acidity of Stomach. *(Sour Stomach.)*

(1) Subnitrate of Bismuth, -   -   -   3 drachms.
     Carbolic Acid,   -   -   -   -   3 grains.
     Mucilage of Gum Arabic,   •   -   1 ounce.
     Peppermint Water,   -   -   -   3 ounces.

Mix and take a tablespoonful 3 or 4 times daily, for an adult.

(2) Bicarbonate of Soda, . -   -   -   2 drachms.
     Aromatic Spirits of Ammonia, -   -   2 drachms.
     Tincture of Ginger,   -   -   -   1 ounce.
     Compound Infusion of Gentian, to make   8 ounces.

Mix and take a tablespoonful or two 3 times daily.

Acne, or Pimples.

     Acetate of Potash,   -   -   -   2 scruples.
     Tincture of Nux Vomica,   -   -   2 drachms.
     Fluid Extract of Yellow Dock, -   -   4 ounces.

Mix and take a teaspoonful in a wineglass of water half an hour before meals.

In addition to this apply the following lotion to the pimples :—

Sulphur,	-	-	-	1 drachm.
Glycerine,	-	-	-	1 ounce.
Rose Water,	-	-	-	7 ounces.

Alcoholism or Drunkenness.

To overcome the desire use :—

(1)
Tincture of Capsicum,	-	-	½ ounce.
Bromide of Potash,	-	-	½ ounce.
Aromatic Spirits of Ammonia,	-	-	3 ounces.
Syrup of Tolu, to make	-	-	6 ounces.

Mix and take a dessertspoonful in water 4 or 5 times daily.

To overcome the sleeplessness due to alcoholism use :—

(2)
Tincture of Nux Vomica.	-	-	1 drachm.
Compound Tincture of Gentian,	-	-	3 drachms.
Spirits of Lemon,	-	-	6 drops.
Spirits of Chloroform,	-	-	1 drachm.
Water to make	-	-	6 ounces.

Mix and take 2 tablespoonsful 3 times daily.

When the stomach is disordered by alcohol use :—

(3)
Compound Tincture of Gentian,	-	-	2 ounces.
Compound Tincture of Calumba,	-	-	2 ounces.
Tincture of Nux Vomica,	-	-	80 drops.

Mix and take a dessertspoonful before each meal.

Dr. Loomis.

Amenorrhœa, or Absence of the Menstrual Flow.

Aqueous Extract of Aloes,	-	-	1 drachm.
Dried Sulphate of Iron,	-	-	2 drachms.
Assafœtida,	-	-	4 drachms.

Make into 100 pills and take one pill after each meal, gradually increased to three.

Dr. Goodell.

Anæmia or Lack of Blood.

(1)
Tincture of the Chloride of Iron,			4 drachms.
Dilute Phosphoric Acid,	-	-	6 drachms.
Spirit of Lemon,	-	-	2 drachms.
Syrup to make	-	-	6 ounces.

Take a dessert spoonful in water after meals.

Dr. Goodell.

(2) Sulphate of Quinine,	-	-	-	20 grains.
Dried Sulphate of Iron,	-	-	-	40 grains.
Sulphate of Strychnia	-	-	-	½ grain.

Divide into 20 pills and take one thrice daily.

Dr. Bartholow.

(3) Dried Sulphate of Iron,	-	-	-	2 drachms.
Carbonate of Potash,	-	-	-	2 drachms.
Syrup, a sufficiency.				

Divide into 48 pills and take one after each meal gradually increased to three.

Aphtha or Sore Mouth.

Solutions of Borax and honey, Borax and Glycerine or of Chlorate of Potash are excellent gargles. Also the following:

Sulphite of Soda,	-	-	-	1 drachm.
Water,	-	-	-	1 ounce.

Mix, and use as a lotion in the mouth.

Sir W. Jenner.

Appetite.

To improve the appetite use :

Gentian,	-	-	-	-	2 drachms.
Quassia,	-	-	-	-	2 drachms.
Cinnamon,	-	-	-	-	2 drachms.

Put into a pint of boiling water, and when cold strain ; take a wineglass-ful thrice daily.

Asthma.

(1) Nitrate of Potash paper,	-	-	One.

Burn in a close room and inhale the fumes.

(2) Bromide of Ammonia,	-	-	-	160 grains.
Chloride of Ammonia,	-	-	-	90 grains.
Tincture of Lobelia,	-	-	-	3 drachms.
Compound Spirits of Ether,	-	-	1 ounce.'
Syrup of Acacia to make	-	-	4 ounces.

Take a dessertspoonful in water every hour or two till the paroxysm has passed.

Dr. Pepper.

(3) Tincture of Lobelia,	-	-	-	1 ounce.
Iodide of Ammonia,	-	-	-	2 drachms.
Bromide of Ammonia,	-	-	-	3 drachms.
Syrup of Tolu,	-	-	-	2 ounces.

Take a teaspoonful every 1, 2, 3 or 4 hours as required to relieve.

Dr. Bartholow.

(4) Compound Spirits of Ether, - - 1 ounce.
    Sulphate of Morphia, - - - 1 grain.
    Water, - - - - 1 ounce.

Mix, and take a teaspoonful every half hour or hour during the paroxysm.

Dr. James.

Biliousness.

(1) Fluid Extract of Stillingia, - - 5 drachms.
    Tincture of Aloes, - - - 2 drachms.
    Tincture of Nux Vomica, - - - 1 drachm.

Mix and take 20 drops in water thrice daily.

Dr. Bartholow.

(2) Blue Mass, - - - 3 grains.
    Compound Extract of Colocynth, - - 3 grains.

Mix and make into 2 pills; take these at once and follow in a few hours with seidlitz powder.

Bladder, Irritability of.

Tincture of Belladonna, - - - 1 ounce.

Take 10 drops, every 3 or 4 hours.

Boils.

Cover the boil with a piece of lint wet in the following :

    Atrophine, - - - - 4 grains.
    Rose Water, - - - - 1 ounce.

And take internally :

    Aromatic Sulphuric Acid, - - - 6 drachms.
    Water to make - - - - 4 ounces.

A teaspoonful in a wine glassful of water 3 or 4 times daily.

Breath, Fetid or Foul.

    Chlorate of Lime - - - - 3 drachms.
    Distilled Water, - - - - 2 drachms.
    Alcohol, - - - - - 2 drachms.
    Oil of Rose, - - - - 4 drops.

Use a teaspoonful in a glassful of water as a gargle for the mouth.

Dr. Bartholow.

Bronchitis, Acute.

(1) Solution of Acetate of Ammonnia, - - ½ drachm.
    Infusion of Senega - - - 3 drachms.
    Water, - - - - - 6 ounces.

Take 2 tablespoonsful thrice daily.

Dr. Niemeyer.

P

(2) Wine of Ipecac, - - - - 2 drachms.
Solution of Citrate of Potash, - - 4 ounces.
Camphorated Tincture of Opium, - 1 ounce.
Syrup of Acacia. - - - - 1 ounce.

Take a tablespoonful thrice daily.

Dr. Da Costa.

Bronchitis, Chronic.

(1) Extract of Eucalyptus, - - - - 1 ounce.
Muriate of Ammonia, - - - 2 drachms.
Extract of Licorice, - - - 2 ounces.
Syrup of Tolu, - - - - 3 ounces.

Take a tablespoonful 4 or 6 times a day.

Dr. Bartholow.

(2) Compound Spirits of Ether, - - 1 ounce.
Syrup of Ipecac, - - - - 1 ounce.
Camphorated Tincture of Opium, - - 1 ounce.
Water, - - - - - 1 ounce.

Take a teaspoonful as required.

Prof. Janeway.

Bronchitis in Children.

Citrate of Potash, - - - - 1 drachm.
Wine of Ipecac, - - - - 1½ drachms.
Compound Tincture of Camphor, - 1 drachm.
Syrup of Tolu, - - - - ½ ounce.
Water to make - - - - 3 ounces.

Dose, for a child two to four years old, one teaspoonful every hour or two.

Dr. Ellis.

Burns and Scalds.

(1) Linseed Oil, - - - - 7 ounces.
Lime Water, - - - - 8 ounces.

Mix, and apply on lint and covered with oiled silk.

(2) Carbonate of Soda, - - - 6 ounces.
Water, - - - - - 3 pints.

Mix, and apply freely for three days on old cloths or lint.

Cancer.

When the cancer is in the stomach, and there is much pain and vomiting, use

Subnitrate of Bismuth, - - - 2 drachms.
Sulphate of Morphia, - - - 1 grain.

Divide into six powders and take thrice daily in milk.

When on the surface of the body use

Arsenious Acid,	-	-	-	2 drachms.
Mucilage of Gum Arabic,		-	-	1 drachm.

Apply as a paste on a cancerous sore, followed after two or three days by bread poultices until the slough separates.

Catarrh, Acute Nasal.

Tincture of Iodine,		-	-	½ ounce.
Carbolic Acid,	-	-	-	-	1 drachm.

Put a small wide-mouthed bottle containing a moistened sponge, in a vessel of hot water ; drop five or ten drops of the above solution on the sponge, and as the iodine vapor ascends in the vapor of the water, inhale it.

Dr. Bartholow.

Catarrh, Chronic Nasal.

Carbolic Acid,	-	-	-	-	40 drops.
Borate of Soda,	-	-	-	-	2 drachms.
Bicarbonate of Soda,		-	-	-	2 drachms.
Glycerine,	-	-	-	-	7 drachms.
Water to make	-		-	-	8 ounces.

Use some of this daily as a spray for the nostrils.

Dr. Dobell.

Chapped Hands and Lips.

(1)
Resin,	-	-	-	-	1 ounce.
White Wax,	-	-	-	-	2 drachms.
Honey,	-	-	-	-	2 ounces.
Oxide of Zinc,	-	-	-	-	7 drachms.

Mix and apply.

(2)
Spermaciti,	-	-	-	-	1 drachm.
Glycerine,	-	-	-	-	2 drachms.
White Wax,	-	-	-	-	15 grains.
Oil of Bitter Almond,	-	-	-	3 drops.

Chilblains.

Tincture of Benzoin,		-	-	-	2 drachms.
Linseed Oil,	-	-	-	-	4 drachms.
Yellow Wax,	-	-	-	-	2 drachms.

Glycerine enough to make an ointment. Apply to the parts.

Cholera Infantum, or Summer Complaint of Children.

(1)
Carbolic Acid,	-	-	-	-	4 grains.
Subnitrate of Bismuth,		-	-	-	2 drachms.
Mucilage of Gum Arabic,	-		-	-	1 ounce.
Peppermint Water,	-	-	-	-	3 ounces.

Take a teaspoonful every **2, 3,** or **4 hours.**

(2) Sulphate of Copper, - - - 1 grain.
    Deodorated Tincture of Opium, - - 8 drops.
    Distilled Water, - - - 4 ounces.

Take a teaspoonful every 2, 3, or 4 hours for a child of one or two years' old.

Dr. Bartholow.

(3) Carbolic Acid, - " - - 24 drops.
    Brandy, - - - - - 24 drops.
    Peppermint Water, - - - 1½ ounces.
    Mucilage of Gum Arabic, - - - 6 drachms.
    Syrup, - - - - - 6 drachms.
    Deodorated Tincture of Opium, - - 10 drops.

Give a teaspoonful every 2 hours.

Chordee.

Powdered Opium, - - - 12 grains.
Powdered Camphor, - - - 24 grains.
White Sugar, - - - - 1 drachm.

Divide into 12 powders and take 1 at bedtime, to be repeated if necessary in 2 hours.

Dr. Sturgis.

Chorea or St. Vitus' Dance.

(1) Tincture of the Chloride of Iron, - - 1 ounce.
    Solution of Arsenious Acid, - - 2 drachms.
    Syrup of Lemon, - - - - ½ ounce.
    Simple Syrup, - - - - 1 ounce.
    Water to make - - - - 4 ounces.

Take half a teaspoonful three times a day, after meals.

(2) Hydrate of Chloral, - - - 4 drachms.
    Tincture of Hyosciamus, - - - 1 ounce.
    Syrup of Lemon, - - - - 1 ounce.
    Cinnamon Water, - - - 4 ounces.

Take a teaspoonful 3 or 4 times a day, according to age.

Colic.

(1) Spirits of Chloroform, - - 2 ounces.
    Compound Tincture of Cardamoms, - 2 ounces.

Take a teaspoonful in water every half hour.

Dr. Bartholow.

(2) Extract of Gentian, - - - 1 drachm.
    Powdered Rhubarb, - - - 1 drachm.

Divide into 20 pills and take one or two pills three times a day.

(3) Powdered Camphor, - - - 2 grains.
     Powdered Capsicum, - - - 2 grains.
     Powdered Ginger, - - - 2 grains.
Divide into 12 pills and take one as required.

Constipation.

(1) Compound Extract of Colocynth, - 12 grains.
     Extract of Belladonna, - - - 2 grains.
     Extract of Gentian, - - - 6 grains.
     Oil of Caraway, - - - 3 drops.
Divide into six pills, and take one at bedtime.

Dr. Goodell.

(2) Castor oil, - - - - 1 drachm.
     Glycerine, - - - - 1 drachm.
     Tincture of Orange, - - 20 drops.
     Tincture of Senega, - - 5 drops.
     Cinnamon Water to make half ounce.
Take the whole at one dose.

(3) Socotrine Aloes, - - - 12 grains.
     Extract of Belladonna, - - - 3 grains.
Divide into 24 pills, and take one or two as required.

Dr Wallace.

(4) Socotrine Aloes, - - - 7 grains.
     Powdered Rhubarb - - - 24 grains.
     Extract of Belladonna, - - - 1 grain.
Divide into 12 pills, and take one or two as required.

Dr. Da Costa.

Convulsions in Children.

(1) Hydrate of Chloral, - - - 5 grains.
     Milk, - - - - 1 ounce.
Inject into the bowel.

(2) Bromide of Soda, - - - 8 grains.
     Bicarbonate of Soda, - - - 8 grains.
     Hydrate of Chloral, - - - 8 grains.
     Water, - - - - 1 ounce.
A teaspoonful every hour until relieved, to a child 4 months old or under.

Cough.

For Adults.

     Spirits of Nitrous Ether, - - 2 drachms.
     Wine of Ipecac, - - - ½ ounce.
     Deodorated Tincture of Opium, - 1 drachm.
     Syrup of Tolu to make - - 2 ounces.
Take a teaspoonful 2 or 3 times daily.

Dr. Bowditch.

For children.

Compound Licorice Mixture,	- -	2 ounces.
Syrup of Wild Cherry,	- -	1 ounce.
Syrup of Tolu,	- - -	1 ounce.

Half a teaspoonful 3 or 4 times a day for infants ; teaspoonful doses at 1 or 2 years ; larger doses for older children.

Dr. Potter.

Croup.

Powdered Blood root,	- - -	20 grains.
Powdered Ipecac,	- - -	5 grains.
Syrup of Ipecac,	- - -	2 ounces.

Take a teaspoonful every quarter of an hour till vomiting occurs ; then half a teaspoonful every hour.

Dr Potter.

Also apply to the throat the following, with a mop or spray.

Lactic Acid,	- - -	3 ½ drachms.
Distilled Water,	- - -	10 ounces.

London Throat Hospital.

Delirium Tremens.

Tincture of Capsicum,	- -	1 ounce.
Fluid Extract of Hops,	- -	1 ounce.
Mucilage of Gum Arabic,	- -	½ ounce.
Cinnamon Water,	- - -	½ ounces.

Take a dessertspoonful as required for the wakefulness and excitement.

Dentition, or Teething.

Bromide of Potash,	. - - -	1 drachm.
Oil of Anise,	- - - -	2 drops.
Mucilage of Gum Arabic, -	- -	1 ounce.
Peppermint Water,	- - -	1 ounce.

A teaspoonful every half hour until relieved.

Diarrhoea.

In adults—when chronic.

(1) Sulphate of Zinc, -	-	-	-	12 grains.
Powdered Opium, -	-	-	-	12 grains.
Powdered Ipecac, -	-	-	-	12 grains.

Divide into 12 pills, and take one three or four times daily.

In Children.

(2) Camphorated Tincture of Opium, -	-	½ ounce.
Aromatic Syrup of Rhubarb,	-	-	½ ounce.
Lime Water,	- - -	-	1 ounce.

A teaspoonful for children.

For Diarrhœa of Teething.

(3) Bromide of Soda, - - - ½ drachm.
    Mucilage of Gum Arabic, - - - 1 ounce.
    Water to make - - - 2 ounces.

A teaspoonful every three hours.

Dr. A. A. Smith.

Diphtheria.

(1) Tincture of the Chloride of Iron, - - 1 ounce.
    Syrup of Tolu, - - - - 1 ounce.
    Solution of the Citrate of Potash, - - 4 ounces.

Take a teaspoonful to a dessertspoonful, according to age, every three hours.

Dr. Anderson.

(2) Tartaric Acid, - - - - 2½ drachms.
    Glycerine, - - - - 4 drachms.
    Peppermint Water, - - 7 drachms.

Apply locally to the patches in the throat every 3 hours.

(3) Thymol, - - - - - 1 drachm.
    Glycerine, - - - - - 1 ounce.
    Water, - - - - - 3 ounces.

Use as a gargle frequently in teaspoonful doses.

Dr. Da Costa.

(4) Sulphite of Soda, - - - - 3 drachms.
    Glycerine, - - - - 2 drachms.
    Water to make - - - - 4 ounces.

Apply in teaspoonful doses locally.

Dr. Da Costa.

(5) Tincture of the Chloride of Iron, - - 1½ drachms.
    Glycerine, - - - - 1 ounce.
    Water, - - - - - 1 ounce.

Take a teaspoonful of this and the next prescription alternately every half hour.

(6) Chlorate of Potash, - - - ½ drachm.
    Glycerine, - - - - ½ ounce.
    Lime Water, - - - - 2½ ounces.

Take a teaspoonful of this and the preceding prescription alternately every half hour.

(7) Muriate of Pilocarpine, - - - ½ grain.
    Pepsin, - - - - - 15 grains.
    Hydrochloric Acid, - - - 2 drops.
    Distilled Water, - - - 8 ounces.

Take a teaspoonful every hour for a child.

A Specific for Diphtheria. Dr. Heer, Health Commissioner at Ratibor, Germany, read a paper before the recent Medical Congress at Berlin, in which he gave particulars concerning the use of brewers' yeast in diphtheria. He has used the article for 35 years in a large hospital, first in scurvy then in diphtheria. In diphtheria he gives children every hour six to eight grains of fluid yeast and causes their mouths and fauces to be mopped at intervals with the same substance mixed with five times the quantity of water. If this is done promptly and energetically the child is saved. Dr. Heer maintains that equally good results follow the use of yeast against all germ diseases. It is efficacious, he says, in preventing the disease. So simple a remedy as this should have a fair trial on this continent, where diphtheria attains a prevalence and fatality unknown elsewhere.

Dropsy.

(1)	Bitartrate of Potash,	-	-	1 ounce.
	Extract of Taraxacum,	-	-	½ drachm.
	Decoction of Taraxacum,		-	8 ounces.

Take half a wine glassful two or three times daily.

(2)	Powdered Squills,	-	-	½ drachm.
	Powdered Digitalis,	-		½ drachm.
	Nitrate of Potash,	-	-	1 drachm.

Divide into thirty pills and take one thrice daily.

Dysentery.

(1)	Acetate of Lead,	•	•	24 grains.
	Ipecac,	•	•	3 grains.
	Powdered Opium,	-	•	3 grains.

Divide into 12 pills and take one every 2 hours until blood ceases, then one every 3 or 4 hours.

Dr. Da Costa.

(2)	Subnitrate of Bismuth,	-	-	2 drachms.
	Aromatic spirits of Ammonia,	-	-	1 drachm.
	Tincture of Opium,	•	-	16 drops.
	Simple Syrup,	•	-	½ ounce.
	Chalk Mixture,	-	-	1 ¼ ounces.

A dessertspoonful to a child over one year; a tablespoonful or more to an adult, every 2 or 3 hours, after emptying the bowels with castor oil.

Dismenorrhœa or Painful Menstrual Flow.

(1)	Extract of Belladonna,	-	•	4 grains.
	Extract of Stramonium,	-	•	5 grains.
	Extract of Hyosciamus,	-	•	5 grains.
	Sulphate of Quinine,	-	-	40 grains.

Divide into 20 pills and take one thrice daily.

(2) Extract of Opium,   -   -   -   5 grains.
     Extract of Cannabis Indicus, -   -   10 grains.
     Extract of Hyosciamus, -   -   -   10 grains.
     Camphor,   -   -   -   -   25 grains.

Divide into 10 pills and take one two or three times daily.

Dr. McLean.

Dyspepsia.

(1) Dilute Hydrochloric Acid,   -   -   1 drachm.
     Tincture of Capsicum, -   -   -   ½ drachm.
     Tincture of Calumba, -   -   -   1½ ounces.
     Wine of Pepsin to make   -   -   4 ounces.

Take a dessertspoonful after meals.

Dr. Pancoast.

(2) Bicarbonate of Soda, -   -   -   2 drachms.
     Aromatic Spirits of Ammonia, -   -   2 drachms.
     Tincture of Ginger,   -   -   -   1 ounce.
     Compound Infusion of Gentian to make -   8 ounces.

Take one or two teaspoonsful thrice daily.

Eczema or Salt Rheum.

     Ointment of Zinc Oxide,   -   -   ½ ounce.
     Ointment of Lead Subacetate, -   -   ½ ounce.
     Chloral, -   -   -   -   -   15 grains.
     Camphor,   -   -   -   -   15 grains.

Apply to the body as an ointment 2 or 3 times daily after bathing with warm water.

Dr. Gross.

Aud take internally the following :

     Citrate of Iron and Ammonia, -   -   1 drachm.
     Citrate of Potash,   -   -   -   2 drachms.
     Fowler's Solution,   -   -   -   1 to 2 drachms.
     Tincture of Nux Vomica, -   -   -   2 drachms.
     Compound Tincture of Cinchona to make   4 ounces.

Take a teaspoonful in water after meals.

Dr. Bulkley.

Epilepsy.

The famous Brown-Séquard remedy has already been given.
The following is also excellent :

     Bromide of Potash,   -   -   -   1 ounce.
     Bromide of Iron, -   -   -   -   4 grains.
     Water, -   -   -   -   -   2 ounces.
     Simple Syrup.   -   -   -   -   6 ounces.

Take a tablespoonful twice daily.

Erysipelas.

Sulphate of Quinine,	-	-	-	1 drachm.
Dilute Sulphuric Acid,	-	-	-	1½ drachm.
Water,	-	-	-	2 ounces.
Tincture of the Chloride of Iron,	-	-	½ ounce.
Spirits of Chloroform	-	-	-	6 drachms.
Glycerine to make	-	-	-	4 ounces.

Take a teaspoonful in water every 2 hours.

Dr. Loomis.

Also apply locally the following :—

Carbolic Acid,	-	-	-	2 drachms.
Alcohol,	-	-	-	-	2 drachms.
Oil of Turpentine	-	-	-	4 drachms.
Tincture of Iodine	-	-	-	2 drachms.
Glycerine,	-	-	-	10 drachms.

Brush over the surface every 2 hours and cover with a thin layer of wadding.

Or the following :—

Acetate of Lead	-	-	-	1 drachm.
Carbonate of Ammonia,	-	-	-	1 drachm.
Rose Water,	-	-	-	8 ounces.

Apply on lint to allay irritation.

Dr. Peart.

Fever, Simple.

(1) Dilute Hydrochloric Acid,	-	-	½ drachm,
Compound Spirit of Ether,	-	-	1½ drachm.
Syrup of Rose,	-	-	-	½ ounce.
Camphor Water to make,	-	-	4 ounces.

Take a teaspoonful to a tablespoonful, according to age every six hours.

(2) Acetate of Potash,	-	-	-	2 drachms.
Spirits of Nitrous Ether,	-	-	4 drachms.
Simple Syrup,	-	-	-	1 ounce.
Solution of Acetate of Ammonia,	-	2 ounces.
Camphor Water to make	-	-	8 ounces.

Take a teaspoonful to a tablespoonful according to age.

Flatulance or Wind.

Gum Camphor,	-	-	-	1 grain.
Powdered Ginger,	-	-	-	1 grain.
Powdered Capsicum,	-	-	-	1 grain.

Divide into six pills and take one as required.

Freckles.

Corrosive Sublimate,	- - -	6 grains.
Dilute Hydrochloric Acid,	- -	1 drachm.
Alcohol, -	- - - -	2 ounces.
Rose Water,	- - - -	2 ounces.
Glycerine, -	- - - -	1 ounce.
Water to make	- - -	8 ounces.

Apply a little to the face at night and wash it off next morning, taking care not to allow any of the wash to enter the eyes or mouth.

Gastralgia or Neuralgia of the Stomach.

(1)
Sulphate of Morphine,	- - -	1 grain.
Carbolic Acid,	- - - -	½ drachm.
Peppermint Water to make	- -	4 ounces.

Take a teaspoonful thrice daily.

Dr. Da Costa.

(2)
Subnitrate of Bismuth,	- - -	16 grains.
Carbonate of Magnesia,	- - -	16 grains.
Dilute Hydrocyanic Acid,	- -	10 drops.
Water,	- - - - -	4 ounces.

Dessertspoonful for child three years old.

Gastritis, Chronic, or Chronic Inflammation of the Stomach.

(1)
Oxide of Silver,	- - - -	12 grains.
Extract of Belladonna,	- - -	3 grains.
Oil of Cloves,	- - - -	20 drops.

Make into 24 pills, and take one pill twice daily.

Dr. DaCos'i.

(2)
Bicarbonate of Soda,	- - -	1½ drachms.
Tincture of Orange peel, -	- -	½ ounce.
Infusion of Calumba, to make	- -	8 ounces.

Take 2 tablespoonsful before each meal.

(3)
Sulphate of Magnesia,	- - -	1 to 2 drachms.
Tartarate of Soda and Potash,	- -	½ to 2 drachms.
Tartaric Acid,	- - - -	20 grains.

Dissolve the whole in a glass of water and drink, an hour before breakfast.

(4)
Alum,	- - - - -	2 drachms.
Extract of Gentian,	- - -	½ drachm.

Divide into 30 pills, and take two pills twice daily.

Gastric Ulcer, or Ulceration of the Stomach.

(1)
Oxide of Silver,	- - - -	10 grains.
Extract of Hyoscianus,	- - -	10 grains.

Divide into 20 pills, and take one thrice daily before meals.

(2) Subcarbonate of Bismuth,   -   -   3 drachms.
     Sulphate of Morphia,   -   -   -   1 to 2 grains.
     Aromatic powder,   -   -   -   1 drachm.
Divide into 12 powders, and take one powder in milk before meals.

(3) Nitrate of Silver,   -   -   -   -   5 grains.
     Extract of Opium,   -   -   -   3 grains.
Divide into 20 pills and take one thrice daily.

Gleet.

(1) Powdered Spanish Fly,   -   -   -   3 grains.
     Oil of Turpentine,   -   -   -   1 drachm.
Divide into 12 pills and take one thrice daily, in obstinate gleet. If stranguary, stop for a day or two.

As an injection use :—

(2) Tannic Acid,   -   -   -   -   10 grains.
     Subnitrate of Bismuth,   -   -   -   2 drachms.
     Rose Water,   -   -   -   -   6 ounces.
Shake, and use a little as an injection thrice daily.

Dr. Maury.

Gonorrhœa or Clap.

For internal use :—

(1) Citrate of Potash,   -   -   -   ½ to 1 ounce.
     Spirits of Lemon,   -   -   -   ½ drachm.
     Simple Syrup,   -   -   -   -   2 ounces.
     Water,   -   -   -   -   1 ounce.
Take a dessertspoonful, largely diluted, 3 or 4 times daily.

Dr. Otis.

(2) Powdered Cubebs,   -   -   -   3 ounces.
     Copaiba,   -   -   -   -   -   1½ ounces.
     Alum,   -   -   -   -   -   2 drachms.
     White Sugar,   -   -   -   -   1 ounce.
     Magnesia,   -   -   -   -   1½ drachms.
     Oil of Cubebs   -   -   -   -   1 drachm.
     Oil of Wintergreen,   -   -   -   1 drachm.
Make into a paste and take a piece the size of a walnut after each meal.

Dr. Otis.

For injection :—

(3) Sulphate of Zinc,   -   -   8 grains.
     Dilute Solution of the Subacetate of Lead,   4 ounces.

Use a little as injection 2 or 3 times daily.

(4) Chloride of Zinc,   -   -   -   1 grain.
     Rose Water,   -   -   -   -   6 to 8 ounces.

Use a little as injection twice daily.

(5) Sulphate of Zinc.   -   -   -   8 grains.
     Acetate of Lead,   -   -   -   15 grains.
     Tincture of Opium,   -   -   -   2 grains.
     Tincture of Catichu,   -   -   - ' 1 drachm.
     Rose Water to make   -   -   -   6 ounces.

Inject some occasionally with urethral syringe.

(6) Powdered Idoform,   -   -   -   ½ ounce.
     Carbolic Acid,   -   -   -   2 drachms.
     Glycerine,   -   -   -   -   2 ounces.
     Distilled water to make   -   -   -   8 ounces.

Use a teaspoonful as an injection twice daily.

Gout.

(1) Wine of Colchicum Root,   -   -   1 ounce.
     Sulphate of Magnesia,   -   -   -   1 ounce.
     Peppermint Water,   -   -   -   10 ounces.

Take a tablespoonful every hour until it operates.

Dr. Scudamore.

(2) Acetic Extract of Colochicum,   -   -   15 grains.
     Aqueous Extract of Opium,   -   -   15 grains.
     Iodide of Potash,   -   -   -   4 drachms.
     Acetate of Potash,   -   -   -   2 drachms.
     Distilled Water,   -   -   3 ½ ounces.
     White Wine,   -   -   -   -   ¼ ounce.

Take 20 drops thrice daily.

Dr. Lallemand.

Hay Fever.

(1) Iodide of Potash,   -   -   -   1 ounce.
     Fowler's Solution,   -   -   -   1 drachm.
     Cinnamon Water,   -   -   -   4 ounces.

Take a teaspoonful every 4 hours.

(2) Extract of Hyoscianus,   -   -   -   12 grains
     Iodide of Potash,   -   -   -   1 drachms.
     Bicarbonate of Potash,   -   -   -   2 drachms.
     Pure Extract of Licorice, -   -   -   4 drachms.
     Anise water,   -   -   -   -   4½ ounces.

Take a dessertspoonful every 4 hours until relieved.

Dr. Weber.

Headache (Bilious or Sick).

Resin of Podophyllin,	-	-	-	2 grains.
Tincture of Ginger,	-	-	-	2 drachms.
Alcohol to make	-	-	-	1 ounce.

Take a teaspoonful in a wineglass of water every night at bedtime; or every second, third or fourth night as required.

Headache (Congestive or Hot).

Spirits of Ammonia,	-	-	-	1 ounce.
Spirits of Camphor,	-	-	-	½ ounce.

Put this into a quart of water in which a handful of common salt has been dissolved; cork tightly and apply to the head on lint or pieces of cloth.

Headache (Nervous).

(1) Extract of Nux Vomica,		-	-	3 grains.
Reduced Iron,	-	-	-	12 grains.
Sulphate of Quinine,	-	-	-	6 grains.

Divide into 12 pills and take one after each meal.

Dr. Hammond.

(2) Phosphide of Zinc,	-	~	-	3 grains.
Extract of Nux Vomica,	-	-	-	10 grains.
Confection of Roses,	-	-	-	a sufficiency.

Divide into 20 pills, and take one after each meal.

Dr. Fordice Barker.

Heart Palpitation.

Powdered Digitalis,	-	-	-	20 grains.
Powdered Colchicum Seeds,	-	-	-	40 grains.
Bicarbonate of Soda,	-	-	-	60 grains.

Divide into 40 pills, take one at first three or four times daily, gradually reduced to one at bedtime.

Dr. Bowditch.

Hemorrhoids or Piles.

(1) Powdered Galls,	-	-	-	20 grains.
Powdered Opium,	-	-	-	10 grains.
Ointment of the Subacetate of Lead,			-	40 grains.
Simple Ointment,	-	-	-	1 drachm.

Make into an ointment, and apply to the piles as necessary.

Dr. Easterlin.

(2) Powdered Galls, - - - - 20 grains.
Powdered Opium, - - - - 20 grains.
Acetate of Lead, - - - - 20 grains.
Tar ointment, - - - - ½ ounce.
Simple Cerate, - - - - ½ ounce.

Make into an ointment and apply night and morning after bathing the parts in cold water.

(3) Extract of Opium, - - - 10 grains.
Powdered Stramonium, - - - 1 drachm.
Powdered Tobacco, - - - ½ drachm.
Simple ointment, - - - - ½ ounce.

Make into an ointment and apply to the piles.

Dr. Shoemaker.

Hemoptysis, or Bleeding of the Lungs.

(1) Fluid Extract of Ergot, - - - 1½ ounce.
Fluid Extract of Ipecac, - - - 2 drachms.
Deodorated Tincture of Opium - - 2 drachms.

Take a teaspoonful every half hour.

(2) Acetate of Lead, - - - - 20 grains.
Powdered Digitalis, - - - 10 grains.
Powdered Opium, - - - 5 grains.

Make into 10 pills and take one every 4 hours.

Dr. Bartholow.

(3) Alum, - - - - - 1 drachm.
White Sugar, - - - - ½ drachm.
Compound Ipecac Powder, - - 20 grains.

Divide in 6 powders and take one every 2 hours.

Dr. Skoda.

Herpes, Zoster, or Shingles.

(1) Phosphide of Zinc, - - - 10 grains.
Extract of Nux Vomica, - - - 10 grains.

Divide into 30 pills. Take one every 2 to 4 hours.

Dr. Bulkeley.

(2) Solution of Hypochlorite of Soda, - 4 ounces,
Water, - - - - - 2 ounces.

Use as a wash for the ulcerated vesicles.

Dr. Fournier.

Hiccough.

Tincture of Nux Vomica,	1½ drachm.
Dilute Nitric Acid,	4 drachms.
Water to make	4 ounces.

Take a dessert spoonful as required.

Dr. Phillips.

Black Pepper in doses of from 2 to 10 grains will also arrest hiccough.

Dr. Phillips.

Hysteria.

(1)

Extract of Ergot,	1 drachm.
Sulphate of Iron,	½ drachm.
Extract of Nux Vomica,	8 grains.
Corrosive Sublimate,	½ grain.

Divide into 30 pills, and take one thrice daily.

Dr. Bartholow.

(2)

Tincture of Assafœtida,	2 drachms.
Ammoniated Tincture of Valerian,	2 drachms.
Tincture of Castor,	2 drachms.
Camphor Water,	7 ounces.

Take a tablespoonful or two every hour.

Impotence.

(1)

Phosphorus,	½ grain.
Extract of Nux Vomica,	6 grains.
Saccharated Carbonate of Iron	40 grains.
Extract of Gentian	30 grains.

Divide into 25 pills and take one, 2 or 3 times a day.

(2)

Tincture of Sanguinaria,	3 drachms.
Fluid Extract of Stillingia,	5 drachms.

Take 15 to 20 drops in water thrice daily.

Inflammation.

(1)

Tincture of Aconite,	1 drachm.
Tincture of Belladonna,	2 drachms.

Take 3 or 4 drops in water every hour.

Dr. Bartholow.

(2)

Tartaric Emetic,	2½ grains.
Sulphate of Magnesia,	2 ounces.
Sulphate of Morphia,	1⅓ grains.
Aromatic Sulphuric Acid,	½ drachm.
Tincture of Veratrum Vivide,	1½ drachms.
Syrup of Ginger,	2 ounces.
Distilled Water,	10 ounces.

Take a tablespoonful every 2, 4, or 6 hours.

Dr. Gross.

Influenza.

(1) Fluid Extract of Ipecac, - - - 2 drachms.
Deodorated Tincture of Opium, - - 4 drachms.
Tincture of Aconite, - - - 1 drachm.

Take 5 to 10 drops every 2 hours.

(2) Fluid Extract of Black Cohosh, - - ½ ounce.
Deodorated Tincture of Opium, - - 1 drachm.
Syrup of Tolu, - - - - 11 drachms.

Take a teaspoonful every four hours.

Insomnia or Sleeplessness.

(1) Hydrate of Chloral, - - - 2 drachms.
Bromide of Potash, - - - 3 drachms.
Tincture of Opium, - - - 1 drachm.
Syrup of Orange Peel, - - - 3 drachms.
Water to make - - - - 2 ounces.

Take a teaspoonful at bed time, repeated if necessary.

Dr. Kane.

(2) Bromide of Potash, - - - ½ ounce.
Hydrate of Chloral, - - - ½ ounce.
Syrup of Wild Cherry, - - - 1 ounce.
Water to make - - - - 2 ounces.

Take a teaspoonful in a wineglass of water; repeat, if necessary, in one hour.

Intermittent Fever, or Fever and Ague.

(1) Sulphate of Quinine, - - - 80 grains.
Dilute Sulphuric Acid, - - - A sufficiency.
Spirits of Nitrous Ether, - - - 4 drachms.
Syrup of Tolu, - - - - 6 drachms.
Water to make - - - - 2 ounces.

Take a teaspoonful every 2 or 3 hours till relieved.

Dr. Da Costa.

(2) Sulphate of Quinine, - - - 30 grains.
Capsicum, - - - - 15 grains.
Powdered Opium, - - - - 3 grains.

Divide into 30 pills and take from 1 to 5 pills as required.

Dr. Piffard.

Should the ague become chronic, use the following:

(3) Carbonate of Iron Mass (Vallit's Mixture), 1 drachm.
Arsenious Acid, - - - - 1 grain.

Divide into 20 pills. Take one thrice daily.

Dr. Bartholow.

For children use the following :

(4) Sulphate of Quinine, - - - ½ drachm.
Powdered Acacia, - - - ½ drachm.
Syrup of Ginger, - - - - 4 ounces.

Take a teaspoonful every 3 or 4 hours.

Lactation, Fever of (Milk Fever).

Tincture of Aconite, - - - 20 drops.
Tartar Emetic, - - - - 2 grains.
Nitrous Spirits of Ether, - - - 1 ounce.
Syrup (S   .), - - - - 1 ounce.
Water o   .ge Flowers, - - - 2 ounces.

Take a teaspoonful in a wineglass of sweetened water every two hours.
Dr. Fordyce Barker.

Leucorrhoea or Whites.

(1) Alum, - - - - - 1 drachm.
Sulphate of Zinc, - - - - ½ drachm.
Borate of Soda, - - - - 4 grains.
Rose Water, - - - - 8 ounces.

Use as injection.
Dr. Bartholow.

(2) Solution of the Subacetate of Lead, - ½ ounce.
Carbolic Acid, - - - 1½ ounces.

One-fourth of this to be added to a pint of water and used as an injection.

Malaria Chronic.

(1) Arsenious Acid, - - - 1 grain.
Carbonate of Iron Mass, - 1 drachm.

Divide into 20 pills and take one thrice daily.

(2) Sulphate of Quinine, - - - 40 grains.
Dried Sulphate of Iron, - - 20 grains.
Arsenious Acid, - - 1 grain.

Divide into 20 pills and take one thrice daily.

Measles.

(1) Carbolic Acid, - - - 1 drachm.
Acetic Acid. - - - 1 drachm.
Deodorated Tincture of Opium, - - 1 drachm.
Spirits of Chloroform, - - - 1 drachm.
Water to make - - - 8 ounces.

Take a teaspoonful every 4 hours till fever abates.
Dr. Keith.

FAVORITE PRESCRIPTIONS. 259

(2) Tincture of Tolu,   -   -   -   2 drachms.
Syrup of Senega,   -   -   -   ½ ounce.
Acetic Acid,   -   -   -   1½ ounces.
Syrup of Wild Cherry to make   -   -   4 ounces.

Take a teaspoonful as required for cough after convalescence.

Menorrhagia or Excessive Menstrual Discharge.

(1) Fluid Extract of Ipecac,   -   -   2 drachms.
Fluid Extract of Ergot,   -   -   -   4 drachms.
Fluid Extract of Digitalis,   -   -   2 drachms.

Take half a teaspoonful to a teaspoonful as required, till vomiting occurs.

Dr. Bartholow.

(2) Extract of Ergot (Squibb's),   -   -   12 grains.
Extract of Cannabis Indicus,   -   -   5 grains.

Divide into 12 pills, and take one every hour until relieved.

Nervous Exhaustion.

Acetate of Strychnia,   -   -   -   1 grain.
Dilute Acetic Acid,   -   -   -   20 drops.
Alcohol,   -   -   -   2 drachms.
Distilled Water,   -   -   -   6 ounces.

Take ten drops thrice daily.

Dr. Marshall Hall.

Nervousness.

Bromide of Potash,   -   -   1 ounce.
Fluid Extract of Guarana,   -   -   1½ ounces.
Syrup of Tolu,   -   -   -   3 ounces.
Water to make   -   -   -   6 ounces.

Take from a teaspoonful to a dessertspoonful 3 or 4 times daily.

Neuralgia.

(1) Chloroform,   -   -   -   -   ½ ounce.
Tincture of Aconite,   -   -   -   ½ ounce.
Soap Liniment,   -   -   -   -   1 ounce.

Apply on flannel and cover with oiled silk.

Internally use the following.

(2) Phosphorus,   -   -   -   ½ to 1 grain.
Alcohol ough to dissolve it.
Spirits of Peppermint,   -   -   -   1 ounce.
Glycerine to make   -   -   -   4 ounces.

Take a teaspoonful after each meal.

Dr. Hamilton.

Nipples, Sore.

Balsam of Peru, - - -	2 drachms.
Oil of Almond, - - - -	1½ drachms.
Mucilage of Gum Arabic, -	2 drachms.
Rose Water, - - -	A sufficiency.

Apply to the nipples after each nursing.

Dr. Phillips.

Paralysis.

(1)
Sulphate of Strychnia, - - -	2 grains.
Distilled Water, - - -	1 ounce.

Inject five drops of this hypodermically.

(2)
Iodide of Ammonia, - - -	1 drachm.
Carbonate of Ammonia, - -	2 drachms.
Solution of Acetate of Ammonia, -	6 ounces.

Take a tablespoonful thrice daily.

Dr. Bartholow.

Perspiration in Consumption.

(1)
Gallic Acid, - - -	30 grains.
Extract of Belladonna, - -	2 grains.

Divide into 10 pills and take two at bedtime.

Dr. Bartholow.

(2)
Tannic Acid, - - -	30 grains.
Powdered Digitalis, - -	15 grains.
Extract of Cinchona, -	A sufficiency.

Divide into 20 pills and take one at bedtime.

Pertussis or Whooping Cough.

(1)
Chloral, - - - - -	1 drachm.
Bromide of Potash, - -	2 drachms.
Syrup of Wild Cherry, - -	1 ounce.
Water, - - - -	1 ounce.

Take a teaspoonful thrice daily for infants.

Dr. Dessau.

(2)
Extract of Aconite, - -	1 grain.
Syrup of Ipecac, - -	45 drops.
Water of Cherry Laural, - -	1 drachm.
Mucilage of Gum Arabic, - -	6½ ounces.

Take a teaspoonful to a tablespoonful according to age every hour.

Dr. Dervieux.

(3) Carbolic Acid, - - - - ½ drachm.
     Chlorate of Potash, - • - 2 drachms.
     Glycerine, - - • - 4 drachms.
     Water to make - - •- - 6 ounces.
Use with a steam atomizer thrice daily.

Dr. J. Lewis Smith.

Pt hisis or Consumption.

(1) *For the Fever.*
     Sulphate of Quinine, - - - 18 grains.
     Powdered Digitalis, - - - 6 grains.
     Powdered Opium, - - - - 3 grains.
Divide into 12 pills. Take one thrice daily.

Dr. Da Costa.

(2) *For the Dyspeptic Symptoms.*
     Jensen's Pepsin, - - - 40 grains.
     Dilute Hydrochloric Acid, - - 2½ drachms.
     Glycerine, - - - - - 5 drachms.
     Juice of Lemon, - - - - 4 drachms.
     Water of Orange Flowers to make - - 4 ounces.
Take a dessertspoonful with meals.

Dr. Hughes.

3 *For the Cough and Restlessness.*
     Tincture of Belladonna, - - - 2 drachms.
     Syrup of Squills, - - - - 2 ounces.
     Sulphate of Morphia, - - - 1 grain.
     Syrup of Tolu to make - - - 4 ounces.
Take a teaspoonful at bedtime and repeat if necessary.

Dr. Gibbons.

Pleurisy.

(1) Acetate of Potash, - - - 1 drachm.
     Infusion of Digitalis to make - - 4 ounces.
Take a teaspoonful every 3 hours for a child 4 or 5 years old, in the second stage.

(2) Mixture of Iron and Acetate of Ammonia, 6 ounces.
Take a teaspoonful to a tablespoonful according to age, in the second stage.

Pneumonia or Inflammation of the Lungs.

(1) Tincture of Viratrum Viridis, - - 40 drops.
     Spirits of Nitrous Ether, - - - 6 drachms.
     Solution of the Citrate of Potash, - - 4½ drachms.
     Syrup of Ginger to make - - - 6 ounces.
Take a tablespoonful every three hours in the early stage.

Dr. Da Costa.

(2) Iodide of Soda,   -   -   -   -   1 ½ drachms.
     Sulphate of Morphia,   -   -   -   ½ grain.
     Simple Elixir,   -   -   -   -   2 ounces.

Take a teaspoonful thrice daily.

Dr. Da Costa.

(3) Carbonate of Ammonia, -   -   -   40 grains.
     Infusion of Serpentaria,   -   -   -   4 ounces.

Take a teaspoonful every 3 hours as a stimulant about the crisis of the disease.

Dr. Bartholow.

Pruritus or Itching.

(1) Carbolic Acid,   -   -   -   -   2 drachms.
     Glycerine,   -   -   -   -   1 ounce.
     Rose water to make   -   -   -   8 ounces.

Apply to the itching part on cloth.

(2) Dilute Hydrocianic Acid,   -   -   ½ to 1 drachm.
     Infusion of Marsh Mallow,   -   -   5 to 8 ounces.

Apply as a lotion.

Dr. Fox.

(3) Cyanide of Potash,   -   -   -   15 grains.
     Water of Cherry Laural,   -   -   -   8 ounces.

Apply as a lotion.

Dr. Anderson.

(4) Prepared Chalk,   -   -   -   -   1 ounce.
     Coal Tar,   -   -   -   -   1 to 2 drachms.
     Linseed Oil,   -   -   -   -   2 ½ ounces.

Make into an ointment and apply.

Dr. Bulkley.

(5) Dilute Hydrocianic Acid,   •   -   2 drachms.
     Borate of Soda,   -   -   •   -   1 drachm.
     Rose Water,   -   •   •   -   8 ounces.

Apply as a lotion.

Dr. Fox.

Rheumatism, Acute.

(1) Oil of Wintergreen,   -   -   -   1 drachm.
     Salicylic Acid,   -   -   -   -   80 grains.
     Borate of Soda,   -   -   •   -   1 drachm.
     Syrup of Tar,   -   -   -   -   2 ounces.
     Anise Water,   -   -   -   •   2 ounces.

Take a dessertspoonful every 2 hours.

(2) Iodide of Potash, - - - - 2 drachms.
Wine of Colchicum Seeds, - - 4 drachms.
Simple Syrup, - - - - 4 drachms.
Peppermint Water, - - - 5 ounces.

Take a tablespoonful every 4 hours.

New Orleans Hospital.

(3) Salicylic Acid, - - - - 2 drachms.
Bicarbonate of Soda, - - - 1 drachm.
Water, - - - - - 2 ounces.

Take a teaspoonful or two every 2 hours.

Dr. Bartholow.

Rheumatism, Chronic.

(1) Etherial Tincture of Guaiacum, - - 1 ounce.
Etherial Tincture of Colchicum, - - 6 drachms.
Etherial Tincture of Cannabis Indicus, - 2 drachms.

Take 25 to 30 drops on sugar every 4 hours for rheumatic and neuralgic symptoms.

Dr. Atlee.

(2) Aconite Liniment, - - - 2 drachms.
Belladonna Liniment, - - - 2 drachms.
Glycerine to make, - - - 2 ounces.

Apply locally over the seat of pain.

Dr. Fothergill.

(3) Oil of Mustard, - - - - ½ drachm.
Oil of Turpentine, - - - - 3 drachms.
Camphor, - - - - - 4 drachms.
Strong Solution of Ammonia, - - 3 drachms.
Tincture of Capsicum, - - - 4 drachms.
Alcohol to make - - - - 6 ounces.

Apply as a liniment.
This is the celebrated Russian Spirit Liniment.

Scabies or Itch.

(1) Corrosive Sublimate, - - - 4 grains.
Alcohol, - - - - - 6 drachms.
Chloride of Ammonia, - - - ½ drachm.
Rose water to make - - - 6 ounces.

Apply as a lotion.

Dr. Fox.

(2) Sublimated Sulphur, - - - 1 drachm.
Peruvian Balsam, - - - - ½ dracham.
Lard, - - - - - 1 ounce.

An ointment for children.

Dr. Duhring.

(3) Sublimated Sulphur, - - - 2 drachms.
Oil of Cade, - - - 2 drachms.
Prepared Chalk, - - - 2½ drachms.
Green Soap, - - - 1 ounce.
Lard, - - - 1 ounce.

Use as an ointment.

Dr. Bulkley.

Scarlatina.

(1) Boric Acid, - - - ½ drachm.
Chlorate of Potash, - - - 2 drachms.
Tincture of the Chloride of Iron, - 2 drachms.
Glycerine, - - 1 ounce,
Syrup, - - - 1 ounce.
Water, - - - 2 ounces.

Give a teaspoonful every 2 hours to a child of 5 years.

Dr. Smith.

(2) Boric Acid, - - - - 3 drachms.
Glycerine, - - - - 4 ounces.

Add to a pint of water and use as a gargle.

Dr. Da Costa.

(3) Carbolic Acid, - - - - ½ to 1 drachm.
Vaseline, - - - - 4 ounces.

Use as an ointment to the entire surface of the body.

Spermatorrhœa or Seminal Loss.

(1) Bromide of Potash, - - - 1 drachm.
Bicarbonate of Soda, - - - 15 grains.
Infusion of Digitalis, - - - ½ ounce.
Sulphate of Atropia, - - - ¹⁄₆₀ grain.

One dose to be taken at bedtime.

Dr. Gross.

(1) Tincture of Spanish fly, - - - 2 drachms.
Tincture of the Chloride of Iron, - 6 drachms.

Take 20 drops in water thrice daily.

Dr. H. C. Wood.

Syphilis.

(1) Red Iodide of Mercury, - - - 3 grains.
Iodide of Potash, - - - 3 to 6 grains.
Tincture of Orange Peel, - - - 1 ounce.
Syrup of Orange Peel, - - - 1 ounce.
Water, - - - - - 8 ounces.

Take a teaspoonful thrice daily after meals.

Dr. O'is.

(2) Red Iodide of Mercury, - - - 2 grains.
Carbonate of Ammonia, - - - 20 grains.
Iodide of Potash, - - - - 3 drachms.
Compound Tincture of Gentian to make - 4 ounces.

Take a teaspoonful in water after meals.

Dr. Fox.

Tonics.

(1) Sulphate of Quinine, - - - 30 grains.
Dilute Sulphuric Acid—a sufficient quantity.
Water, - - - - - 2 ounces.
Tincture of the Chloride of Iron, - - ½ ounce.
Spirits of Chloroform, - - - 6 drachms.
Glycerine to make - - - 4 ounces.

Take a teaspoonful after meals.

Dr. Loomis.

(2) Citrate of Iron and Quinine, - - 1 drachm.
Carbonate of Ammonia, - - - 30 grains.
Compound Tincture of Gentian, - - 2 ounces.
Tincture of Quassia, - - - 2 ounces.
Syrup, - - - - - 1½ ounces.
Water to make - - - - 8 ounces.

Take a dessertspoonful thrice daily.

Belleview Hospital, N.Y.

(3) Tincture of the Chloride of Iron, - - 10 drops.
Tincture of Nux Vomica, - - 10 drops.
Water to make - - - - 1 drachm.

One dose to be taken thrice daily after meals.

Blackwell's Asylum.

Typhoid Fever.

(1) Tartar Emetic, - - - - 1 to 2 grains.
Sulphate of Morphine, - - - 1½ grains.
Water of Cherry Laural, - - - 1 ounce.

Take a teaspoonful every 2, 3 or 4 hours.

Dr. Bartholow.

(2) Sulphurous Acid, - - - - 2 drachms.
Dilute Sulphuric Acid, - - - 2 drachms.
Deodorated Tincture of Opium, - - 20 drops.
Syrup of Orange, - - - - 3 drachms.
Water to make - - - - 6 ounces.

Take one-sixth of this every 4 hours.

Dr. Wilks.

(3) Sulphate of Quinine,   -   -   -   10 grains.
     Dilute Hydrochloric Acid, -   -   -   ½ ounce.
     Syrup of Orange,   -   -   -   1 ounce.
     Caraway Water to make,   -   -   -   6 ounces.

Take a tablespoonful in an equal amount of water every 3 or 4 hours.

Dr. Murchison.

Ulcers and Sores.

(1) Iodoform, -   -   -   -   -   2 drachms,
     Mucilage of Gum Arabic, -   -   -   15 drops.
     Oil of Peppermint, -   -   -   -   2 drops.
     Glycerine, -   -   -   -   -   20 drops,

Use on unhealthy ulcers.

Dr. Bronson.

(2) Corrosive Sublimate,   -   -   -   15 grains.
     Carbolic Acid,   -   -   -   -   30 drops.
     Water to make   -   -   -   -   4 ounces.

Use on syphilitic ulcers, with cotton, and renew once a day.

Dr. Fox.

Urination, Painful, in Women.

     Fluid Extract of Belladonna,   -   -   ½ to 1 drachm.
     Fluid Extract of Buchu, -   -   -   1 ounce.
     Nitrous Spirits of Ether,   -   -   -   2 ounces.

Take a teaspoonful 3 or 4 times daily.

Vomiting.

(1) Dilute Hydrocianic Acid, -   -   -   1 drachm.
     Water of Cherry Laural,   -   -   -   2 ounces.

Take a teaspoonful every 2 hours till relieved.

(2) Hydrobromic Acid,   -   -   -   2 ounces.

Take half a teaspoonful in a wineglass of water 4 times daily.

Vomiting in Pregnancy.

(1) Oxalate of Cereum,   -   -   -   12 grains.
     Ipecac,   -   -   -   -   -   15 grains.
     Creosote, -   .   -   -   -   24 grains.

Divide into 12 pills, and take one every hour.

Dr. Goodell.

(2) Subnitrate of Bismuth,   -   -   -   160 grains.
     Oxalate of Cerium,   -   -   -   40 grains.
     Sulphate of Morphia,   -   -   -   1½ grains.
     Syrup,   -   -   -   -   -   2 ounces.

Take a teaspoonful every hour till vomiting ceases.

Dr. Van Valzah.

Worms.

(1) Fluid Extract of Pink Root, - - 1 ounce.
     Fluid Extract of Senna, - - - ½ ounce.

Give a teaspoonful to a child between 3 and 5 years.

Dr. Smith.

(2) Pomegranate-root Bark, - - - 2 ounces.

Make a tea of this and take before 11 a.m., and after two hours take the following :

Castor Oil, - - - - 3 ounces.
Turpentine, - - - - 1 drachm.
Ethereal Extract of Male Fern, - - 1 drachm.

Take at a dose. *Dr. Wilde.*
The above is the adult dose.

Yellow Fever.

(1) Carbonate of Potash, - - - 20 grains.
     Camphorated Tincture of Opium, - 1 drachm.
     Water, - - - - - 8 ounces.

Take 2 tablespoonsful every hour or two to produce sweating.

Dr. Dickson.

(2) Chloroform, - - - - ½ ounce.
     Tincture of Camphor, - - - ½ ounce.

Take 2 drops as required for the vomiting.

CHAPTER XIV.

DIETETIC RECEIPTS FOR THE SICK ROOM.

Importance of the subject—Rules for preparing and serving food for the sick—*Nutritious, cooling, and soothing drinks :* ¡Lemonade—Effervescing lemonade—Barley water—Linseed tea—Arrowroot drink—Milk punch—Wine whey—Egg and sherry—Ice—Toast and water—Nutritious ʔcoffee—Milk and Isinglass—A soothing drink—Milk and cinnamon drink—Caudle—Apple water—Chocolate—Chocolate milk—The Invalid's tea—Rose tea —Sage tea—Oatmeal tea. *Gruels :* Water gruel—Milk gruel—Flour gruel—Rice gruel —Barley gruel. *Broths and soups :* Chicken broth—Mutton broth—Whole beef tea— Quickly-made beef tea—Vegetable soup—Bread soup—Spinach soup—Beef and hen broth. *Meats and Vegetables for Invalids:* Table, in order of digestibility, of some articles of animal food—Boiled pigeon or partridge—Bread sauce—Relish for fish—Minced fowl and egg—Fowl and rice—Stewed oysters—The invalid's cutlet—The invalid's mashed potato—Potato surprise. *Jellies for invalids:* Isinglass jelly—Strengthening jelly— Mutton jelly—Bread jelly—Rice blanc-mange—Arrowroot blanc-mange—Sago jelly— Tapioca jelly—Panada—Calves' feet jelly—Currant jelly—Irish moss blanc-mange— Gelatine blanc-mange, *Puddings for Invalids:* Rice pudding—Bread pudding—Batter pudding—Milk for puddings or stewed fruit—Rice and apple—Vermicelli pudding.

In many, if not in all diseases, the choice and the preparation of the articles for the patient's table are of the utmost importance. Food is often the best medicine, and the cook may frequently be of more service than the druggist. But not uncommonly, the dishes served the invalid, like the drugs administered to him, fail of their effect because of their faulty preparation. Those who cook for the sick share the responsibility of treatment with the physician and pharmacist.

The character and amount of the food required by the sick vary, of course, with the nature and stage of the illness. Those sick of a *fever* need to be nourished by frequent supplies of nutritious, easily digested food, in a fluid form, for solid food is, ordinarily, then rejected by the stomach, because of the loathing it excites. The weakest stomach, in such case, will take a small wineglassful of milk or beef-tea, every hour or two. If the milk clot in the stomach, and cheesy lumps be thrown up, this can be readily guarded against by adding a tablespoonful of lime water to each wineglassful of milk. As no regular meals can be taken, they should not be attempted. Small quantities of fluid food at short intervals, will furnish, during the twenty-four hours, a large amount of nourishment, which the irritable stomach, when thus approached, will absorb unconsciously. In inflammatory rheumatism, meat in any form, solid or fluid, is injurious ; the patient must be put upon preparations of **rice, potatoes, bread, arrowroot,** gruel, vegetable or meatless soups, and

268

jellies. In dyspepsia and weak digestion, the invalid requires frequent small meals, at which he should drink very sparingly, and not all at the beginning of the repast. Persons subject to hysterics need a generous meat diet, and must avoid all spirituous or fermented liquors. An excellent drink for them is a mixture of equal parts of soda water and milk. For consumptives, milk and suet are excellent articles of diet, but the best of all is cod-liver oil, which is the most readily digested fat of which we have any knowledge. In disease of the heart, a dry diet is most conducive to the comfort of the invalid, as liquids are absorbed very slowly by the stomach ; the table should be generous, from which stimulants, however, are to be excluded, as they readily excite the heart's action.

We cannot better preface the varied receipts of this chapter, than by quoting the words of one whose eminence in the profession, and whose large and long experience, give them the weight of authority. Professor Gross says : " The diet of the sick room has slain its thousands and tens of thousands. Broths, and slops, and jellies, and custards, and ptisans, are usually as disgusting as they are pernicious. Men worn out by disease and injury must have nutritious and concentrated food. The ordinary preparations for the sick are, in general, not only not nutritious, but insipid and flatulent. Animal soups are among the most efficient supporters of the exhausted system, and every medical man should know how to give directions for their preparation. The life of a man is his food. Solid articles are, of course, withheld in acute diseases, in their "er stages ; but when the patient begins to convalesce, they are frequently borne with impunity, and greatly promote recovery. All animal soups should be made of lean meat ; and their nutritious properties, as well as their flavor, may be much increased by the addition of some vegetable substance, as rice or barley. If the stomach is very weak, they may be diluted, or seasoned with pepper."

The following rules must be observed in preparing, cooking, and serving food for the sick :—

All the utensils employed should be scrupulously clean.

Never make a large quantity of one thing at a time.

Serve everything in as tempting and elegant a form as possible.

Put only a small quantity of an article on a dish at a time.

Keep milk and other delicacies on ice in warm weather.

Never leave food about a sick room.

Never offer beef tea or broth with the smallest particle of fat or grease on it, nor milk that is sour, no meat or soup that is turned, nor an egg that is bad, nor vegetables that are underdone.

NUTRITIOUS, COOLING AND SOOTHING DRINKS.

Lemonade. Take of :—
        Sugar, two or three lumps.
        Lemon, one.

Well rub the sugar on the rind of the lemon, squeeze out the juice, and add to it half a pint or a pint of cold or iced water ; or, better still, one or two bottles of soda water.

Another Lemonade. Pare the ring of three lemons as thin as possible, add a quart of boiling water and a quarter of an ounce of isinglass. Let them stand till next day, covered, then squeeze the juice of eight lemons upon half a pound of lump sugar ; when the sugar is dissolved, pour the lemon and water upon it, mix all well together, strain it, and it is ready for use.

Effervescing Lemonade. Squeeze two large lemons, and add a pint of spring water to the juice, and then four or five lumps of white sugar. When required for use, pour half of it into a tumbler, and add half a teaspoonful of baking soda ; stir and drink while foaming.

Barley Water. Take of :—
> Pearl barley, half a quarter of a pound.

Wash with cold water. Boil for five minutes in some fresh water, and then throw both waters away. Then pour out two quarts of *boiling water*, and boil it down to a quart. Flavor with thinly-cut *lemon rind*, and sugar to taste, but do not strain, unless at the sick person's special request.

This is an excellent receipt for making hard water more digestible.

Linseed Tea. Take of :—
> Whole linseed,
> White sugar, each one ounce.
> Licorice root, half an ounce.
> Lemon juice, four tablespoonsful.

Pour on the materials two pints of *boiling water*, let them stand in a hot place four hours, and then strain off the liquor.

This makes an admirable soothing drink, which acts also upon the kidneys.

Arrowroot Drink. Take of :—
> Arrowroot, two tablespoonsful.
> Cold water, three tablespoonsful.

Mix together and pour in about half a pint of boiling water. When well mixed, add, by degrees, half a pint of cold water, stirring all the time, so as to make it perfectly smooth. It should be about the consistance of cream ; if too thick, a little more water may be added. Then pour in two wineglassfuls of sherry, or one of brandy, add sugar to taste, and give it to the invalid in a tumbler. A lump of ice may be added.

Milk-Punch. Take of :—
> Good brandy, two tablespoonsful.
> Cold, fresh milk, one tumblerful.

Mix with sugar and nutmeg to taste.

This is a useful drink when a stimulant is required in conjunction with a nutriment. It is a medicinal drink, and must not be given indiscriminately.

Wine Whey. Take of :—
> Fresh milk, one pint.

Boil it, and so soon as the boiling point is reached add as much good maderia or sherry as will coagulate it. Strain, and sweeten or flavor for use.

This preparation, when nicely made, renders great service to the sick in proper cases.

Eggs and Sherry. Beat up, with a fork, an egg till it froths, add a lump of sugar and two tablespoonsful of water. Mix well. Then pour in a wineglassful of sherry, and serve before it gets flat. Half the quantity of brandy may be used instead of sherry,

This is a valuable preparation in cases of great prostration, when stimulants and concentrated nutriment are required.

Ice. It has been found by experiments on the gastric juice that low temperature does not exercise any deleterious influence upon it, though it is quite spoiled by heat. The supply of the juices necessary to digestion is arrested by feverishness of the system and in hot weather and in hot rooms. It cannot, therefore, but be beneficial to the stomach to reduce the unusual temperature to which it has been brought by the overheated blood. Hence, ice makes a most valuable addition to the tables of both sick and well. It is very injurious during the exhaustion following violent exercise, or the real cooling attending excessive perspiration. Lake ice is much superior to pond ice or snow.

Toast and Water. Take of :—
> Bread, one slice, from a stale loaf.
> Boiling water, one quart.

Toast the slice of stale bread (a piece of hard crust is better than anything else for the purpose) to a nice brown on each side, but do not allow it to burn or blacken. Put it into a jug, pour the boiling water over it, cover it closely, and let it remain until cold. When strained it will be ready for use.

Toast and water should always be made a short time before it is required, to enable it to get cold ; if drank in a tepid or lukewarm state, it is an exceedingly disagreeable beverage. If, as is sometimes the case, this drink is wanted in a hurry, put the toasted bread into a jug, and only just cover it with the boiling water ; when this is cool, cold water may be added in the proportion required, and the toast and water strained. It will then be ready for use, and is more expeditiously prepared than by the above method.

Nutritious Coffee. Dissolve a little isinglass in water, then put half an ounce of freshly ground coffee into a saucepan, with one pint of new milk, which should be nearly boiling before the coffee is added. Boil both together for three minutes. Clear it by pouring some of it in a cup and dashing it back again. Add the isinglass and leave it to settle before the fire for a few minutes. Beat up an egg in a breakfast cup and pour the coffee into it; or, if preferred, drink without the egg.

Milk and Isinglass. Take of :—

> Isinglass, a pinch or two.
> Milk, a tumblerful.

Mix well and boil. Serve with or without sugar, as preferred.

A Soothing Drink. Take of :—

> Isinglass, a pinch.
> New milk, a tumblerful.
> Bruised sweet almonds, half a dozen.
> Sugar, three lumps.
> Boil together.

Milk and Cinnamon Drink. Boil, in a pint of new milk, sufficient cinnamon to flavor it pleasantly, and sweeten with white sugar.

This may be taken cold, with a teaspoonful of brandy, and is very good in cases of diarrhœa. Children may take it milk-warm, without the brandy.

Caudle. Beat up an egg to a froth, add a wineglassful of sherry, and half a pint of gruel; flavor with lemon-peel and nutmeg, and sweeten to taste.

Apple Water. Slice two or three ripe apples, without paring, into a pitcher, pour out a quart of scalding water, let it stand till cool, and sweeten with sugar.

Chocolate. Put milk and water on to boil. Scrape the chocolate fine, one or two squares to a pint, as will best suit the stomach. When the mixture of milk and water boils, take it off the fire, throw the chocolate into it, mix it well, and serve it up with the froth. The sugar may be mixed with the scraped chocolate, or added afterwards. It should never be made before it is wanted, as heating it again injures the flavor, and causes a separation of the oil.

Chocolate Milk. Dissolve an ounce of scraped chocolate in a pint of boiling new milk.

The Invalid's Tea. Pour into a small china or earthenware teapot a cup of quite boiling water; empty it out, and while the teapot is still hot and steaming, put in the tea. Add enough boiling water to wet the tea thoroughly, and set it close to the fire to steam, for five or six minutes. Then pour in the quantity of boiling water required, from the kettle, and it is ready for use.

Rose Tea. Take of :—

> Red rosebuds (the white heels being taken off), half an ounce.
> White wine vinegar, three tablespoonsful.
> White sugar candy, one ounce.

Put them in two pints of boiling water, and let them stand near a fire for two hours, then strain.

Similar sour drinks may be made of apple jelly, syrup of gooseberries, etc. A variety is always agreeable.

Sage Tea. Take of :—

> Green sage leaves, plucked from the stalks and washed clean, half an ounce.
> Sugar, one ounce.
> Outer rind of lemon-peel, finely pared from the white, quarter of an ounce.

Put them in two pints of boiling water, let them stand near the fire for half an hour, then strain.

When the sage is dried, it must be used in rather less quantity than above mentioned.

In the same manner, teas may be made of rosemary, balm, southern wood, etc., and are convenient to prevent a thirsty invalid taking too much tea and coffee when not good for him.

Oatmeal Tea. Take of :—

> Oatmeal, a handful.
> Barley water, a gallon.

Mix in a deep vessel. Let the oatmeal subside, which it does in half an hour, and pour off the tea. Hard water may be made digestible in this manner.

GRUELS.

Gruels should be thick, but not too thick ; thin, but not too thin. Served in a tumbler they are more appetizing than when served in a basin or cup and saucer.

Water Gruel. Take of :—

> Fine Oatmeal, a dessertspoonful.
> Cold water, a tablespoonful.

Mix. Add a pint of boiling water, and boil it ten minutes, keeping it stirred.

Milk Gruel. Take of :—

> Fine oatmeal, four tablespoonsful.
> Milk, a quart.

R

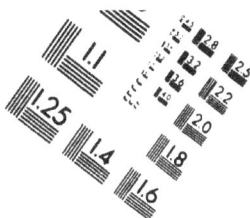

IMAGE EVALUATION
TEST TARGET (MT-3)

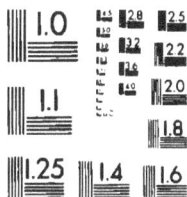

Stir the oatmeal smoothly into the milk. Then stir it quickly into a quart of boiling water, and boil up a few minutes, till it is thickened. Sweeten with sugar.

Flour Gruel. Take of :—

> Flour, a tablespoonful.
> Water, half a tumblerful.

Mix smoothly. Set on the fire, in a saucepan, half a tumblerful of new milk, sweeten it and, when it boils, add the flour and water. Simmer and stir them together for a quarter of an hour.

Rice Gruel. Take of :—

> Fine rice, two tablespoonsful.

Soak for half an hour in cold water. Pour off the water, and to the rice add a pint, or rather more, of new milk. Simmer gently till the rice is tender, then press through a sieve and mix with the milk. Heat over the fire, add a little more milk gradually, pour off to cool, and flavor with salt or sugar.

Barley Gruel. Take of :—

> Pearl barley, two ounces.
> Port wine, a tumblerful.
> Rind of lemon, one.
> Water, one quart and a pint.
> Sugar to taste.

After well washing the barley, boil it in a tumblerful of water for fifteen minutes. Then pour this water away. Put to the barley the quart of fresh boiling water, and let it boil until the liquid is reduced to half; then strain it off. Add the wine, sugar, and lemon peel. Simmer for five minutes and put it away in a clean jug. It can be warmed from time to time, as required.

BROTHS AND SOUPS.

Broths, soups and beef tea should not be kept hot, but heated up as required. Neither should they ever be made in the sick room.

Chicken Broth.

Skin, and chop up small, a small chicken, or half a large fowl, and boil it, bones and all, with a blade of mace or sprig of parsley and a crust of bread, in a quart of water, for an hour, skimming it from time to time. Strain through a coarse colander.

Chicken broth, poured on thin pieces of bread laid on the bottom of the dish, makes a good sauce for boiled chicken or partridge, when the invalid is well enough to be allowed solid food.

Mutton Broth. Take of :—

> Lean loin of mutton, one pound, exclusive of bone.
> Water, three pints.

Boil gently till very tender, throwing in a little salt and onion, according to taste. Pour out the broth into a basin, and when it is cold skim off all the fat. It can be warmed up as wanted.

If barley or rice is added, as is desirable during recovery from sickness, it must be boiled first, separately, till quite soft, and put in when the broth is heated for use.

Whole Beef Tea.
The virtue of beef tea is to contain all the contents and flavors of lean beef in a liquid form. Its *vices* are, to be sticky and strong, and to set in a hard jelly when cold.

Take half a pound of fresh-killed beef for every pint of tea required, and remove all fat, sinews, veins and bones. Cut up into pieces under half an inch square, and soak for twelve hours, in one-third of the water. Take it out and simmer for two hours in the remaining two-thirds of the water, the quantity lost by evaporation being replaced from time to time. Then pour the boiling water on the cold liquor in which the meat was soaked. Dry the solid meat, pound it in a mortar, freed from all stringy parts, and mix with the rest.

When the beef tea is made daily, it is convenient to use one day's boiled meat for the next day's tea, and thus it has time to dry and is easier pounded.

A wholesome flavoring for beef tea is fresh tomato. A piece of green celery stalk, or a small onion and a few cloves, may also be boiled in it. Leeks give it a fusty flavor, and mushroom ketchup, sometimes introduced, is of doubtful composition.

While this is cooking, some more hastily prepared, in the following way, may be used.

Quickly-made Beef Tea.
Take one pound of raw beef, minced, for each pint of water. Stir up cold and let it stand for one hour. Then place the vessel in which they are mixed in a pan of water, and heat for another hour, over a slow fire, being careful not to boil, as then the preparation becomes gluey, and is not equally nutritious or digestible. Run the tea through a coarse strainer, and flavor at discretion.

Vegetable Soup.
Take of butter, half a pound. Put it in a deep stew-pan, place it on a gentle fire till it melts, shake it about, and let it stand till it has done making a noise. Have ready six medium-sized onions, peeled and cut small; throw them in and shake them about. Take a bunch of celery, cut in pieces about an inch long, a large handful of spinach, cut small, and a little bundle of parsley, chopped fine; sprinkle these into the pan, and shake them about for a quarter of an hour; then sprinkle in a little flour and stir it up. Pour into the pan two quarts of boiling water, and add a handful of dry bread crust, broken

in pieces, a teaspoonful of pepper, three blades of mace, beaten fine ; boil gently another half hour. Then beat up the yolks of two eggs, with a teaspoonful of vinegar, and stir them in, and the soup is ready.

The order in which the ingredients are added is very important.

Bread Soup. Take the crust of a stale roll, cut it in pieces, and boil it well in a pint of water, with a piece of butter as big as a walnut, stirring and beating them till the bread is raised. Season with celery and salt.

Spinach Soup. Pick all the stalks from one and a half pounds of fresh spinach ; wash it and clip it ; pour it in a three-quart stew-pan, with a quarter of a pound of butter ; stir it over the fire for five minutes ; add an ounce of flour, and stir again for three or four minutes, then stir in two quarts of chicken broth till it boils. Simmer it on a coal stove for half an hour, and add a small teaspoonful of cream. Serve with it some fried or baked bread.

Endive or lettuce soup may be prepared in the same way.

Beef and Hen Fruit. Take of :—

> Lean beef, one pound.
> Hen, one-half, boned.

Pound together in a mortar ; add salt ; put in a stew pan with two and a half pints of water, and stir over the fire till boiling. Then add carrots, onions, leeks, and celery, cut fine. Boil for half an hour. Strain and serve.

MEATS AND VEGETABLES FOR INVALIDS.

TABLE, IN ORDER OF DIGESTIBILITY, OF SOME ARTICLES OF ANIMAL FOOD.

Sweet bread.
Boiled chicken.
Venison.
Lightly boiled eggs, new toasted cheese.
Roast fowl, turkey, partridge and pheasant.
Lamb, wild duck.
Oysters.
Boiled haddock, trout, perch.
Roast beef.
Boiled beef.
Rump Steak.
Roast veal.
Boiled veal, rabbit.
Salmon, mackerel, herring.
Hard-boiled and fried eggs.

Wood pigeon, hare.
Tame pigeon, tame duck, geese.
Fried fish.
Roast and boiled pork.
Heart, liver, lights and kidneys of ox, swine and sheep.
Lobsters.
Smoked, dried, salt and pickled fish.
Crab.
Ripe old cheese.

Boiled Pigeon or Partridge. Clean and season, inclose it in a puff paste, and boil. Serve in its own gravy, supplimented by the liver rubbed up with some stock, and do not forget the bread sauce.

Bread Sauce. Take of :—

> The crumbs of a French roll.
> Water, a tumblerful.
> Black pepper, six to eight corns.
> Onion, a small piece.
> Salt, to taste.

Boil till smooth ; then add a piece of butter about as big as a walnut, and mix for use. It is good, hot, with hot birds, cold, with cold birds, and is an excellent food for the sick.

Relish for Fish. Fish is made more digestible, and has its flavor brought out, by a few drops of lemon juice squeezed over it.

Minced Fowl and Egg. Take of :—

> Cold roast fowl, one.
> Hard-boiled egg, one.
> New milk or cream, three tablespoonsful.
> Butter, half an ounce.
> Flour, one tablespoonful.
> Salt, pepper, or cayenne, to-taste.
> Lemon juice, one teaspoonful.

Mince the fowl and remove all skin and bones. Put the bones, skin and trimmings into a stewpan, with one small onion, if agreeable to the patient, and nearly half a pint of water. Let this stir for an hour, then strain the liquor, chop the egg small, mince with the fowl, add salt and pepper, put in the other ingredients, let the whole just boil, and serve with thin slices of toasted bread.

Fowl and Rice. Take of :—

> Rice, one quarter of a pound.
> Broth, one pint.
> Butter, one ounce and a half.
> Minced fowl, egg and bread crumbs.

Put the rice into the broth, let it boil very gently for half an hour, then add the butter, and simmer it until quite dry and soft. When cold, make it into balls, hollow out the inside, and fill them with mince made according to the foregoing receipt, but a little stiffer. Cover with rice, dip the balls into egg, sprinkle with bread crumbs, and fry a nice brown. A little cream stirred into the rice before it cools improves it very much.

Stewed Oysters. Take of :—

> Oysters, half a pint.
> Butter, half an ounce.
> Cream, one-third of a pint.
> Flour, cayenne and salt, to taste.

Scald the oysters in their own liquor. Take them out, beard them, and strain the liquor. Put the butter into a stewpan, dredge in sufficient flour to dry it up, add the oyster liquor, and stir it over a sharp fire with a wooden spoon. When it boils, add the cream, oysters and seasoning, and simmer for one or two minutes, but not longer, or the oysters will harden. Serve on a hot dish, with thin slices of toasted bread. A quarter of a pint of oysters, the other ingredients being in proportion, makes a dish large enough for one person.

The Invalid's Cutlet. Take of :—

> Nice cutlet, from loin or neck of mutton, one.
> Water, two teacupsful.
> Celery, one very small stick.
> Pepper and salt, to taste.

Have the cutlet cut from a very nice loin or neck of mutton. Take off all the fat, put it into a stewpan, with the other ingredients ; stew very gently indeed for nearly two hours, and skim off every particle of fat that may rise to the surface from time to time. The celery should be cut into thin slices before it is added to the meat, and care must be taken not to put in too much of this in gredient or the dish will not be good. If the water is allowed to boil fast the cutlet will be hard. Time, two hours, very gentle stewing.

The Invalid's Mashed Potato. Boil one pound of potatoes with their skins on, till they are tender or brittle. Peel them and rub them through a fine sieve. When cool, add a small teacupful of fresh cream and a little salt, beating up lightly until the whole is quite smooth. Warm up gently for use.

Potato Surprise. Scoop out the inside of a sound potato, leaving the skin attached, on one side, to the hole, as a lid. Mince up fine the lean of a juicy mutton chop, with a little salt and pepper ; put it in the potato, pin down the lid, and bake or roast. Before serving (in the skin) add a little hot gravy, if the mince seems too dry.

JELLIES FOR INVALIDS.

Isinglass Jelly. Boil an ounce of isinglass and a dozen cloves (if liked), in a quart of water, down to a pint. Strain, hot, through a flannel bag on two ounces of sugar candy, and flavor.

Strengthening Jelly. Simmer, in two quarts of soft water, one ounce of pearl barley, one ounce of sago, one ounce of rice, till reduced to one quart. Take a teacupful, in milk, morning, noon and night.

Mutton Jelly. Take of:—

> Shanks of mutton, six.
> Lean beef, half a pound.
> Water, three pints.
> Crust of bread, toasted brown.
> Pepper and salt, to taste.

Soak the shanks in water several hours, and scrub them well. Put the shanks, the beef and other ingredients into a saucepan, with the water, and let them simmer, say, gently, for five hours. Strain it, and when cold, take off the fat. Warm up as much as is wanted at a time.

Bread Jelly. Take the crumb of a loaf, break it up, pour boiling water over it, and leave it to soak for three hours. Then strain off the water containing all the noxious matters with which the bread may be adulterated, and add fresh. Place the mixture on the fire and let it boil till it is perfectly smooth. Take it off, and after pouring out the water, flavor with anything agreeable. Put it into a mould, and turn it out when required for use.

Rice Blanc-mange. Take of:—

> Ground rice, one-quarter of a pound.
> Loaf sugar, two ounces.
> Butter, one ounce.
> Milk, one quart.
> Flavoring, of lemon peel.

Mix the rice to a smooth batter, with a little milk, and put the remainder into a saucepan, with the butter, sugar and lemon peel. Bring the milk to boiling point, stir in the rice. Let it boil for ten minutes, or till it comes away from the saucepan. Grease a mould with salad oil, pour in the rice, let it get perfectly cold and turn out.

Arrowroot Blanc-mange. Take of:—

> Arrrowroot, two tablespoonsful.
> Milk, three-quarters of a pint.
> Lemon and sugar, to taste.

Mix the arrowroot, with a little milk, to a smooth batter; put the rest of the milk on the fire and let it boil. Sweeten and flavor it, stirring all the time, till it thickens sufficiently to come from the saucepan. Put it into a mould till quite cold.

Sago Jelly. Take of —

> Sago, two tablespoonsful.
> Water, one pint.

Boil gently, until it thickens, frequently stirring. Wine, sugar and water may be added, according to circumstances.

Tapioca Jelly. Take of :—

> Tapioca, two tablespoonsful.
> Water, one pint.

Boil it gently for an hour, or until it assumes a jelly-like appearance. Add sugar, wine and nutmeg, with lemon juice to suit the taste of the patient and the character of the ailment.

Panada. Take of :—

> Bread crumbs, one ounce.
> Mace, one blade.
> Water, one pint.

Boil, without stirring, till they mix and turn smooth. Then add a grate of nutmeg, a small piece of butter, sugar according to taste.

Calves'-feet Jelly. Take two calves' feet and add to them one gallon of water, and boil down to one quart. Strain, and, when cold, remove all fat. Then add the whites of six or eight eggs, well beaten, (a pint of wine, if desirable), half a pound of loaf sugar, and the juice of four lemons, and mix well. Boil for a few minutes, constantly stirring. Then strain through a flannel bag.

Currant Jelly. Boil together equal weights of white sugar and the juice of white currants, until the mixture solidifies by cooling, as shown by dropping a few drops on a cold plate. Remove the scum, and form the jelly in suitable vessels.

A tablespoonful of this jelly in a tumbler of cold water makes a delightful acid drink, very grateful to many invalids.

Irish Moss Blanc-mange. Take of :—

> Irish moss, half an ounce.
> Fresh milk, one pint and a half.

Boil these down to such a consistency as to retain a form when cold. Remove any sediment by filtering, and then add the requisite quantity of sugar, with lemon juice or peach water, to give an agreeable flavor.

The moss, before being used, must be well washed in cold water, to remove its saltish taste.

Gelatine Blanc-mange. Boil one ounce of shred gelatine in a quart of milk for ten minutes, stirring constantly. Sweeten to the taste, flavor with peach water or essence of vanilla, and strain into a mould.

PUDDINGS FOR INVALIDS.

Rice Pudding. Boil two ounces of rice in a pint of milk, assiduously stirring till it thickens. Take it off and let it cool. Then well mix in two ounces of butter, a quarter of a nutmeg, grated, and sugar in moderation, according to taste. Pour it into a buttered dish and bake.

Bread Pudding. Pour over a French roll half a pint of boiling milk, cover it close, and let it stand till it has soaked up the milk. Tie it up tightly in a cloth, and let it boil for a quarter of an hour. Turn it out on a plate and sprinkle a little white sugar over it. The addition of burnt sugar or tincture of saffron, will give it the established yellow color.

Batter Pudding. Take of :--

Flour, three teaspoonsful.
Milk, one pint.
Salt, a pinch.
Powdered ginger.
Nutmeg.
Tincture of saffron, each a teaspoonful.
Boil.

It will be observed in these three receipts eggs are avoided, as when baked, or even when boiled, so long as it is necessary to boil puddings, they are quite insoluble in a weak stomach.

Milk for Puddings or Stewed Fruit. Boil a strip of lemon and two cloves in a pint of milk. Mix half a teaspoonful of arrowroot in a little cold milk, and add it to the boiling milk Stir it till about the consistency of cream. Have ready the yolks of three eggs, beaten up well in a little milk. Take the hot milk off the fire, and as it cools, add the eggs and a tablespoonful of orange flower water, stirring in constantly till quite cool. Keep it in a very cool place till required for use.

Rice and Apple.—Boil about three tablespoonsful of rice in a pint and a half of new milk, and simmer, stirring it from time to time, till the rice is quite tender. Have ready some apples, peeled, cored, and stewed to a pulp, and sweetened with a very little loaf sugar. Put the rice round a plate, and the apple in the middle, and serve with a little of the above preparation of milk, if liked.

Vermicelli Pudding. Take of :—

Vermicelli, two ounces.
Milk, three quarters of a pint.
Cream, one-quarter of a pint.
Butter, one ounce and a half.
Eggs, two.
Sugar, one ounce and a half.

Boil the vermicelli in the milk till it is tender, then stir in the remaining ingredients (omitting the cream if that be not obtainable).   Butter a small tart dish, line with puff paste, put in the pudding and bake.

We conclude this chapter with the following judicious counsels from the pen of Prof. CHAMBERS, of London : " When a patient cannot be raised in bed without risk of exhaustion, a crockery or glass feeder is a convenience, but the same vessel, or even one of the same appearance, should not be used for food and for medicine.   If the patient's mouth be foul, as in small-pox or putrid fever, it should be cleansed when he is fed.   The administration of nutriment should then be so frequent that it is not allowed to become again foul.   Food should, as a rule, be as near the natural temperature of the body as possible.   But when the febrile heat is very high, or there is much nausea, some of it may be iced, with advantage.   When life seems passing away under their eyes, the friends will often shrink from tormenting (as it seems to them) the sick man with food.   Let them not despair ; many a one has recovered after the doctor has taken his leave with a sad shake of the head, and without making a fresh appointment.   And let them also be stimulated by this fact, namely, that the pains of death are aggravated, if not mainly caused, by the failure of nutrition.   Even when apparently insensible, the dying suffer much increased distress from want of food, though they cannot express their sufferings."

CATHARTICS OR PURGATIVES.

Cathartics or purgatives are medicines which loosen the bowels. Those which act violently are called drastics, those which act mildly, aperients or laxative.

Remarks in regard to their use. The habitual employment of purgatives is a practice productive of great injury, causing dyspepsia and many other troubles of the stomach and bowels. Purgatives should not be given so that their operation will interfere with the regular hours of rest. They should not be taken immediately after a full meal. The action of every purgative is followed by a greater or less amount of costiveness. This is especially true of rhubarb, least so of castor oil. In cases of great debility cathartics should be avoided.

Effervescing Cream of Tartar. Take of :—

> Cream of tartar.
> Carbonate of soda, each three drachms.
> Water, a tumblerful.

Put the whole into a stone jug or bottle, and attach the cork firmly. To be taken in the morning, before eating.

Magnesia and Rhubarb. Take of :—

> Magnesia, one ounce.
> Rhubarb, two drachms and a half.
> Powdered ginger, two scruples.

Mix and divide into eight powders. Take one or two in the evening, at bed-time, to obtain a laxative effect in the morning.

Rhubarb and Epsom Salts. Take of :—

> Powdered rhubarb, one drachm.
> Epsom salts, one ounce.
> Spirits of peppermint, two drops.
> Water, a tumblerful.

One or two tablespoonsful will produce a laxative effect.

283

May Apple, or Mandrake. Take of :—

Powder : resin of May apple (Podophyllin), one grain.
Powdered hyoscyamus leaves, eight grains.
Powdered ginger, twelve grains.

Mix and divide into four powders. One or two at bedtime in torpor of the liver and bilious disorders. A much better and safer pill than blue mass or other mercurials, so frequently employed indiscriminately in these cases.

Jalap and Cream of Tartar. Take of :—

Powdered jalap, one drachm.
Cream of tartar, six drachms.

Mix and divide into six powders. Dose, one, in molasses.

Calcined Magnesia. Take of :—

Husbands' or Ellis' magnesia, thirty grains,

And dissolve it in a little milk or water for one dose. This is an excellent cooling laxative. Its operation is promoted by the drinking of lemonade.

Seidlitz Powders. These are to be obtained of any druggist. Two

powders are given together, a white and a blue one ; each of which is to be dissolved separately, in a tumbler one-third full of water, and the two solutions then mixed and drank while foaming, in the morning, before breakfast. This is a very popular, gentle laxative, and well borne by the stomach, when other medicines of the kind disagree.

Purgative Mineral Water. Take from a bottle of the solution

of the citrate of magnesia, to be had of any druggist, a teacupful every two hours, until it operates. For a child five years old, a wineglassful is the proper dose. This preparation, which tastes like lemonade, is one of the most agreeable of laxatives.

CLYSTERS OR INJECTIONS.

Clysters or injections are solutions thrown into the lower bowel, in order to act as purgatives, as astringents to check diarrhœa, or stop bleeding, or as nutrients to nourish the patient in those exhausted conditions when food cannot be given by the mouth.

A Purgative Injection. Take of :—

Epsom salts, one ounce.
Sweet oil, two tablespoonsful.
Starch water, one pint.

To be given when a purgative is required.

Common salt and molasses also make an excellent purgative injection ; a tablespoonful of each in a pint of water, with or without the addition of a little soap.

An Astringent Injection. Take of :—

> Subnitrate of bismuth, twenty grains.
> Tincture of catechu, a teaspoonful.
> Milk, a wineglassful.

For one injection, to be repeated in twelve hours. Useful in checking the purging of consumption, fevers, etc.

Nutritive Injections. Life can be prolonged, and even, in many cases, preserved, by the persistent use of nutritive injections when, in ailments like ulceration of the stomach, it is ' possible to give food by the mouth, as it is at once rejected by the stomach. Nutritive injections are made of strong beef tea, milk, raw eggs, cod liver oil, and, even in extreme cases, of diluted brandy.

COLD, MODE OF APPLYING.

Cold has been employed in the treatment of disease from the earliest times. It is applied in various ways, by cold baths, by streams of cold water, by cold moist sponges and cloths, by bladders filled with ice, and by the evaporation of ether.

Cold Application. Take of :—

> Nitre, half an ounce.
> Sal ammoniac, two drachms.
> Vinegar, three tablespoonsful.
> Water, a pint.

Mix. This solution, applied, by means of sponges or cloths, to the head, and elsewhere, where intense cold is desired, produces a more powerful effect than cold water or pounded ice.

Cold Without Moisture. When it is desired to apply a freezing mixture to a small portion of the skin, it may be readily done by putting a mixture of ice and salt in a bladder, or a tumbler, or a lamp glass covered with a piece of bladder.

Hydropathic Belts. A hydropathic belt consists of a belt, five or six inches wide and long enough to pass two or three times round the body. It is dipped into cold water, carefully wrung out, wound round the trunk, and covered by a wider and longer dry band. About every hour, or as often as it becomes dry, it is to be changed. A bandage may be applied in the same manner upon various parts of the body, and particularly over the joints attacked by rheumatism. An eruption of the skin is usually produced by this application, which is frequently of service.

COUNTER-IRRITANTS.

Counter-irritants are applications intended to irritate the parts to which they are applied, and by exciting artificial congestion or inflammation, to modify disease existing in a distant part.   How this curative power is exerted, it is difficult to say.   Professor STILLE remarks: " It is a familiar fact that the body is an assemblage of organs, which are constantly exerting a reciprocal influence upon one another, so that all are more or less involved in the derangements of each.   Examples of sympathy between remote parts, and exerted through the organ of the mind, are innumerable.   Emotions of pleasure or shame suffuse the face with blushes ; while fear and the depressing passions blanch the cheeks, chill the extremities, and bedew the skin with a cold sweat.   So, emotions of pity or tenderness make the tears flow ; the odor or sight of agreeable food, or even the thought of it, makes the mouth water ; while disgusting objects turn the stomach, and alarming ones suspend digestion or destroy the appetite.   Obstinate constipation has been overcome by causing the patient to stand upon a wet marble pavement ; consumption has often supervened upon the suppression of an issue or other habitual discharge ; and still more frequently, apoplexy and other internal congestions have followed the same causes.   The translation of gout and rheumatism to the brain, heart, stomach, etc., when suppressed in the extremities, is familiar to every   practitioner.   The coryzas, and sore throats, and pulmonary c⸱ ⸱rrhs diarrhœas, and other affections which arise from merely getting ⸱⸱⸱ ⸱ ⸱t wet, are matters of daily experience, whose reality cannot be deni   nor explained away.   They are neither more nor less intelligible the   he effects of counter-irritation, and both must be accepted because th   ⸱ are facts."

Shakspeare, whose wonderful acquaintance with the actions of the body has furnished us with more than one illustration, speaks, in " Romeo and Juliet," of the principal of counter-irritants, as follows :—

> " Tut, man, one fire burns out another's burning,
> One pain is lessened by another's anguish. . . .
> Take thou some new infection to thy eye,
> And the rank poison of the old will die."

Counter-irritation is effected by the application of various substances which redden or blister the skin.

Croton-Oil Liniment.   Take of :—

> Croton oil, thirty drops.
> Sweet oil, two tablespoonful.

Mix.   Produces, when rubbed on, redness and eruption of the skin. A useful application to the chest in ' eginning consumption.

Iodine Paint. Take of :—

> Tincture of Iodine,
> Alcohol, equal parts.

To be applied with a camel's-hair brush.   Useful in many cases of persistent pains in the joints and limbs (See, also, Blisters and Cups).

CUPPING.

Physicians employ two sorts of cups, known as wet cups and dry cups. The former are for the purpose of extracting blood, the latter for counter-irritation by reddening the skin.   A dry cup can be readily applied by any one.   All that is necessary is a tumbler or wine-glass, and a little piece of cotton or paper, which is to be wet with spirits of wine, set on fire, thrown into the glass, which should then at once be firmly pressed down over the skin, when the fire will be quickly extinguished (without causing any pain), and the skin drawn up forcibly into the glass.   The same object may be accomplished by holding the tumbler over a light until the air within is well heated, and then applying it quickly and closely to the skin.

Cupping is of benefit in rheumatic ailments, and in many affections of the chest and of the large joints.

DISINFECTANTS.

Disinfectants are substances which possess the power of destroying poisons capable of producing disease, and of removing disagreeable odors and gases by decomposing both them and the bodies from which they proceed.

The principal disinfectants are carbolic acid, coal tar, creosote, charcoal, chlorine, permanganate of potash, quicklime, sulphate of iron (copperas or green vitriol), sulphur and fresh earth.

Carbolic Acid as a Disinfectant. Take of :—

> Impure carbolic acid, one ounce.
> Water, one gallon.

Mix.   Sprinkle over the floors of privies, about sinks, etc.

Charcoal as a Disinfectant. Powder some wood charcoal and expose it, in open pans, in the place to be disinfected.   It has the advantage over lime preparations, being without odor.

Chlorine as a Disinfectant. Chlorine water, to be obtained from any druggist, is a useful agent for correcting stenches, and, diluted with water, for washing foul sores.

Permanganate of Potash. Take of :—

> Permanganate of potash, a teaspoonful.
> Water, a quart.

Expose, in saucers, in the sick room.   Useful for musty closets and foul cellers.   It has no odor itself.

Green Vitriol.   Sulphate of iron, commonly called green vitriol, or copperas, in powder, alone, or mixed with lime, is an excellent disinfectant for privy-wells, slaughter-houses, ditches, etc.

Fresh Earth.   Fine dry earth, sprinkled over offensive matters, is an admirable disinfectant.   A knowledge of this fact has led to the construction of earth closets.   A box of dust from the road, and a tin cup, kept at the side of the closet or chamber vessel, so that the earth may be thrown immediately upon the dejection, will serve as a complete deodorizer, and answer the purpose almost as well as the elaborately-constructed patent earth closets now in the market.

Sulphur.   Take of :—

Milk of sulphur, a teaspoonful.
Water, one pint.

Mix.   Sprinkle over clothes to be disinfected, and iron with a hot flatiron.

The fumes of burning sulphur may be employed for disinfecting outhouses, closets, carriages, etc.

To Quickly Remove a Bad Smell.   An unpleasant odor may be quickly removed from the sick room by burning in it dried lavender or cascarilla bark, with the window open.

EMETICS.

Emetics are medicines which cause vomiting, They are used to remove from the stomach poisons or crude indigestible matters, to dislodge things lodged in the throat or air-passages, and to excite the action of the skin and of the liver. They are not, except, of course, in cases of poisoning, to be given when there is disease of the heart, great irritability of the stomach, or much general debility. They are well borne by children, and of much service in many of the ailments of infancy and childhood. Their action is promoted by drinking plentifully of warm water, and by tickling the throat with a feather. When the vomiting produced is too violent or too long continued, it may be checked by a few drops of laudanum, or by applying a mustard plaster over the pit of the stomach.

Mustard Emetic.   Take a teaspoonful of mustard in a teacupful of warm water, every ten minutes, until vomiting is produced.

This is an efficient, quick, and safe emetic.

Alum Emetic.   Take a teaspoonful of powdered alum in a little honey, syrup, or molasses, every fifteen minutes, until vomiting is produced.

Common Salt Emetic. Add one or two teaspoonsful of salt to a teacupful of warm water. Take every ten or fifteen minutes, till vomiting is produced.

Salt and Mustard Emetic. Mix a teaspoonful each of salt and mustard in a teacupful of warm water. Repeat every ten minutes, until free vomiting is brought on.

EYE-WASHES OR WATERS.

Eye-waters, or collyria, as they are called by physicians, are solutions applied directly to the eye or eyelids.

Alum Eye-wash. Take of :—

> Alum, one grain.
> Pure water, two tablespoonsful.

Mix. A useful wash, night and morning, for inflamed eyes.

Brandy Eye-wash. A teaspoonful of brandy to two tablespoonsful of water makes a serviceable eye-water when a stimulant is wanted.

Arnica Eye-wash. Take of :—

> Tincture of arnica, five drops.
> Pure water, two tablespoonsful.

Mix. Often of benefit in weak or sore eyes.

Tea Eye-wash. Ordinary tea, when cold, makes a valuable eye-water in many cases.

FOMENTATIONS, OR STUPES AND STEAMINGS.

Fomentation is the application of warmth and moisture to the surface of the body by means of a flannel or soft cloth. Steaming consists in exposing a part to the vapors arising from a piece of flannel wrung out in boiling water; it is often employed in affections of the eyes.

An Ordinary Fomentation. Immerse a piece of flannel in boiling water, remove it and put in a wringer made by attaching stout toweling to two rods. The wringer is twisted around the flannel very strongly, till as much as possible of the water is pressed away. The wringer is useful, as the flannel is too hot, when first removed from the boiling water, to be grasped by the hand. When wrung as dry as possible, fomentations prepared in this way may be applied very hot, without fear of scalding or blistering the skin. The flannel, when applied to the part, should be covered with a piece of oiled silk or rubber cloth, and changed before it becomes cold. On the removal of the fomentation the skin should be at once gently dried and covered with a piece of dry flannel.

S

If the precaution of covering the fomentation with oiled silk, muslin, or paper, or a rubber cloth, be neglected, the warm, comforting flannels will be converted, in a few minutes, into cold, clammy, wet ones, disagreeable and hurtful to the patient.

Turpentine Fomentations. Steep a piece of lint or linen in oil of turpentine, place it over the part, and immediately apply over it flannel, heated as hot as it can be borne.

This is, frequently, more effectual than a mustard plaster.

Another Turpentine Fomentation. Sprinkle the flannel, wrung out of hot water in the manner described, with a tablespoonful of turpentine.

This will act as a counter irritant, rapidly reddening the skin and relieving pain in many cases.

Opium Fomentation. Instead of turpentine, employ laudanum, as directed in the preceding receipt. Used to relieve pain.

Mustard Fomentation. Add a quarter of a pound of mustard to a pint of boiling water. Wring the flannel cloth out in this solution, in the manner above directed.

This fomentation quickly reddens the skin, and is frequently useful in allaying pain.

GARGLES.

Gargles, to be of benefit, must be frequently repeated, and their use persevered in.

Gargle of Brandy. A mixture of equal parts of brandy and water makes a useful gargle in some cases of sore throat.

Gargle of Alum. Take of :—

> Alum, two teaspoonsful.
> Water, a tumblerful.

Mix. Used to remove offensive breath depending upon inflamed throat.

Gargle of Lime-Water. Pour upon a quarter of a pound of fresh unslackened lime two quarts of hot water. After standing several hours carefully decant the clear liquid, without shaking up the lime. This is a valuable gargle in diphtheria and croup.

Gargle of Chlorate of Potash. Take of :—

> Chlorate of Potash, a teaspoonful.
> Water, a tumblerful.

Mix. An excellent gargle for ordinary sore throat.

Gargle of Sage and Linseed. Take of :—

> Sage, two ounces.
> Linseed, one ounce.
> Boiling water, one pint.

Mix. To be used cold in the early stages of inflamed throat.

HEAT, MODE OF APPLYING.

Moist heat is applied by means of fomentations and poultices, which see. Dry heat may be applied in various ways. Flannel, highly heated in an oven, or before the fire, may be employed, but it cools quickly. Hot sand, though heavy, and, therefore, for many purposes improper, retains its heat for a long time. It should be heated over the fire in an iron pan, and put in a warm linen bag of the proper shape for the object in view. Chamomile flowers are lighter than sand, but more quickly lose their warmth. They are to be heated, and placed in a linen bag, in the same manner as the sand. Hot salt, in a bag, is a ready method of applying heat in many cases, as, for instance, to the back of the neck, at night, to relieve headache. A thin piece of flat-tile, when it can be procured, can often be used with advantage. It is lighter than sand, and, when heated in an oven, and wrapped in a flannel, retains its warmth for a considerable time. A heated dinner-plate, or a hot brick, wrapped in flannel, may sometimes be employed, as may also bottles of hot water, well corked.

LINIMENTS.

Liniments are used for the double purpose of causing the removal of swellings and for reddenning the skin, and so act as counter-irritants. They are applied by rubbing, either with the bare hand or with the hand covered by a piece of flannel, oiled silk or muslin, or a piece of bladder.

Camphor Liniment. Take of :—

> Camphor, one ounce.
> Olive oil, four ounces.
> Rub up the camphor in the oil.

Hartshorn and Oil. Take one part of hartshorn to two parts of oil ; mix.

Useful for stiff neck and lumbago.

Opodeldoc. Take of :—

> Hard white soap, three ounces.
> Camphor, one ounce.

Put them in a bottle, and add a tumblerful of spirits of wine, or brandy or any other spirit, and as much water. Shake the bottle from day to

day, till the soap and camphor are dissolved, when the liniment is ready
for use.

A mixture of two tablespoonsful of this liniment with a teaspoonful of
laudanum, is very valuable to lull violent rheumatic pain.

Mustard Liniment, is, for stimulating the surface, one of the
best, as it is very manageable, and may be made to act either very slightly, or so severely as to take the skin off, according to quantity used and
the time the rubbing is kept up.   The best guide as to the quantity required is the feelings of the person rubbed.   At first, there is a pleasant
sensation of heat, then a little pricking, and next a positive smarting ;
when this is produced, leave off, for if the rubbing be continued, it will
soon flay the skin, and, as a consequence, prevent its being rubbed again
for three or four days.   An ounce of fresh flour of mustard put into a
bottle with a pint of spirits of turpentine, and shaken daily for two or
three days, make this liniment.   The mustard will settle to the bottom,
and the clear fluid should be then poured off.   Do not leave the mustard
in and shake the bottle up before using ; if so, it will give the skin a
coating of mustard, and render the application unnecessarily severe.   It
is excellent for lumbago and chilblains.

Sometimes it is necessary to keep up irritation on the skin for length
of time without disturbing the constitution, which some irritants will do.
The best application for this purpose is—

Croton Oil. Of which ten or a dozen drops are to be rubbed in
lightly with the fingers, guarded with a piece of oiled silk, for two or
three nights.   Generally on the second day the surface is red and puffy,
and on the third day a large crop of little blisters, about the size of hemp
seed, cover the skin.   When these appear, the rubbing must be stopped.
In the course of a few hours, the fluid in the blisters changes to matter,
and these pustules begin to tingle and itch furiously.   As soon as this
happens, prick each with the point of a needle, and press out the matter
with a handkerchief.   In the course of a week the skin has been completely reproduced, and then the croton oil may be used again ; but it
does not blister quite so quickly as when first applied.   The croton may
be used for months, and is a most excellent mild irritant.

Lime Liniment. Take of :—

> Lime water,
> Flaxseed oil, each a tumblerful.
> Mix them.

Chloroform Liniment. Take of :—

> Pure chloroform, a wineglassful.
> Olive oil, two wineglassesful.
> Mix them.

A useful liniment in many painful rheumatic and neuralgic affections.

Lead Ointment.   Take of :—
> Solution of sugar of lead, two ounces.
> Olive oil, two wineglassesful.
> Mix them.

Turpentine Liniment.   Take of :
> Rosin cerate (to be had of the druggist), three-quarters of
> a pound.
> Oil of turpentine, a tumblerful.

Add the oil to the cerate, previously melted, and mix them.

LOTIONS OR WASHES.

Washes are employed either for soothing and cooling inflamed parts, for stimulating sluggish sores to heal, or for drying and absorbing discharges.

In applying a cooling lotion or wash, a single piece of linen should be wet with it and laid on the part, which should not be wrapped up nor covered with the bed-clothes. So soon as the cloth dries it should be again dipped in the lotion, or wet by squeezing a spongeful of the wash over it.

A stimulating lotion is applied by dipping lint or rag into it, putting it on the sore and confining it by a bandage.

Lint, for use for this and so many other purposes, in dressing wounds and sores, is made by unraveling old linen, soft from use and washing. It may be prepared by scraping tightly-stretched linen with a sharp knife. A "patent lint" is sold at all the drug stores, in rolls or sheets, which is more compact than loose lint, one side being fleecy and the other smooth. Charpie is an excellent sort of lint, much employed. It is made as follows : Cut a piece of lint into small pieces, a few inches square, and completely unravel it thread by thread. The coarser kind may be made of old tablecloths. Old linen is much better than new for making charpie.

Drying lotions are applied, by means of lint or rag, to cracked skin, and to scalds, burns, and sores which weep or discharge very freely.

Cold Water Wash is as good a wash as any, to produce evaporation, if care be taken to have the wet linen well exposed to the air ; and it has the further advantage of being almost always at hand.

Spirit Wash. Half a quarter of a pint of spirits of wine, or a quarter of a pint of brandy, or any other good spirit, added to a pint of water, make this wash.

Vinegar Wash is made by mixing one-fourth of vinegar to three-fourths of water.

To a pint of either of the former washes half a tablespoonful of laudanum may be added, if the pain suffered be very severe.

Lead Wash, or Goulard-Water or White Wash, as it is often called, for common purposes, may be made by dissolving one drachm of sugar of lead in a pint of soft water. Some persons are very fond of using this wash, with the addition of spirits of wine as an evaporant.

Lime Water Wash. A very simple application ; is one of the best, and very easily made. Take half a pound of unslaked lime, and three-quarters of a pint of water. Put the lime into an earthen pot, and pour a little of the water upon it, and as the lime slakes pour the water on by little and little, and stir up with a stick. The water must be added very slowly, otherwise the water will fly about in all directions, and the great heat suddenly produced will crack or break the vessel which contains it. After three or four hours, when the slaked lime has sunk to the bottom, the clear fluid may be poured off, and put in a stoppled bottle, away from the light.

Oxide of Zinc Wash is made by putting four drachms of oxide of zinc into a pint of lime water, which does not, however, dissolve, but merely suspends it. It is, therefore, always necessary to shake the bottle well up, so that the linen may entangle the proper quantity of oxide.

Sal Ammoniac Wash. Take of :—

> Sal ammoniac, one drachm.
> Water, a tumblerful.

Mix, with or without the addition of a teaspoonful of laudanum. Useful for painful, sluggish sores.

Chlorate of Potash Wash. Take of :—

> Chlorate of potash, a teaspoonful.
> Water, a tumblerful.

A useful application to bad sores, and for chapped and cracked hands.

Borax Wash. Take of :—

> Borax, a teaspoonful.
> Glycerine, a tablespoonful.
> Water, a tumblerful.
> Mix. An agreeable soothing lotion.

An Absorbent Wash. Take of :—

> Oxide of zinc, two drachms.
> Water, a tumblerful.
> Mix.

Arnica Lotion. Take of :—

> Tincture of arnica, a tablespoonful.
> Water, a tumblerful.
> Mix. A useful lotion for sprains, bruises and burns.

OINTMENTS OR SALVES.

Ointments or salves are usually prepared by rubbing of a medicine with lard. They should not be kept on hand too long, as they become rancid and unfit for use.

Sulphur Ointment.   Take of :—

> Flowers of sulphur, half a pound.
> Lard, one pound.
> Oil of bergamot, two teaspoonsful.
> Mix up together.   A curative ointment for itch.

Common Ointment.   Take of :—

> Yellow wax, one ounce.
> Olive oil, four ounces.
> Lard, one ounce.
> Mix. A soothing salve.

Tannin Ointment.   Take of :—

> Tannin, one drachm.
> Lard, one ounce.
> Mix. An excellent astringent salve.

Camphor Ointment.   Take of :—

> Camphor, ten grains.
> Spirits of wine, a few drops.
> Lard, one ounce.
> Mix. Useful in some skin diseases.

Oxide of Zinc Ointment.   Take of :—

> Oxide of zinc, one drachm.
> Lard, two ounces.
> Mix. Useful for chapped skin and sores.

PAIN REMOVERS, OR ANODYNES.

The boldest doubter, he who, in rude health, is a most defiant and sarcastic skeptic of the power of medicine, is forced, by the terror and victory of pain, to confess the beneficence of the substances given us by a merciful Creator for the subduing and the destroying of that pain.

> " Pain is perfect misery, the worst
> Of evils, and, excessive, overturns
> All patience."

The most powerful means for relieving pain, such as opium by the mouth, morphia injected under the skin, and chloroform by inhalation, are such potent agents for evil, as well at good, that their use can only be trusted to those trained to employ them. But there are other means which can be handled, with greater or less success, by non-medical persons. Many of them we have mentioned in speaking of painful ailments. but we group, in this place, a number of receipts for the relief of pain, which are of no little value, and which can be employed, with the exercise of ordinary care and judgment, without the risk of doing harm.

Quinine Powders. Take of quinine, twenty grains, and divide it into eight powders. One of them twice a day is an excellent remedy in many forms of neuralgia (particularly in that occurring in persons living in districts where chills and fever prevail), and sick headache.

Sal Ammoniac. Take of sal ammoniac, twenty grains, for one dose ; in a half tumbler of water. Repeat the dose at the end of every hour, until four doses are taken, when, if no relief is had, the medicine is not appropriate to the case, and it is needless to continue. If it afford relief, as it does frequently, in an almost magical manner, in neuralgia of the face, and other parts, it should be taken three times a day, for a week or two, after the attack.

Camphor. Half a teaspoonful of the spirits of camphor in water, is an excellent remedy for the pains of colic and diarrhœa.

Painful joints are also relieved by the rubbing in over them of spirits of camphor, by itself or mixed with a little laudanum.

Chloral. Take of :—

> Chloral, two drachms.
> Sweetened water, half a tumblerful.

Take a tablespoonful for a dose. Useful to procure ease and sleep in many ailments, particularly rheumatic pains and pains arising from burns.

Coffee. Squeeze the juice of a lemon in a small cup of strong black coffee. This will often afford immediate relief in neuralgic headache.

Tea ordinarily increases neuralgic pain, and ought not to be used by persons affected with it.

Iodide of Potassium. Take of :—

> Iodide of potassium, one drachm.
> Sweetened water, half a tumblerful.
> Mix. Take a tablespoonful three times a day, in pains of the joints and bones that are worse at night.

External Applications. Warm and hot baths are admirable remedies for pain. So, also, are poultices and hot fomentations, for which receipts are given under their heads in this chapter.

Hygienic Means. Whatever improves the general health, saves and relieves pain. Pure air is, therefore, an anodyne.

> " Ye, who amid this feverish world would wear
> A body free from pain, of cares a mind,
> Fly the rank city, shun its turbid air."

Light rooms ; warm clothing ; regulated gymnastic exercise, out-door recreation ; proper amount of sleep ; the avoidance of fatigue of body and mind ; and of intemperance in alcohol and tobacco, and in everything which impairs the nervous force ; together with change of air and climate as restoratives, are all means by which pain may be escaped or its edge blunted.

Many forms of neuralgia are relieved by change of air, as well as various constitutional disorders causing pain. Sea air is particularly useful in numerous instances.

POULTICES.

" Poultices are blessings or curses, as they are well or ill made," was a saying of the celebrated Dr. Abernethy. They should be spread thickly, as a general rule, otherwise they dry quickly, and irritate the part they are intended to soothe.

All poultices should be covered over by a piece of oiled silk, muslin, or paper, to retain the heat and moisture.

Bread and Water Poultice. Scald out a basin, for, in order to make a good poultice perfectly, boiling water is necessary ; then, having put in some hot water, throw in coarsely crumbled bread, and cover it with a plate. When the bread has soaked up as much water as it will imbibe, drain off the remaining water, and there will be left a light pulp. Spread it, a third of an inch thick, on folded linen, and apply it when of the temperature of a warm bath.

Or, carefully pare away the hard, brown crust from a slice round the loaf of stale bread, dip it into hot water, lift it out at once, and apply immediately, if not too hot.

Flaxseed Meal Poultice. The celebrated Dr. Abernethy gave the following directions for making this poultice :—

" Get some linseed powder, not the common stuff full of grit and sand. Scald out a basin ; pour in some perfectly boiling water ; throw in the powder, stir it round with a stick, till well incorporated ; add a little more water and a little more meal ; stir again, and when it is about two-thirds the consistence you wish it to be, beat it up with the blade of a knife till all the lumps are removed. If properly made, it is so well worked together, that you might throw it up to the ceiling, and it would come down again without falling to pieces ; it is, in fact, like a pan-cake. Then take it out, lay it on a piece of soft linen, spread it the fourth of

an inch thick, and as wide as will cover the whole inflamed part ; put a
bit of hog's lard in the centre of it, and when it begins to melt, draw
the edge of the knife lightly over and grease the surface of the poultice.
When made in this way, oh ! it is beautifully smooth ; it is delightfully
soft ; it is warm and comfortable to the feelings of the patient."

The Bran Poultice is a sort of "entire," or half-and-half, partly
poultice, partly fomentation, and is a very good application for setting up
and keeping up perspiration on a part ; but it requires to be often
changed, for it very quickly becomes sour and then has not the most
agreeable smell.   It merely consists of bran moistened, but not made
wet, with hot water ; and enough of it should be put into a flannel bag,
sufficiently large to cover the part, to fill it about one-third ; if more
bran be put in, the bag becomes unpleasantly heavy.   It must then be
held before the fire, and the bran turned about again and again, till it
is thoroughly heated.   Thus warmed, it must be quickly applied, and
the bran should be gently spread, so as to cover the whole extent of the
bag.

The mixture is to be heated in a pot, carefully stirred, to prevent
burning, and, when sufficiently warm, must be spread on linen, like any
other poultice.

Starch Poultice.   Add a little cold water to the starch, and blend
the two into a pap ; then add sufficient boiling water to make a poultice
of the required consistence, which must be spread on linen.

Useful in skin eruptions attended with much heat and pain, and, in
general, when a soothing application is required.

Charcoal Poultice.   The charcoal may either be mixed with the
ingredients of the poultice or sprinkled over the part and covered with
a simple poultice, or the following receipt may be used :   Take of :—

> Wood charcoal, in powder, a tablespoonful.
> Bread, three or four slices.
> Flaxseed-meal, three tablespoonsful.
> Boiling water, one tumblerful.

Mix.   A useful application to offensive wounds and sores.

Carrot Poultice.   Boil the carrots till they become quite soft,
mash them with a fork, and spread the pulp on linen, in the ordinary
way.

A turnip poultice may be made in the same manner.

Alum Poultice.   Composed of the white of two eggs and a teaspoonful of powdered alum.   An excellent astringent.

Mush Poultice.   Stir Indian-meal, in small quantities, into water
kept boiling in a pan, until the whole has acquired the proper degree of
thickness.

Slippery Elm Poultice. Made by moistening, with hot water, the inner bark of slippery elm, ground into a fine powder.

Arrowroot Poultice. Add enough boiling water to arrowroot, previously mixed with cold water into a smooth paste, as will make it of the required thickness for spreading. A pleasant soothing poultice.

Onion Poultice. Mash some partially roasted onions, and spread them upon folds of muslin. Applied to the chest, useful in the croup and catarrh of children ; applied to the arms and legs, useful to prevent the fits of children.

Mustard Poultice. For a mustard poultice, a sufficient quantity of powdered mustard should be taken to make a thin paste the required size. This should be mixed with boiling water, with a small quantity of vinegar added, if a very strong poultice is required, and spread on brown paper, with a piece of thin muslin over it.

A mustard poultice should generally be kept on from ten to twenty minutes, but some skins will bear it much longer than others. If the skin is very irritable afterwards, a little flour should be sprinkled over it. This will remove the burning sensation.

Bread and Milk Poultice. Upon the crumbs of stale wheat bread, in a basin, pour boiling milk, stirring with the back of a spoon until the mixture has the thickness of mush. Spread and apply.

TONICS.

Tonic medicines are those which give strength to the system. They act slowly, and must be persevered in to obtain their full effects. Excellent results are frequently obtained by changing from one to another, when the first tried fails, or has ceased to do good.

Quinine in Powder. Take of :—

Quinine, two scruples.

Divide into twenty powders. Take one three times a day, well covered in a little scraped apple. The apple disguises, completely, the bitter taste of the medicine.

This is an admirable tonic in nervous and other forms of debility, and in loss of appetite.

Iron Powder. Take of :—

Reduced iron, two scruples.
White sugar, in powder, two teaspoonsful.

Mix, and divide into twenty powders. Dose, one powder, in molasses, syrup or preserves, three times a day.

Reduced iron, which is a tasteless powder of an iron gray color, may be obtained from any druggist, and is one of the very best preparations of iron which can be taken by pale, thin-blooded people.

In using this, as well as the other preparations of iron, it is necessary to persevere for several months, to reap the fullest results.

Potassio-tartrate of Iron. Take of :—

Potassio-tartrate of iron, one drachm.

Divide into twelve powders. Take one, in syrup or preserves, three times a day. An admirable tonic in dyspepsia, and loss of appetite, and debility. It has the advantage over many other preparations of iron, of not constipating the bowels, and of being rapidly digested.

Half an ounce of the potassio-tartrate of iron, added to a pint of cherry after solution, may be used instead of the powders, in tablespoonful doses.

The best time for taking this, and other preparations of iron, is with the meals, and not on an empty stomach. The juice in the stomach during digestion readily dissolves the iron, which, if taken while fasting, may cause pain and uneasiness.

Columba and Ginger. Take of :—

Bruised columba, one ounce.
Bruised ginger, a quarter of an ounce.
Boiling water, one pint.
Mix and strain.

A wineglassful four or five times a day is a useful tonic in persistent diarrhœa.

RECEIPTS FOR THE HYGIENE OF THE PERSON.

Under this head we group a variety of useful receipts for the care of the hair, the skin and the teeth. They are all harmless, as well as efficient, which, too frequently, is not the case with the many perilous compounds widely advertised, and sold as tooth powders, hair tonics, mouth and skin washes, and lip salves, under high-sounding names. Any druggist can put them up, and at a less price than is asked for the dangerous secret preparations they are designed to replace.

TOOTH POWDER.

Take of :—

Powdered camphor,
Powdered orris root, each two drachms.
Precipitated chalk, half an ounce.

Mix thoroughly. The chalk employed should be the "precipitated," and not the "prepared" chalk of the druggist.

Another excellent tooth powder is the following :—
Take of :—

> Freshly-prepared willow charcoal, or
> Freshly-prepared areca-nut charcoal, one ounce.
> Keep it in a tightly-corked-bottle.

Another powder especially valuable when the teeth have been stained by taking iron :—
Take of :—

> Tannic acid, quarter of an ounce.
> Sugar of milk, two ounces.
> Red lake, half a drachm.
> Oil of teaberry, or cloves, a few drops.
> Mix with care.

The following tooth powder is of benefit when the gums are sore and spongy :—
Take of :—

> Powdered myrrh, quarter of an ounce.
> Powdered borax, half an ounce.
> Precipitated chalk, one ounce.
> Powdered orris root, quarter of an ounce.
> Mix.

The following tooth powder is excellent when the saliva is acrid and the breath sour.
Take of :—

> Bicarbonate of soda,
> Powdered talc, each half an ounce.
> Oil of anise, a few drops.

Improper tooth powders are powdered pumice stone, which rapidly wears off the enamel, the protecting cover of the teeth (pumice stone is a frequent ingredient in secret tooth powders), cigar ashes, cream of tartar, or any other acid.

MOUTH WASHES.

These are useful when the gums are tender, or the breath offensive, or the teeth rapidly decaying.

Camphor Mouth Wash. Take of :—

> Spirits of camphor, half a teaspoonful.
> Milk-warm water, a wineglassful.
> To be used several times a day, and at bedtime.

Honey Mouth Wash. Take of :—

> Honey of rose, half a teaspoonful.
> Milk-warm water, a wineglassful.
> Mix, and use as above directed.

Myrrh and Cinchona Mouth Wash. Take of :—

> Tincture of myrrh,
> Compound tincture of cinchona, each half a teaspoonful.
> Milk-warm water, a wineglassful.
> Mix, and use as above directed.

Brandy Mouth Wash. Take of :—

> Pure French brandy, a teaspoonful.
> Milk-warm water, a wineglassful.
> Mix, and use as above directed.

Permanganate of Potash Mouth Wash. Take of :—

> Permanganate of potash, four grains.
> Rose water, four fluid ounces.
> Oil of peppermint, a few drops.

This is an excellent mouth wash for foul breath, caused by bad teeth or disordered secretions of the mouth It slightly stains the teeth, but does them no injury; on the contrary, being an excellent preservative, and a valuable remedy for preventing and curing toothache. The discoloration may be easily taken off by a tooth brush or sponge.

Chlorate of Potash Mouth Wash. Take of :

> Chlorate of potash, two or three teaspoonsful.
> Water, a tumblerful.
> Oil of teaberry, a few drops.
> To be used several times a day.

When offensive breath comes from a foul stomach, twenty grains of bisulphite of soda, in half a tumbler of water, with a little essence of peppermint, twice a day, is an excellent remedy. Or three grains of chlorinated lime, known also as chloride of lime, in a wineglassful of water, several times a day, may be taken. Charcoal internally, is also of use in such cases.

For masking the scent of onions, and other disagreeable acquired odors, freshly roasted coffee grains are useful, or a small portion of Canada snakeroot, chewed.

HAIR TONICS.

Bark Hair Tonic. Take of:—

Red cinchona tea, a tumblerful.
Brandy, a wineglassful.
Pure glycerine, a tablespoonful.

Mix. Apply, night and morning, for scurf and falling of the hair.

Ammonia Hair Tonic. Take of :—

Stronger water of ammonia,
Castor oil, each one ounce.
Old brandy, two ounces.
Rose water, six ounces.

Mix. This mixture must not be employed oftener than every other day.

Ointment for Dandruff. Take of :—

Powdered borax, twenty grains.
Lead water (diluted solution of subacetate of lead), two
  drachms.
Fresh lard, one ounce.
Attar of roses, a few drops.

Mix. to be rubbed on the scaly patches on the scalp every morning, after the skin has been cleaned by soap and water.
Or, the same ingredients may be used in a wash as follows :

Wash for Dandruff. Take of :—

Powdered borax, twenty grains.
Lead water, two drachms.
Rain water, half a tumblerful.
Pure glycerine, a tablespoonful.
Mix. Use once or twice a day.

LIP SALVES, LOTIONS, ETC.

Lip Salve. Take of :—

Oxide of zinc, thirty grains.
Spermaceti ointment, half an ounce.
Attar of roses, one drop.

Mix. To be applied, night and morning, for irritated and cracked lips

Lotion for the Hands. Add a tablespoonful of pure glycerine to a pint of water. This well rubbed in, but not wiped off, will soften and whiten the hands, and protect them from the air.

Ointment for Fetid Feet. Take of :—

> Crystalized carbolic acid, five grains.
> Ointment of oxide of zinc, one ounce.

Mix. Apply, morning and evening, after washing the feet in cold water, with which a few teaspoonsful of alum have been mixed. In such cases, the wearing of a thin sole of felt inside the shoe, which should be removed several times a week, wet in a solution of permanganate of potash (twenty grains to the ounce of water), dried, and reinserted, is of benefit. The stockings should be of wool, and changed once a day, and the same pair of shoes should not be worn on two consecutive days.

WASHES FOR THE FACE.

Take of :—

> Powdered borax, half an ounce.
> Pure glycerine, one ounce.
> Camphor water, one quart.

Mix. Wet the face with this, morning and evening, allow it to remain for several minutes, and wash in rain water. An excellent lotion to prevent chapped skin, to remove sunburn, and cleanse the pores of the skin.

Take of :—

> Fresh lemon juice, a wineglassful.
> Rainwater, one pint.
> Attar of roses, a few drops.

Mix, and put in a well-corked bottle. Wash the face and hands with this several times a day (letting it stay on for several minutes before drying with the towel). This preparation is highly recommended by the celebrated Dr. Wilson, of London, for clearing the complexion of "muddiness."

Take of :—

> Juice of the cucumber, pressed from the fruit, a sufficient quantity.

Boil it over a quick fire, cool rapidly, and bottle. Apply a tablespoonful, diluted with two tablespoonsful of water, night and morning. This preparation is much employed in France, for clearing the complexion.

In some parts of England the following wash is much employed for removing sunburn and whitening the skin.
Take of :—

> Fresh horseradish root, one ounce.
> Cold buttermilk, one pint.
> Put aside for four hours.

Or, take of :—

> The juice of horseradish, one part.
> Cidar vinegar, two parts.

Mix. Has the same uses as the above receipt, and is also recommended for removing freckles.
Take of :—

> Benzoin, two ounces.
> Pure alcohol, one pint.

Mix. A tablespoonful of this, added to a tumbler of water, turns it white, and makes a most agreeable wash for clearing the complexion.

Ointment for Sunburn. Take of :—

> Spermaceti,
> Oil of almond, each two ounces.
> Honey, one teaspoonful.
> Attar of roses (or any scent), a few drops.

Melt the spermaceti in a pipkin, then add the oil of almonds, and, when they are thoroughly mixed, stir in the honey. Take the pipkin off the fire, and stir, constantly, until it is cool, adding the scent.

Apply at night, after washing the skin, and allow it to remain until morning. It relieves the irritated burning skin, and lessens the redness. Cold, fresh cream smeared over the affected parts sometimes does good.

As lunar caustics and tincture of iodine are medicines in common use, and stain the skin and body linen, the following directions for removing these stains will be found useful :—

To Remove Stains of Lunar Caustic (Nitrate of Silver) from the Skin. Wash the parts discolored by handling a stick of solution of lunar caustic in a solution of iodide of potassium in water. The water will turn the discolorations from brown to dead white. Then wash in a solution of spirits of hartshorn.

To Remove Stains of Lunar Caustic from Linen. The best way to remove stains of nitrate of silver from linen is to moisten the stain with a few drops of a solution of one drachm of cyanide of potassium in a wineglassful of water, to which a few drops of tincture of iodine has been added immediately before using the solution. The linen should

T

then be well rinsed in clear water. This plan removes even the oldest stains of nitrate of silver, provided the operation be carried on in a moderately lighted room. The cyanide of potassium may be obtained of the druggist, but is a violent poison, which should be kept and handled with great care.

Another Way of Removing Nitrate of Silver Stains from the Fingers. Moisten the part with tincture of iodine; immediately apply spirits of hartshorn, and wash off.

To Remove Iodine Stains from Linen. Dissolve two drachms of the hyposulphite of soda in half a tumbler of water, and soak the stain in the solution, then wash in water.

CHAPTER XVI.

THE TOILET:

NEW, SIMPLE, AND EFFICIENT PREPARATIONS.

The Hair. A thick, handsome head of hair is generally acknowledged to be a good thing to have. Many do possess it, and many others might, but from sheer neglect in the care of this "divine ornament." The scalp should be kept clean and free from scurf dandruff, and the dirty accumulations caused by the use of oily hair preparations, &c. Every body knows this, and yet how generally neglected is the care of the hair and scalp.

Hair Wash. An unequalled hair wash for cleansing the scalp and hair from all impurities is a tablespoonful of hartshorn in a pint of water. Rub thoroughly into the hair and over the scalp, and then wash the head with clear water.

The use of this wash once or twice a week, renders the hair and beard soft and glossy, and greatly promotes its growth.

Dandruff. *A Cure.*—After the scalp and hair have been purified of dirt and dandruff by use of the above wash, future annoyance from dandruff can be prevented by dampening the scalp three or four times per week (oftener if necessary) with sulphur-water, made by putting one half ounce of flour of sulphur into a pint of water, shaking occasionally for two days, and then pouring off into a clean bottle. This "cure for dandruff" is new and of great value. The neglect of keeping the scalp free from those scaly particles known as dandruff shows a lack of personal cleanliness not particularly commendable. A lady with her hair powdered with scurf is not particularly angelic ; and a gentleman with his coat-collar whitened with dandruff is surely not a fascinating object.

Hair Preparations. As the hair of but few persons is sufficiently oily of itself to be of desirable appearance, and to keep in a proper position, the use of various hair preparations is almost universal. A fine article is made as follows :—

Strong Alcohol, - - -	1 pint.
Castor Oil, - - -	3 ounces.
Tincture of Spanish Flies, - -	2 drachms.
Oils Bergamot and Lavender, of each -	10 drops.

Nothing superior to the above was ever made. It is not too greasy to be objectionable. It renders the hair glossy and silken, and is sufficiently stimulating to prevent the hair from falling out, and often induces an unusually fine growth, and not the least, it is cheap, and quickly made.

Glycerine Washes. Glycerine washes, for the hair, are objectionable. They render the hair disagreeably sticky, causing it to "catch dirt" to such an extent that the scalp must be washed every few days, in order to keep it in a wholesome condition.

Pomatums Pomatums are liable to nearly the same objections. Besides, they are almost without exception made of lard, which is liable to become rancid and acrid, thus irritating the scalp, and not unfrequently causing sores.

Grey Hair. *Hair-dyes.*—From various causes, generally unknown, the hair turns prematurely grey. Nothing will restore it to its original natural color, but it can be artificially colored so as to be quite of satisfactory appearance.

The quantity of hair-dye used in this country almost exceeds belief. The very best, because producing an almost perfectly natural color, and at one time the most popular, was the well-known sulphur and sugar of lead dye. In 1866 it was sold under more than four hundred and sixty names, by actual computation. As one dollar was charged per bottle, and as the entire cost was about five and one half-cents per bottle, including the bottle and wrapper, almost every druggist and barber in the country put up the stuff under some attractive tit e, and proclaimed its virtues in the most positive manner. It was always "not a dye," and it would "restore the hair to its original color." Rich firms advertised it enormously, and the result was that for a number of years it seemed as if every third person was using the abominable dye. The odor it gave to the person using it was truly disgusting. The sulphur and lead were quickly absorbed through the pores of the scalp and the head, and the entire body became stenchful indeed from the fœtid, sulphur-laden perspiration.

Presently physicians in every part of the country found that cases of paralysis (formerly an uncommon disease) were becoming alarmingly frequent, and on due investigation it was found that the cause was from the use of the lead and sulphur hair-dye.

Dr. G. H. Taylor, who treats a large number of paralytics every year, at his well-known "Movement-cure Institute" in New York found this class of patients increasing to an unusual extent. Investigation showed that in many recent cases the cause was from the use of the hair-dye in question. An eyelid or one side of the face or neck is usually first paralyzed, then an arm, finally an entire side,—generally the left. In 1869 a lady was taken to Dr. Taylor for consultation who had used the hair-dye so long and so freely, that her grey hairs had become jet black. The odor from her person was sickening. She was entirely paralyzed—could only move her head, and speak. She died after lingering a few weeks in this condition, at her home in Brooklyn, N.Y.

A year or two since an editor in a western city died from the effects of this poisonous dye. The newspapers immediately made a hue and cry; chemists analyzed and warned; doctors cautioned their patients; the result has been that the use of the preparation has decreased largely. Still it is extensively used, and it is to save many from possible disease and death that we have given so much space to the notice of this most dangerous article.

"But I am not old and my hair is turning grey,—what shall I do?" Well if you *will* color your hair, use only such dyes as are known to be harmless—which are the "silver dyes," almost the only kind in the market.

The following simple "hair stain" has no superior. Put into a perfectly clean bottle :—

Nitrate of Silver,	-	-	-	-	75 grains.
Ammonia,	-	-	-	-	7 ounces,
Alcohol,	-	-	-	-	3 ounces.
Water,	-	-	-	-	9 ounces.

Keep the bottle well corked and in a dark place. The hair should be well cleaned with soap and warm water, and then wiped dry, before each application. The stain is used by combing the hair thoroughly with a fine comb dipped in the preparation; no washing is required after the operation. The first application gives a reddish-brown color. The second a brown, and the third a black color. After each application expose the head to the sunlight for fifteen or twenty minutes, if convenient, then oil and keep well oiled.

Gentlemen will find this a most satisfactory article for obliterating white hairs among their whiskers, and for "touching up" the beard or moustache to any shade desired.

To give "hair dye receipts" here, we think, would be superfluous, as numbers of them can be found in any of the one or more "receipt books" that almost every family possesses.

Curling the Hair. Preparations are being continua'ly advertised that are claimed will cause the straightest, stiffest hair to curl in wavy, massive ringlets. A genuine swindle. Nothing but mechanical means —the curling iron, &c., will cause the hair to curl, or crimp, in the slightest degree.

The insane belief of a certain class of young persons, that the hair can be curled by the use of "curling fluids," &c., was recently rather expensively illustrated. A western sharper, fully alive to the weakness of young folks on the subject of curly hair, advertised in several papers of large circulation, a "magnetic curling-comb," warranted to cause the stiffest hair to curl beautifully. In short time this genius received over thirteen thousand orders at one dollar each—the price of the comb. A common ten cent horn comb, with a bit of copper wire twisted about it, was sent at first, but presently nothing whatever.

Hair artificially curled is not becoming, while the twisting process around the wood or iron breaks the fibres of the hair and seriously injures it.

Bandolines. Bandolines for the hair, are preparations entirely free from grease for keeping the hair in place, giving it a glossy appearance, &c. One of the best, and most simple, is made by pouring a teacup of boiling water, on ten or fifteen quince-seeds; strain, pour into a bottle, and add five drops of the oil of cloves or cinnamon.

Two Hair-brushes. Every person should be provided with two hair-brushes—an ordinary brush of good stiff hair, the other a friction or shampoo-brush of stiff, black uneven hair. Before dressing the hair in the morning, the scalp should be thoroughly " polished' with the shampoo-brush. Nothing gives such tone and vigor to the scalp, and prevents a tendency to fall out. It is an excellent sanitary operation in other respects. A person who of a morning gets up drowsy and unrefreshed, after giving his scalp a brisk rubbing with the stiff brush, will experience a surprising change in his feelings for the better. A brisk friction of the scalp in a like manner will often cure headache.

Baldness. Baldness, unless caused by fever or a similar cause, is generally incurable. One of the most successful remedies ever used, and we believe now published for the first time, is prepared as follows:—

Glycerine,	-	-	-	-	4 ounces.
Tannin,	-	-	-	-	2 ounces.
Tincture of cantharides,	-	-	-	2 drachms.
Oil of capsicum,	-	-	-	-	10 drops.

Apply to the bald spots morning and evening.

Character as Indicated by the Hair. Stiff, straight and abundant hair and beard are combined with a character which is straightforward, unyielding, strong, and rather bluff.

Fine hair and dark skin, show purity, goodness, and strong mind.

Black hair, a dark skin, and bilious temperament are usually found together. There is strength of character and sensuality.

Fine brown hair indicates exquisite sensibility, with strong will for what is good and right, if not perverted.

If the hair is coarse, black, and sticks up, there is not much talking, and the person is apt to be stubborn, sour, and harsh.

White hair, as a general rule, indicates a good, easy, lazy fellow. There is animosity in coarse red hair, with unusual firmness of purpose and strength of character. Hasty, impetuous, rash people, have curly and crispy hair. Red hair indicates a fiery temperament, passion and devotion. Auburn hair, which is hair of a golden-hue, having a yellowish tinge, with a florid face, gives purity, and great capacity for enjoyment and suffering. Wavy hair is pliable, yielding, accommodating. The dark-haired races—the Spaniard, the Malay, the Mexican, the Indian, and the Negro, have physical strength, endurance, robustness in body. The light-haired races are the thinkers of the world, the poets, and the artists. Dark-brown hair combines the two, and is the most desirable.

THE SKIN, COMPLEXION, &c.

Pimples, Flesh-worms, or " Blackheads." The skin is a marvellous piece of lace-work, through the interstices or pores of which there is constantly escaping insensible perspiration and an oily secretion.

In a torpid condition of the skin there appears on the face, nose, and lower part of the forehead of many persons, what are called "black heads," "flesh worms,' &c.,—often in great numbers, and causing an exceedingly disagreeable appearance. When the skin is pressed with the finger-nails they come out in vermicular form, having a black point or head, which gives rise to the name grubs, or flesh-worms, which they are, being real, living worms, which has been ascertained by microscopical examination.

They can generally be permanently removed by pressing them out with the finger-nails, and then bathing the parts with mild salt water. The skin should receive gentle friction daily with a coarse towel.

Pimples, Red and Matterated. If these are the result of an inactive liver, disordered digestion, or constipation, proper means must be employed to remove these ailments. Nothing will accomplish this more effectually and harmlessly than the use of the following pills :—

Strychnia, -	-	-	-	1 grain.
Quinine, -	-	-	-	1 scruple.
Leptandrin,	-	-	-	1 drachm.
Hydrastrin,	-	-		15 grains.

Make into thirty pills, one to be taken immediately after breakfast, daily.

Pimples are generally the result of an inactive and unhealthy condition of the skin. A cure is generally accomplished by the use of the following :—

Corrosive sublimate,	-	-		8 grains.
Muriate of ammonia,	-	-	-	30 grains.
Water,	-	-	-	-	1 gill.

Bathe the pimples twice daily with this wash, using a bit of soft cotton cloth. Carbolic soap or pine-tar soap, to be bought in almost any drug store, dipped in water and rubbed over the pimply portion of the face, usually proves an effectual cure.

Freckles, Brown Spots, Moth Spots. A peculiar chemical

combination of iron and oxygen in the blood of many persons produces that appearance of the skin known as freckles, moth-patches, brown spots, etc. It is no question of health—those most freckled being generally blessed with most vigorous health. Those with very light or red hair (which is caused by red-colored oil, more strongly impregnated with iron than others), are most liable to freckles, etc.

The proper means of removing freckles and moth-patches is in the use of those chemicals which will dissolve and dissipate the existing combination situated in the second or middle membrane of the skin. As those with freckly faces have very naturally an intense desire for a clear skin, it gives us pleasure to be able to give here a new and positive cure for freckles, etc.

Sulpho-Carbolate of zinc,		·	2 parts.
Glycerine,	·	·	25 parts.
Rose-Water,	·	·	25 parts.
Alcohol,	·	·	5 parts.

Apply twice daily, and let remain from one-half-hour to one hour. Then wash with cold water.

To improve the complexion—

Corrosive sublimate,	·	·	10 grains.
Oxide of zinc,	·	·	½ drachm.
Muriate of ammonia,	·	·	¼ drachm.
Soft water,	·	·	½ pint.

Use a piece of soft cotton cloth, and apply once or twice, daily, slightly moistening the skin. The above has been extensively used for the removal of freckles, and other discolorations of the skin. A simple and excellent cosmetic for softening, whitening, and beautifying the complexion and hands, is made by mixing four parts of the yellow of an egg and five parts of glycerine, and rubbing them well together with a pestle. Apply to the face on going to bed. Rub well into the hands after each washing. It will keep for years, and is an admirable preparation for all bruises of the skin.

The use of paints and powders, however harmless they may be of themselves—as powdered starch—clogs the pores of the skin, and it becomes rough, sallow, and wrinkled, and often pimply ; and a painted or powdered woman is such an unlovely object !

The Teeth. Probably less attention is given to the care of the teeth than to any other part of the body. Yet the teeth are of the first importance ! But comparatively few rinse the mouth regularly after eating, or use a tooth-pick habitually. And food clogged between the teeth is such a disgusting sight ! It ferments—rots, giving the breath a bad smell, and causing the teeth to decay.

Besides the use of a tooth-pick, and rinsing the mouth after eating, the teeth should be thoroughly scrubbed with a good stiff tooth-brush at least once a day—on retiring for the night is the best time, using cold water. The teeth of many persons can not be kept clean and white without the use of a powder. The following is good, cheap, and safe :—

Prepared chalk,	-	-	-	4 ounces.
Orris-root (powdered),	-	-	-	2 ounces.
Green myrrh (powdered),		-	-	¼ ounce.
Oil of cinnamon,		-	-	20 drops.

Keep in a corked bottle.

Charcoal irritates the gums, and no preparation containing it should be used. Soap and soap preparations give the teeth a yellowish appearance, and hence are decidedly objectionable.

Ill-Smelling Breath. If caused by filthy teeth, or a decaying tooth, the remedy is plainly to clean the teeth, and to have the offending tooth filled or taken out.

If the cause is from a disordered stomach, of course the breath will be offensive till the digestion becomes good again, by due attention to diet, and proper medication.

But the breath of many persons is ill-smelling from no apparent cause. Such persons can render the breath sweet temporarily at any time by chewing a kernel of roasted coffee.

Wash your Feet. In cities, facilities for taking a hot bath are so many and convenient, that every respectable person feels himself bemeaned, if he does not occasionally give himself a good scrubbing with soap, brush and hot water. But in villages and the country proper, things are different. To attempt a bath in a washtub placed in the barn or back kitchen, can not be considered particularly convenient ; and the result is bathing in the country is not the general practice. One consequence is a good many feet that give forth exceeding bad smells !

The pores of the soles of the feet are much larger than any on any other parts of the body, and they pour out a large amount of perspiration, which condenses, mixes with dust and dirt, and forms that hard scaley crust that almost every person has noticed at some time on the soles of their feet. Wearing tight boots or shoes, the perspiration is confined—there is no "ventilation," and the result is about the feet of many persons, especially men, a most rottensome smell. Soak the feet ten or fifteen minutes, at least once a week, in a pail of hot water, well soaped, and then giving them a thorough scraping with a stiff-bladed knife. This will prevent all danger of bad smells from the feet, and add much to your feeling of personal comfort, at the same time lessening the liability of taking cold. Those who keep their feet scrupulously clean are much less liable to cold .et—have warmer feet in winter—than those who do not.

Item. The feet can be kept comfortably warm on an exceedingly cold day by wrapping a piece of newspaper about the feet over the **stockings, and then drawing on the boots.**

CHAPTER XVII.

GENERAL OBSERVATIONS REGARDING THE YOUTH OF BOTH SEXES, AND THE DUTY OF PARENTS AND GUARDIANS.

When the famous Alexander Pope, of Twickenham, gave utterance to the celebrated aphorism, "The proper study of Mankind is Man," he was doubtless fully cognizant of the fact, that the laconism embraces everything appertaining to the sexes as such, as well as to the human family generally. To a mind so astute and analytical as his, it must have been obvious that most if not all of the defects, mental and physical, peculiar to any generation or people, were attributable solely to the imperfect training of its youth, or to the indulgence of such inharmonious and incompatible marriage relations as disfigure the annals of the present day, and as have marked so frequently those of past ages.

As in the vegetable kingdom, the selection of proper seed and soil is indispensable to the production of a perfect plant, so in the animal is the enlightened and judicious blending of the sexes a *sine qua non* to the production of a being representing all the excellence of its species. This is an axiom the most unassailable ; and hence the vital necessity of accepting it in all its integrity, and of never transgressing it in any respect upon the exalted plane of human existence.

The sentiment of love, in its highest and most divine acceptation, can obtain between the sexes only. Although far from antagonistic to that of friendship or affection, it differs widely from it ; inasmuch as it 'has more important ends to attain, and can never exist between individuals of the same sex. Friendship or affection for one another may characterize the intercourse of men, or of women ; but love, in its truest sense, never. This latter is the golden link which unites us at once to our opposites and to heaven, and that culminates in that holy and mysterious compact which results in the propagation of our species, and the accomplishments of our mission in this direction.

While in pursuit of the study of this question, however, we must be careful not to confound or confuse the love under consideration with the mere animal passion that so often steals its guise to gratify the cravings of lust, and that so constantly betrays the youth of both sexes in excesses that terminate, on one side, at least, in years of misery or shame. And here we would address ourselves more especially to the inexperienced maiden whose guileless heart is too often open to the deceitful blandishments of some cruel suitor who has but one object to attain, or to the sincere and ardent professions of thoughtless youth, who, without pausing to analyze the motives which actuate him or the stability of his in-

314

tentions, accomplishes her ruin, and leaves her to learn, alas ! too late, that, save before the altar, no woman is justifiable in placing her character and happiness in the keeping of any man.   In such instances, deceit and sincerity being alike at fault, the only safe course for the maiden who would escape the Scylla of the one or the Charybdis of the other, is to keep watch and ward on the battlements of her prudence and virtue, and, no matter how impassioned and sincere the pleadings of any individual upon whom she may have bestowed her affections, preserve both intact, as the only means of retaining his love and respect, should he be a true man, and of keeping herself unsullied in the eyes of society and of the world generally.

Although delicate and difficult, the task of whispering some truths into the ears of a young maiden arrived at the years of discretion, yet, so necessary to her well-being and happiness in every possible relation is it that she should be made thoroughly aware of the untoward influences which so constantly obtrude themselves into even the purest atmosphere, we venture, although with some hesitancy, to assume the serious undertaking.   And here we may observe, in the first place, that the primary elements of all that makes life worth a single hour's purchase, are to be found in a thorough recognition of what we owe to the Creator, to ourselves, and to society ; and the possession of a mind free from the taint which disfigures some of the literature of the day, and from those low desires and loose ideas, which, with scarce a single exception, result from its perusal.   Nothing can be more dangerous to the youthful mind than even a passing glance at the works of any of those authors who appeal to the animal passions in a manner so insidious and ruinous and who, before a young maiden is aware of it, destroy all her sense of delicacy, and often, alas ! betray her into those dreadful excesses which, although kept the profoundest secret from every living soul save herself, invariably end in the total loss of innocent purity and the utter destruction of all physical beauty.   We need not be more explicit upon this subject, but may summon on the witness-stand in proof of what we here state, the sallow and lifeless features, the dim eyes, and desponding gait, which are significant to the astute medical man, and which are to be encountered so frequently in what is termed the very best society.   Any violation of the laws of the Creator in this, as in every other connection, is sure to be visited upon the aggressor ; and when we come to consider that the first offence in the particular relation now alluded to, leads quickly to another, and yet another, until transgressions crowd upon each other thick and fast, and seize upon the whole being, we shall be able to perceive at once how vital the necessity for every young maiden to eschew with prayerful diligence the source of such terrible dangers, and to be in a position to feel within herself, at the period when she may be called upon to give her hand at the altar to some one worthy of all the love and affection that could be bestowed upon him, that she enters the marriage state as pure in mind and body as the veriest child, and that, in this relation, not a cloud or a regret can obscure the sunshine of her after years.

In the observance of the course which is here suggested or implied, lies the corner-stone of all the bliss that attends upon the sacred compact into which two young souls enter for life. And here we would observe, that in this direction, a serious and solemn duty devolves upon mothers, to instruct, at the proper moment, their daughters as to the dangers that beset them, both mentally and physically, at a certain age. In this relation there should be no false delicacy felt. The truth must be told, and in a manner the most unmistakable. A life of happiness or of misery hangs on the issue, and there should be therefore no mincing of the matter. The crime of self-abuse, if we must say it, is not confined to the sterner sex only. Some of the most fearful examples of it amongst females are to be met with terrible frequency, and of a character so hopeless as to embarrass all medical interference, and to seal the doom of those who had so fallen from their natural birthright and high estate.

How indispensable, then, the proper education of young girls, and how necessary to surround them with a mental and moral atmosphere the most pure and desirable. Let them begin aright, as children, and learn to think well and soberly as they advance in years, and there is nothing to be feared. Let their minds and dispositions be formed on the models of the virtuous fireside rather than upon those of the gaily decorated and frivolous saloon or drawing-room. Let them be taught to respect their own persons, as a sacred trust from heaven, and to feel that any violation of the laws appertaining to their physical being, in the sense under consideration, can not fail to be visited with the direst results. This knowledge the judicious mother can impart by degrees, and in her own way. She has at her command various modes and opportunities of approaching the subject successfully, which do not obtain in the case of any other individual whatever. Consequently, she is to a great extent accountable for the future happiness or misery of those who lie nearest her heart in this matter ; for it must be obvious, that she, above all, can influence their conduct and habits of thought before they arrive at that period of life when they are presumed to think and act for themselves.

It is surprising how often children of unusually tender years are led by bad companionship to familiarize themselves with the abominable practice which we need not again pause to particularize. It is within our own knowledge, that not long since, in the city of New York, a young girl of great prominence, and most respectably connected, became a hopeless victim to this awful infatuation. And this was the more lamentable, as she promised to be as lovely as the day, and was possessed of a form and figure that were of exceptional beauty. At first her mother was unaware of the cause of her gradual transformation ; but when made sensible of the truth, she found, perhaps too late, that to her own criminal neglect the disaster was to be mainly attributed ; for, before her unfortunate child was eleven years of age, she had learned to make war upon her mind and body in a manner so effectual that it was pitiable to look upon her when she arrived at the age of thirteen.   What has become of her since we are unable to say ; but this much we know, it will take a strong hand to rescue her from the most terrible of fates, or to restore her to any degree of health or strength.   To the companionship of books and children of questionable morals this whole disaster may in reality be attributed, for the parents of this wretched being were themselves of morals the most irreproachable, but not wise and watchful in their day and generation.

The necessity, then, of the strictest caution on the part of mothers in the selection of books or playmates for their children becomes obvious at a glance.   The child is the marble from which the woman is sculptured ; and if the youthful block is disfigured, fractured, or broken, where may we find the moral, mental, or physical chisel that shall obliterate or remove the damning defects ?

An eminent physician has justly observed on this head :—

" We now approach a part of our subject which we would gladly omit, did not constant experience admonish us of our duty to speak of it in no uncertain tone.   We refer to the disastrous consequences on soul and body to which young girls expose themselves, by exciting and indulging the morbid passions. . Years ago Catherine E. Beecher sounded a note of warning to the mothers of America on this secret vice, which leads their daughters to the grave, the madhouse, or, worse yet, the brothel.

" Gladly would we believe that her timely admonition had done away with the necessity of its repetition.   But current medical literature, and our own observation, convince us that the habit of self-abuse has increased rather than diminished   Surgeons have recently been forced to devise painful operations to hinder young girls from ruining themselves, and we must confess that, in its worst form, it is absolutely incurable.

" The results of the constant nervous excitement which this habit produces are bodily weakness, loss of memory, low spirits, distressing nervousness, a capricious appetite, dislike of company and of study, and, finally, paralysis, imbecility, or insanity.   Let it not be supposed that there are many who suffer thus severely ; but, on the other hand, let it be clearly understood that any indulgence whatever in these evil courses is attended with bad effects, especially because they create impure desires and thoughts which will prepare the girl to be a willing victim to the arts of profligacy.   There is no more solemn duty resting on those who have the charge of young females than to protect them against this vice.

"But, it is exclaimed, is it not dangerous to tell them anything about it? Such a course is unnecessary. Teach them that any handling of the parts, any indecent language, any impure thought, is degrading and hurtful. See that the servants, nurses, and companions with whom they associate, are not debased ; and recommend scrupulous cleanliness.

"If the habit is discovered, do not scold or whip the child. It is often a result of disease, and induced by a disagreeable itching. Sometimes this is connected with a disorder of the womb, and very frequently with worms in the bowels. Let the case be submitted to a judicious, skilful medical adviser, and the girl will yet be saved. But do not shut your eyes, and refuse to see this fact when it exists. Mothers are too often unwilling to entertain for a moment the thought that their daughters are addicted to such a vice, when it is only too plain to the physician."

We have it on high authority that modesty is the chief quality in the adornment of women ; and in no case is it more grateful and becoming than in that of a young maiden who has arrived at that important and interesting period, when she may be wooed and won and made a wife. What the age of puberty is has been made a careful study by medical men. In the temperate zone, fourteen years and six months is the average period of its first appearance in healthy girls. If it occurs six months earlier or later, then there is probably something wrong. There is sometimes a wider deviation for the age here stated than this, and without any serious meaning ; but at no time is such a deviation to be neglected. In a vast majority of cases it is owing to some defect in constitution, health, or formation, and should be seen to and corrected at once, otherwise years of hopeless misery may be the result. "Mothers, teachers," observes the author just quoted, "it is with you this responsibility rests. The thousands of miserable wives who owe their wretchedness to the absence of proper attention at the turning point of their lives, warn you how serious is the responsibility."

The foundation of old age, observes a celebrated author, is laid in childhood, but the health of middle life depends upon puberty. This maxim is invaluable. The two years which change the girl into the woman, frequently seal forever her happiness or misery in this life. They decide whether she is to become a healthy, cheerful wife and mother, or the reverse,—to whom "marriage is a curse, children an affliction, and life a burden." Both sexes mature more early in hot climates than in temperate or cold ones. Within the tropics, marriages are usual at twelve or fourteen years of age. Such precocity, however, is the precursor of early decay ; for a short childhood portends a premature old age, and *vice versa.*

It is not a favorable symptom to experience any indication of puberty before the usual average time, as it betokens a weakly and excitable frame. Let us therefore enumerate the principal causes which incline to hasten it unduly.   Idleness of body, highly-seasoned food, stimulants, such as beer, wine, liquors, and in some degree, coffee, tea, and irregular habits of sleep.   The mental causes are, however, still more potent in tending to premature development.   What stimulates the emotions leads to unnaturally early sexual life.   Late hours, children's parties, sensational novels, questionable pictorial illustrations, love stories, the drama, the ball-room, talks of love and marriage, &c., all hasten the event which transforms the girl into the woman.   This becomes obvious when we compare the average of puberty in large cities and country districts, it being clearly ascertained that the females in the former mature from six to eight months sooner than those in the latter ; and the result may be seen more plainly in the well-preserved farm-wife of thirty when compared with the languid and faded city lady of the same age.

During the two short years, then, that transform " the awkward and angular girl of fourteen " into the graceful maiden of "sweet sixteen," the utmost caution is to be observed in every relation, moral and physical. The magic wand of the fairy is at work, and a new creature, as it were, is being released from her chrysalis state, with sentiments and responsibilities that must be kept well in hand.   The transformation goes on until at last the system acquires the requisite strength, and furnishes itself with reserved forces, when the monthly periods commence.

A writer of great judgment and experience on this subject, asserts that one of the most frequent causes of disease about the age of puberty is starvation.   He avers that many a girl is starved to death, from the fact of food of an improper quality being given to her, or from the circumstance of sustenance being administered to her in insufficient quantities, or at improper hours.   Hence, from the want of proper nourishment, the system becomes enfeebled and subject to attacks of disease, and especially to those of consumption.   The food at such periods should be abundant, varied, and simply prepared.   Good fresh milk should be used daily, while tea and coffee should be thrown aside totally.   Fat meats and vegetable oils, so generally disliked by girls at this age, are exactly what they require at this juncture of their lives.

All kinds of exercise proper for a young lady, and especially those which lead into the pure open air and sunshine, are also beneficial at this momentous crisis ; and a particular kind is to be recommended for those whose chests are narrow, whose shoulders stoop, and who have a hereditary predisposition to consumption.   If it is systematically practised along with other means of health, we would guarantee any child, no matter how many relatives have died of this disease, against invasion. It is voluntary inspiration.   Nothing is more simple   Let her stand erect, throw the shoulders well back, and the hands behind ; then let her slowly inhale pure air to the full capacity of the lungs, and retain it for a few seconds *by an increased effort ;* then it may be slowly exhaled. After one or two natural inspirations, let her repeat the act, and **so on** for ten or fifteen minutes, twice daily. Not only is this simple procedure a safeguard against consumption, but, in the opinion of some learned physicians, it can even cure it when it has already commenced.

At first the monthly loss of blood exhausts the system. Therefore, plenty of food, plenty of rest, plenty of sleep, are required. That ancient prejudice in favor of early rising should be discarded now, and the girl should retire early, and, if she will, sleep late. Hard study, care, or anxiety, should be spared her. This is not the time for rigid discipline.

Clothing is a matter of importance, and, if we were at all sure of attention, there is much we would say about it. The thought seriously troubles us, that, so long as Canadian women consent to deform themselves, and sacrifice their health to false ideas of beauty, it is almost hopeless to urge their fitness for, and their right to, a higher life than they now enjoy.

With thoroughly healthy girls, what is usually termed the monthly period, continues to recur at regular intervals, from twenty-five to thirty days apart. This is true of something like three out of every four. In others, a long interval, occasionally six months, occurs between the first and second sickness. This latter, *if the general health is perfect*, need excite no apprehension; but, under the slightest mental or physical derangement, the case must at once receive intelligent treatment. Perfectly healthy young women have, on the other hand, been known to have been unwell for sixteen days, while others again experienced this change every thirty-five or thirty-six days only. This appears not easy of explanation, and may, perhaps, be attributed to some inherited peculiarity of constitution. In this relation, climate seems to play a prominent part; as travellers tell us, that in Lapland this phase of woman's physical life occurs but three or four times a year.

"At this critical period," observed an able physician, "the seeds of hereditary and constitutional diseases manifest themselves. They draw fresh malignancy from the new activity of the system. The first symptoms of tubercular consumption, of scrofula, of obstinate and disfiguring skin diseases, of hereditary insanity, of congenital epilepsy, of a hundred terrible maladies, which from birth have lurked in the child, biding the opportunity of attack, suddenly spring from their lairs, and hurry her to the grave or madhouse. If we ask why so many fair girls of eighteen or twenty are followed by weeping friends to an early tomb, the answer is, chiefly from diseases which have their origin at the period of puberty. It is impossible for us to rehearse here all the minute symptoms, each almost trifling in itself, which warn the practised physician of the approach of one of these fearful foes in time to allow him to make a defence. We can do little more than iterate the warning, that, whenever at this momentous epoch any disquieting change appears, be it physical or mental, let not a day be lost in summoning *skilled, competent* medical advice."

From what has been now said, it is impossible for any mother of ordinary good understanding to mistake for a moment the shoals and quicksands, both moral and physical, that surround her daughters, whether as mere girls or grown-up maidens. A careful supervision of the company they keep and the books they read—their determined exclusion from the society of either men or women of lax conversation or morals, and the careful inculcation of self-respect, which can alone be based upon proper pride and purity, will go far to obviate the dangers that beset their path, and so ground them, ultimately, in the principles of virtue and a correct demeanor, as to put evil thoughts to flight on their first approach.

While a handsome person and excellent acquirements are always desirable on the part of any individual who pays his addresses to a young maiden, yet there are other and more important considerations which ought to overshadow mere physical beauty or mental attainments ; and these are a high sense of honor, and a thorough and practical conception of the duty we owe to God and man. These latter constitute the imperishable part of our nature when properly moulded, and are the staff upon which we can lean with confidence when our mere physical being loses all its brightness, and totters to its fall. Hence the necessity of warning the young and untutored heart against mere outward appearance, and directing it towards a recognition of those attributes and features on the part of the sterner sex, upon which a life of true happiness can alone be founded.

This is a matter of such paramount importance that we feel the necessity of impressing it, to the utmost of our ability, upon those who have arrived at the years of understanding, and whose hearts may yet be free, or partially so ; and who may be induced to pause ere they commit their happiness for life to the hands of those who may not only be unworthy of the woman's love, but who seek to obtain it under the false pretences of a comely exterior, or the adventitious matter of dress, while every fibre of their nature may be selfish beyond measure, and set only upon the momentary gratification of a passion that when once satisfied turns aside from the hapless and unsuspecting object that has inspired it, and leaves her to mourn in the silence of her lonely chamber, the fate she might have avoided, but that is now beyond repair.

How warm soever her feelings and sentiments, every step taken in the paths of courtship by a marriageable maiden should be well observed and guarded. In her conversation or conduct there should be nothing of thoughtless levity, or anything that would warrant a familiarity on the part of her suitor which might not be taken in the presence of some dear friend or relative. To observe a proper and well-considered course in this relation, is to insure the increasing admiration of the being upon whom she has bestowed her affections, if he be worthy her love. And here we may observe that this is no mere speculation, but a fact as firm-

U

ly established as any in human experience. For although the tide of passion may run high in an unguarded moment, and set in against heaven and society, yet the terrible and painful ebb follows as surely as effect follows cause, and leaves at least one of the thoughtless culprits stranded forever on the bleak and barren shore of her earthly existence.

There is, therefore, nothing so desirable as firmness and caution on the part of the young maiden in her intercourse with her accepted lover; and both can be observed without wounding his susceptibilities, or impressing him with the idea of either prudery or coldness on her part. Her sentiments in this relation can be conveyed through a thousand different channels, and with such force and effect as to impress and influence to the proper extent any individual possessed of correct feelings, or of the mental and moral requisites to make a wife happy.

Let us glance for a moment at the fate of many beautiful and warm-hearted maidens whose happiness has been wrecked, even in this city, through the fiendish machinations of perfidious suitors. Scarce a house of ill-fame in our midst but has one or more inmates of this character—poor, thoughtless, and confiding creatures, that would sooner had thought an angel of light capable of deceit than those who had betrayed and ruined them. But they would not be warned, or had not been advised until the die was cast; and hence, without a hope, their wretched downward career began apace, until at last, with their ears familiarized to the ribald song and jest, they sought refuge from the upbraidings of conscience in the intoxicating cup, which so completed and rendered hideous the work of debauch that their persons, once beautiful perhaps to intensity, became a loathsome mass, that provoked both horror and disgust in even the coarsest nature.

And some of these unfortunates had been raised in the lap of luxury, while others had been the idols of respectable and loving households; but the education of both classes had been neglected, as neither had been brought up by strictly moral parents, who had attended upon their footsteps with pious care while they were yet children, and who, as they approached the period of maidenhood, had not excluded every moral taint from the atmosphere they breathed, or taught them true allegiance to the divine laws and those which sustain our great social fabric. Had their guardians inculcated those moral perceptions and principles, without which a woman is the darkest stain on humanity, all would have been well; and perhaps the false suitor, regenerated, or rather transformed, through the persuasive influence of such goodness and piety might have been induced to turn from the evil of his ways and have led to the altar a happy and beloved bride, the very being who had opened up the approaches to her total ruin by overstepping the boundaries of prudence or those of modesty, without pausing to consider that a step once taken in this latter direction is never recovered.

Although the mental and physical tendencies of mere girls may vary

in no small degree, we are of the firm belief, that, under even the most
unfavorable circumstances, both may, through judicious and proper
treatment, be brought to harmonize with the great objects of creation.
In view of the accomplishment of this vast desideratum, then, the early
inculcation of proper religious principles, and the ensample of healthy
conversation and moral excellence in the family circle are of paramount
importance. And this is quite compatible with the freedom necessary to
the happiness and well-being of even the lightest heart and most joyous
disposition. There is not an innocent amusement or pleasure incident
to the life of a young girl, that may not be heightened and sanctified, in
a measure, through the adroit and loving guidance of a mother of sound
observation and an ordinary well-trained mind. Cruel and unjust as it
may be, and is in many cases, the axiom is a safe one, that the animal
passions of men lead them into the blindest excesses, and that in the attempt to gratify them they too frequently lose all sight of the consequences, and pause only to count the cost when a keen sense of the
frailty of their victim prevents them from making the only possible adequate atonement in relation to one already so fallen in their eyes. This
is the true state of the case, and the results already glanced at, are, as we
have seen, the most lamentable. Mothers and guardians should therefore bestir themselves, if they would do a noble and abiding work in this
connection ; and never relax their vigilance until those under their charge
have attained the age of maturity and understanding, in the fullest sense.
Here the parent plays a most important part, and must, if she would see
her daughter a happy wife and mother, train her in all the paths of virtue and correct thought. The surveillance may be gentle and loving,
but it must at the same time be constant and inflexible. Every rock and
shoal must be pointed out, and dwelt upon with force and clearness, and
the guiding lights of self-respect, purity of speech, and careful demeanor,
held constantly aloft and in full view. No other course can possibly
succeed, or add, in such cases as now command our attention, to the sum
of human happiness. Consequently both mother and daughter should
understand each other upon an issue so vital ; the younger and more inexperienced looking, with full confidence, for counsel and advice to her
truest friend and rightful preceptor, and forming no acquaintance or
friendship, with a view to matrimony, without her sanction and approval.

In this relation the quick wit and keen eye of the sober and thought-
ful matron will be seldom at fault. She will be able to determine with
something like unerring accuracy, and speedily, the character of the
suitor who may seek to win the affections of her child, and who must
not be judged on mere external appearance, or be taken at his own esti-
mate. Here an honest heart, industrious habits, and a good record, are
of the first importance, and more desirable than gold itself. In saying
so much on this head, however, it is not to be supposed that mere world-
ly wealth is to be disregarded, as an element in any compact between

two hearts that would become one, and spend together a life of usefulness and independence. On the contrary, we hold it indispensable, that no young maiden of sound judgment permit her affections or the solicitations of her lover to betray her into the cares and responsibilities of matrimony without the possession or prospect of sufficient means to render her home comfortable, and remove her beyond the probability of want. In this, both caution and prudence should be observed ; for to step into poverty and its consequent domestic embarrassments, is often to step out of love ; and then, alas ! for the future of both parties.

The vital necessity, then, of worthy male companionship, in the first instance, for any young girl whatever, must be obvious to even the most commonplace intelligence. If those who surround her are pure and good, and the uncompromising enemies of the free-love taint and principles which are now so rife in certain quarters, her selection of a proper companion for life will be the less difficult, as fewer chances present themselves for bestowing her affections unworthily. She can take no more important step between the cradle and the grave than that which leads her to the altar. It should therefore be well considered and guarded, as once it is taken the die is cast forever.

Let us then hope that all those most deeply concerned will ponder well the facts we have laid before them on this all important subject, and let no mother relax for a single moment the vigilance that should wall out from her daughters the dangerous books and companionship to which we have made such distinct reference. In addition, let every young maiden who is approaching the interesting and critical period already named, look well to her footsteps, and beware of allowing her affections to be captured by a pleasing exterior only on the part of one of the opposite sex. However agreeable an attractive face and form, these do not comprise all that is necessary to the most abiding and exalted manhood, and are not unfrequently a delusion and a snare. There is no absolute manliness without manly principles ; and no true happiness without moral rectitude and a proper sense of our duty towards heaven. These are the attributes and sentiments that tend to make earth a paradise, and that survive all mere physical excellence, inasmuch as they belong to our immortal part. Let them therefore be sought after assiduously by both mother and daughter in the person of any suitor for the hand of the latter ; and let there be no uncertain sound in the premises. Let those desirable features be so prominent in the acts and the demeanor of the man as to be obvious at a glance, and let them not be simulated on certain occasions. True virtue and uprightness of thought and conduct soon manifest themselves in those possessed of such inestimable treasures ; and whenever or wherever we see their absence verified either by word or act, we may rest assured that the transgressor is unworthy the love of any woman who values her own happiness, or who hopes to make the marriage state, under heaven, all that it ought to be, both morally and physically.

FOR THE ESPECIAL PERUSAL OF YOUTHS OF UNDERSTANDING, AS
WELL AS FOR THAT OF FATHERS AND GUARDIANS.

If man is " the noblest work of God," most assuredly every word
thought, and act of his, ought to tend towards the elevation of his mental
and physical being ; because this implied perfection is to be regarded in
a measure as the result of his acquiescence in all the laws, moral and
otherwise, appertaining to his nature. That is, if he would assume and
maintain the high position accorded to him, he must not violate any of
these divine precepts or rules laid down for his guidance ; but from the
first moment of his responsibility to the last of his earthly existence, walk
in the way of godliness, virtue, and truth, and never transgress any of
the provisions relative to the true development and important mission of
his animal structure.

The licence accorded to boys when compared with that allowed to
girls of the same age, ought to engender in them a chivalrous respect for
the gentler sex, and never urge them into anything savoring of egotism
or tyranny. The apparent superiority is but simply the result of greater
physical strength, and the freedom with which the one sex is permitted
to move through the world compared with that accorded to the other.
There is in reality no mental superiority in the one over the other ; for
in this respect it has been shown that the impress of man has been left
mentally on the age more than that of woman ; because, from some ill-
judged laws or rules of society, she has been subject to restrictions which
circumscribe to an unwarrantable extent her sphere of action.

But while laying it down as an axiom that there is perfect mental
equality between the sexes, we can not refuse to entertain the idea that
woman is the weaker vessel physically, and that her dependence upon
man, and her claims to his love and protection, arise to some extent from
this cause, although the sublime mystery of her being appeals to him in
a higher and more abiding sense. This taken for granted, then, and
perceiving, as even the most unphilosophical can, that both sexes were
designed to harmonize with each other in every possible relation, and
that the weaker and more beautiful is obviously entitled to greater consideration than the more robust, it behooves the latter to look well to his
manhood and the perfection of his mental and physical status, for a
defect in either is not only destructive of anything like love in the female
bosom, but of a character which, with but few exceptions, build up an
insurmountable barrier between the man and the woman towards whom
he may be attracted.

We say mental and physical status here, because the body is seldom
injured through excesses of any description, without the mind suffering
commensurately. And as the foundations of a healthy physical existence are laid in childhood, we would observe that this fact, in relation
to boys, as we are now treating on them more especially, should be

recognized to the fullest extent by fathers and guardians, as upon their judgment, affection, and fidelity, the character of the rising male generation must mainly depend.

Although, as already observed, a greater latitude is allowed to boys than girls, in almost every relation, yet this latitude must be circumscribed and confined to certain well-defined, healthy bounds. And here we would again dwell upon the vital necessity of good companionship and good books, where example and precept harmonize with the exalted ends to be attained in after years. Of course, the family circle is to be regarded as the true starting point, whence the earliest lessons in vice or virtue are derived, and should this prove to be lax in any of its teachings, moral or religious, the very germs of success are embarrassed or destroyed at once.

It is astonishing how quick young lads, not much more than half way to their teens, acquire bad habits and principles from impure associates, or the unguarded conversations which sometimes occur at their own fireside. From both these sources the worst consequences are to be apprehended, as they gradually undermine every principle of good, and so familiarize the tender ear and understanding with what is most pernicious, because of the almost indelible and fatal impress that is left upon the unreasoning susceptibilities. Whatever may be said to the contrary, there are unmistakable traces of the early hearthstone to be found in the lives of most men ; and such being the case, how indispensable it is that the atmosphere that surrounds it should be free from taint, and that the greatest caution should be observed that nothing transpires within its sacred limits that might have the slightest tendency to mar the man in the child, or thwart the beneficial designs of nature regarding him.

In view, then, of the influences of the family circle, and that of the father upon the son, while yet a mere youth or child, that circle should be made as attractive as possible, and on a plane thoroughly comprehensible to the intelligence to which it appeals. If parents would shape their   en to the noblest ends, they must gain their confidence and   by becoming children themselves in a measure. They cannot   or teach to any purpose from a reserved or exalted pedestal, or   ugh the instrumentality of the sober long-faced truths which are applicable to grave years only. The atmosphere of youth is, in the natural order of things, bright and happy ; and if we would influence, by precept or example, those still surrounded by it, we must assume to breathe it ourselves for the time being, and sow our earliest good seeds in its tender light. Once the kind and judicious father has won the heart and confidence of his little son, the road to the fullest success lies wide open ; and feet that under other circumstances would assuredly have turned aside under less benign influence from the paths of rectitude and truth are now easily directed into that glorious upward and onward course which never fails to culminate in happiness here and hereafter.

Some of the most terrible evils that beset the path of so many of the young of the sterner sex result from association with children of their age who have learned to debauch their own persons, and to indulge in those secret habits which have been long noted as fraught with disaster and death. So early the period at which unsuspecting little ones are led into this terrible snare by their seniors of two or three years, that few will be inclined to believe that this first essay in crime has been attempted at the age of eight, and continued until an actual drain on the young energies and vital forces of the system commenced. It is, therefore, obvious that the lynx eye of the father or guardian, or of both parents, should at this tender period of youth be brought constantly to bear upon all the acts and tendencies of their young ward or progeny, and that they should seek to inculcate those pure ideas and aspirations which are the only security against this danger. Let the playmates, the toys, and the rudimentary books of the child be carefully selected, and let there be for him an abiding attraction about his own door and fireside, so that he may not be necessitated to look abroad for any of those innocent pastimes or recreations that have such charms for the young. The great error of many parents, in this direction, lies in the dry and solemn homilies, which they conceive ought to be forever mumbled into the ears of their children, or the long religious services to which they regard their subjection as wise and indispensable. This is a fatal mistake. The medium through which the young are to be taught successfully their dependence upon heaven and their duty towards God and man, should rather consist of the green fields, fruits, and flowers,—of sunlit skies, running brooks and balmy winds,—the song of birds, the changing seasons, and the summer woods. These and the beneficent design of the Creator in calling them into existence for the benefit of man in a pre-eminent degree, should form the earliest pages presented for the study of the child ; and if each of the beauties and truths they contain be carefully explained upon the basis of that higher information which may be sought through the " revealed word" as the young student advances in years, the result will be most happy, and culminate in all the perfection possible of attainment in this earthly sphere.

But while inculcating these salutary lessons, there must be no false delicacy on the part of the father, relative to pointing out, in the clearest possible manner, the dangers with which his inexperienced and susceptible child is assuredly surrounded. The parent is but ill-versed in his duty, or the prevalence of the crime of self-abuse in the young, who fancies that by keeping his youthful offspring in ignorance, so far as he knows, of the character and manner of this frightful offence, that the child must necessarily escape its taint. Let there be no misconception on this head ; for it may be accepted as a leading fact, that nineteen boys out of every twenty learn something of it at a very tender age ; and from sources, too, but badly qualified to warn them of its terrible results.

And be it further remembered, that no matter how innocent and unconscious of guilt the first attempt at its commission on the part of a poor
unsuspecting young creature, there is something almost fatal in tamper-
ing with even the undeveloped organs of generation, or in endeavoring
to excite them to undue or premature action. This fact must be dwelt
upon in the plainest manner, and so impressed upon the youthful mind
as to satisfy it that a single move made in this direction tends to speedy
destruction and death to both body and soul. So soon, then, as the
child is capable, in any degree, of comprehending advice and instruction,
this subject must be broached in the best and most impressive manner
known to the judicious father who would acquit himself fully in the sight
of God a' 'man ; for to leave the matter to mere chance, in the hope
that escape was possible through a fortuitous combination of circumstances, would be to be guilty of a crime the most heinous, and scarcely
second to that of murder.

To the youth of understanding, however, we may address ourselves
more directly ; assuming that he may not be altogether free from the
taint of this sin.   In the first place, then, let us lay it down as an inexorable fact, that so terrific and contaminating this practice in any degree,
that it not only tends to destroy every particle of physical beauty and
manhood, but, if persisted in, results in absolute idiocy, or a premature
and most horrible death.   This is no overdrawn picture, but may be
taken as an absolute fact, in connection with the quite as inexorable
truth, that the youth addicted to this prevailing vice is as certain to render himself incapable of propagating his species or consummating the
holy sacrament of matrimony, as that the sun is in the heavens at mid-
day.   In the unnatural excitement which saps his whole being, he may
fancy, at the time of self abuse, that this is not so ; but should the crisis
ever arrive when a loving and confiding wife is betrayed into his arms,
then comes that terrible humiliation which is worse than death.   There
is no escape from the consequences of this monstrous offence should it
be indulged in to a certain point ; for then it seizes upon the whole being, and like the infatuation of the arsenic eaters, whether continued or
abandoned, ends alike in the most appalling mental and physical sui-
cide.

What youth, then, of the slightest manly feeling, or intelligence, would
so make war upon his physical being, as to render him disgusting to the
pure and good of the opposite sex, and even to those who had fallen
from virtue, and taken refuge in the lowest brothel ?   Can it be possible
that any individual who sets the slightest store by the love of a beautiful woman, or the possession of every manly attribute and sentiment,
will, after the perusal of these startling and inexorable facts, approach
the edge of this shuddering abyss, or, if partially engulfed in it, will not
struggle to extricate himself at once.   There may yet be time to retrace
his steps, and escape the awful depths that yawn beneath him ; while

the commission of a single offence more may hurl him irrevocably to his dreadful doom. Oh ! could we but depict in adequate language the fearful and hideous wrecks that have resulted from this crying sin against nature, each particular hair of the youthful aggressor would stand on end, like the quills on " the fretful porcupine." But so thickly strewn around us, under a thousand loathsome forms, are the evidence of this most damning crime, that those who run may read if they only will.

And now that we have dwelt at some length on this part of our topic, let us turn for a moment to those youths who have sufficient virtue, manliness, and strength, to eschew this great evil, and say to them, that although they have escaped it, their path is still beset with other snares which lead to discomfort and misery in a lesser degree only. If not prostrated at this disgusting shrine, we have known the manhood of many to fall a victim to unholy and misguided passion, and who, by giving full rein to their unbridled lust, or falling into the snare of some casual circumstance, have brought woe to true and unsuspecting hearts that loved " not wisely, but too well." And here we would observe, that when a respect and highest consideration for the opposite sex are not entertained by a man whatever his condition, he is unworthy the recognition of society or the favor of heaven. There is something of infinite treachery and cowardice on the part of a suitor, sincere or pretended, who steals into the affections of a young maiden, with a view to betraying her, or who having once gained them on an honorable basis, takes, in an unguarded moment, advantage of the love she bears him, to humiliate her in her own eyes, if not to accomplish her ruin beyond redemption. No true and chivalrous man or youth who respects the person or the memory of the mother who bore him has ever been guilty of doing such dastardly violence to the sex, or of treading ruthlessly beneath his feet what God has made so confiding and beautiful, and what is in every relation the counterpart of his own sister. There is something here which demands attention, and which should be subjected to the strictest analysis and scrutiny. To the professional profligate, who knows no law, human or divine, we do not address ourselves. We leave him in the hands of the Living God, who is sure to call him to account when he least expects it. To such, however, as are not vitiated in this relation, and whose inexperience is beset by warm passions and susceptibilities, we would give a few words of caution and advice ; hoping to enlist their sympathies and attention regarding a matter which affects so vitally their interest and happiness, not only in this world but in the world to come.

To be succinct, then, when a youth finds himself approaching the threshold of manhood, or that period of life which succeeds his mere boyish days, he in most cases enters upon a new and charming phase of his existence, which is expressed mainly in a desire for female society, and generally for that of one being beyond all others, who in his admiring eyes appears to be the best and most beautiful of her sex. Now this is nothing more nor less than the first dawning of love, and before he would have its partial victim commit himself irrevocably to the overpowering passion of any individual case, we would implore him to pause on the verge of the charmed circle, and ere he become hopelessly entangled in its delicious mazes, seek advice from his natural guardian, and analyze for himself all the circumstances surrounding the being who has awakened such strange sensations in his bosom, and the possible result of attempting to unite his destiny to hers forever.

To this end, he must endeavor to look upon things as they really exist, and not permit his enthusiasm or admiration to present them in any romantic or fanciful light. As a primary step in this direction, he must measure his own pecuniary circumstances and prospects, and see how far justified he might be in endeavoring to win the affections of any young maiden with a view to making her his wife. We are, of course, aware that but few inexperienced youths pause thus practically upon the threshold of new hopes and aspirations ; but then this does not nullify the wisdom and necessity of doing so. No honorable or prudent person will commit himself to a step so serious as that of marriage, without seeing, to some extent, the road before him ; nor will he tamper with the love or affections of any woman whatever, whom he considers unworthy to become his partner for life. In the fullest manliness on this point there is a chivalry the most noble and exalted. Everything like deceit must be discarded totally, and if it is found that the over-sanguine fair one has misinterpreted any word or act of kindness on the part of him who may regard her as a friend only, she must be undeceived, and at once. The safest rule to follow in this relation is for the young man to be cautious, and never inspire any hopes or confidences in any of the opposite sex, that he does not wish to exist. There are a thousand channels through which the pleasantest intercourse may flow securely, without entering upon this vital one ; and these are familiar to every person of ordinary good understanding. Where there is no intention of awaking in the female bosom a sentiment of love, there should be neither act nor word calculated to provoke it ; and if, as is often the case, one of the softer sex, overstepping the bounds of prudence if not of modesty, makes advances on her part, then the only honorable and correct course of the object of her affections, is to withdraw himself totally from her society.

After becoming satisfied that he is in a position to support a wife, and being free in person from such physical blemishes as have been already alluded to, the first care of the expectant bachelor should be to select from among his acquaintances a maiden of comely looks, industrious habits, and sound and pure morals. These are essentials the most important, and must underlie all the accomplishments which render a woman fascinating, if she is ever to become the light of her own household. There can, of course, be no objection to the possession of those ornamental acquirements which render a drawing-room so attractive at times, such as music, bright conversational powers, and all the agreeable phases of a polite education ; but these, desirable as they undoubtedly are, must not be accepted as the true constituents of happiness ; but rather as the agreeable guise that the more serious and abiding attributes assume, until the moment f r their more active agency presents itself again. No man of moderate means has ever dined off a piano solo, or supped off a dish of fashionable gossip ; and hence the ne cessity of looking for something more substantial in the person to whom he might be induced to pay his addresses, with a view to matrimony, and of eschewing every female, no matter what her attractions, who has not within herself the knowledge and elements that constitute a good housewife. The freaks of fortune are often both sad and surprising ; and hence it is of vital importance that parents educate their children in some trade or calling that might stand their friend in the hour of adversity, when they might find themselves deprived of the last shilling. And here we may refer, briefly, to the helpless condition of some of the fashionable youths who are depending solely upon circumstances for a life of ease and pleasure to which they devote themselves, and who, through a single turn of the wheel of fortune, might be reduced to beggary, from the fact of their having no positive means at the ends of their own fingers of earning their daily bread. On this point we would urge the knowledge of some useful employment on the part of the young of both sexes, no matter what wealth may surround them for the time being ; because by its acquisition they are, in a measure, secured against fate, and have an inner and more satisfactory sense of independence than flows from the possession of mere perishable riches.

But, now presuming that our young suitor is fairly on the carpet, and that, with the sanction of his parents and guardian, which is indispensable, he has determined to win, if possible, the heart of some fair one supposed to be possessed of all the attributes, and the germs of all the qualifications, adverted to, it must be a' parent that his hour of danger and difficulty has arrived, and that, in dealing with it, the greatest caution and prudence ought to be observed. With a view to the fullest success, then, and in furtherance of the great object upon the holiest and highest basis, the moment he perceives a being worthy all his love and affection, she must at once become sacred in his eyes, and never be the subject of any familiarity that might shock her ears or her sense of propriety in any degree, or that might tend to lower her in her own estimation. The more chaste and considerate his intercourse with her, to even the most trifling word or whisper, the higher her sense of his nobility, and, consequently, the warmer and the more profound her sentiments of love towards him. It is a mistaken idea that an occasional questionable jest or brilliant *double-entendre* has any charm for the ear of a young maiden of correct perceptions ; for, although the circumstances of her position may often extort a smile from her as a foil for her pain

and discomfiture, the shadow of the cruel, though unintentional, offence
does not easily pass away from her, inasmuch as it is calculated to awake
in her chaste bosom doubts as to the morality and purity of the bein
whom she may love dearly.   Let there be then, on the part of the suitor
a noble consideration for the woman who has given her whole heart to
him ; and let him feel that the bonds which she is willing to assume, ca
be only made holy and happy when forged in a sense of true delicac
and the highest moral obligations.   One impure, indelicate, or low word
uttered in the ear of a truly chaste and virtuous woman may be destruc
tive of her true happiness for all time to come ; while a single trifling ac
savoring of the libertine could not fail to estrange her from the trans
gressor forever, if faithful to her pride or sex, or else so humiliate her in
her own eyes, as to cause her to feel that the love she brings to the altar
is not so worthy, so fresh, or so sanctified, as it would have been had it
not been soiled and dishonored, in a measure, by him who should have
guarded it more jealously.

The demeanor, then, to be observed by a young man, in relation to the
maiden of his choice, must, while open, generous, and warm, be carefully studied, elevated, and free from the slightest taint of immorality.
The step which he premeditates is the most serious that could possibly
be undertaken by him, and, as it involves a partnership for life
with a being whom he is to pledge himself before God and man to love
and cherish until death, it behoves him to make the compact one of the
most chaste and sacred, so that it may never pall upon his sensibilities,
but always, under heaven, bear the impress of unfading youth.

And now that we have been so explicit on this point, we must go
farther, and warn the ardent and well-meaning suitor of another precipice
that besets his path from the period of his declaration to his appearance
before the altar, and that is, the great danger that attends the warm
embraces which a lover sometimes bestows upon his affianced, when
alone, without presuming for an instant that his passions may outstrip
his reason, and, in a moment of intense excitement, hurry him into an
excess that would destroy the purity and self-respect of the being who
confided so implicitly in his honor, until she stood before him a guilty
and injured woman.   This is a matter for most serious consideration ;
for, notwithstanding that the aggressor makes every reparation in his
power and still redeems his plighted vows, the memory of this great
indiscretion or crime is likely to overshadow all his subsequent married
life.   Let it then be distinctly understood, that even the ordinary
embrace so frequent between two young lovers before they become
united in marriage, must be indulged in with caution and reserve, where
no eye but that of heaven is upon them.   The passions, like tinder,
often take fire from the slightest spark, and it therefore becomes a
matter of the last importance, that until a man and a woman become
one according to the laws of society as well as those of the Creator, the

stronger vessel must keep his desires in hand with a bit and bridle the most inflexible and stern ; otherwise, all the charm and brightness which properly belong to the marriage state can not fail to be tarnished, or, perhaps, ultimately extinguished in gloom.

When entering upon that phase of his life which tends towards matrimony, the young and ardent suitor, once that his heart is truly engaged, should treat the object of his affections with such loving respect and consideration, as could not fail to ennoble him in her eyes, and secure her heart upon a true and abiding basis. Every maiden of modesty and womanly instincts is thoroughly conversant with what is due to her, and alive to every circumstance however trifling, bearing upon the character of her intended. The more profound her love the keener her discrimination, and the more jealous her eye and ear. Not a single expression or act appertaining to his intercourse with her or others but is weighed unconsciously, and subjected to that subtile process of analysis which is almost instantaneous in its results. It is therefore ungenerous, and as dangerous as it is unjust and reprehensible, to wound her susceptibilities in any relation ; for notwithstanding that she may permit, without actual censure, or apparent chagrin, any slight dereliction of duty to pass in this relation, she can not fail to feel his unworthiness, and the shock of her esteem and affections which it must necessarily entail.

But if we have dwelt upon what may be termed the two leading evils which beset the path of youth, there are yet others, scarcely less danger-ous, which require the most careful consideration ; and one of these is the sin of intemperance. We need not travel out of every day common record for evidence to establish the dread consequences of this dire infatuation and its general prevalence. Like most other criminal practices, it steals through the first stages of its progress by slow and imperceptible degrees ; but at each unconscious step, so surely does it entangle its victim in its meshes, that not unfrequently he passes beyond the final point of redemption before he is thoroughly alive to his lost and hopeless state. And here, again, the influence and habits of the home circle become of the first importance. If unswerving temperance be the inexorable rule of the household—if the seductive cup, in even the most harmless aspect, be banished from it with firmness and persistency, there is every hope that the dangerous out-door influences which so throng the path of the inexperienced, may be met and neutralized.

But if the household be tainted, to any extent, with the vice —if indul-gent fathers and mothers will tempt their children with an occasional sweet spoonful of the poison, they do neither more nor less than set fire to one end of a slow fuse which is almost sure to result, one day or other, in the most fearful destruction.

But as we are speaking, as it were, to a youth of understanding, we must appeal to his own manhood and sense of right. And here we

would observe, that next to the appalling crime of self-abuse, that of
habitual intemperance is most destructive of pure love, and of the
physical capacity or power to realize from the chaste passion all that
heaven designed it should accord.   No woman of correct feeling or
judgment has ever bestowed her heart upon a habitual toper, or enjoyed
his society for a moment, not to speak of his maudlin embrace.   This is
true beyond question, and to an extent so dreadful, that no inconsider-
able portion of the infidelity which a neglected and disgusted wife visits
upon her husband to-day may be traced simply to the vice of habitual
drunkenness on his part.   No matter how warm the affections of the
maiden, or how sincere the love of the wife, this curse tramples out both
alike ; and hence the necessity of the utmost vigilance on the part of any
youth who would win and retain the heart of a pure and beautiful
maiden, or preserve his health and manhood intact until both, at a ripe
old age, declined in the natural order of things.   It may be laid down
as an axiom, then, that no true love can exist between a good and pure
woman and the man who renders his person disgusting to her, and who
in a measure emasculates himself through the constant use of intoxicating
liquors : and when we come to dwell on the fact, that the prostration of
his high mission and manhood arises from the indulgence of a loathsome
vice, whose inception is to be traced to the first fatal and seductive glass
that is thoughtlessly raised to his lips, surely the individual who is yet
free from the taint of this curse, or the man who is its partial victim only,
should, if these lines happen to meet his eye, never approach—or dash
from him forever—the cup that has been so fraught with some of the
direst crimes, heart-aches, and miseries, known to the human family.
When, therefore, it is a fact beyond contradiction, that a persistent use
of stimulants of any description tends to generate morbid and adventitious sexual desires, that are succeeded by a reaction the most depressing,
inconvenient and dangerous, and that frequently result in a permanent
injury, what shall we say in denunciation of the habitual use of those
fiery potations that not only lead to rags and beggary, but corrupt the
blood, disfigure the features, and trail the last particle of manhood in the
dust ?

    To all within reach of our warning voice we would then say, beware of
the first glass, and those inebriates or thoughtless persons who would
tempt you to pollute your lips with it.   There is danger in even looking
upon it or in breathing the atmosphere inhaled by those who have
passed its fearful Rubicon, without either the wish or the power to retrace their steps.   Avoid such, and their social meetings or orgies, with
all the strength and decision of which you are capable ; for in this relation also "wide is the gate and broad is the way that leadeth to
destruction."

    The vice of intemperance in youth is not unfrequently associated with
that of gambling, and is almost invariably wedded to that of promiscuous

sexual intercourse ; and here we find another dangerous pitfall besetting the paths of the young on their way towards matrimony, and the attainment of the fullest and most perfect manhood. Fatal even as the crime of gambling, *per se*, may be, it is less terrible in its effects than the physical and moral destruction which results from the loathsome and contaminating embraces of the lost and lewd woman who sells her person to every passer by, and so poisons the life blood of her unsuspecting young victims as to entail upon them a life of disease l misery horrible to contemplate. A man may lose his money at the gaming table, and suffer the pangs of remorse and the beggary that it involves, but criminal as the passion for play is, and frightful as have been the domestic disasters that have resulted and do still result from it, yet if he have escaped the vice of drink and the taint of impure and loathsome women —if his physical structure and mind is not wholly poisoned and debauched—there is hope for him, as a woman's love can survive the one, although it dies out into absolute hatred and disgust under the other.

There is little more to be said on this part of the subject. We have touched all its vital points, and trust that the seeds we have endeavored to plant, in all honesty and good will, may be found to bear the sweet and abiding fruit they are so pre-eminently calculated to yield. In summing up the whole case, then, we would say to fathers and guardians, form the thoughts and habits of your wards or children by a joyous and virtuous fireside. Let their earliest perceptions of right and wrong be based on no uncertain foundation. Set them examples of morality, and inculcate that religion in them which is filled with innocent sunshine and which alone is calculated to inspire their tender hearts with sentiments of true love towards their Creator and their fellow men. Warn them, in the plainest possible manner, of the danger of bad company, loose habits, and any and all of the evils and infamous practices that lie in wait for them, and when they grow up to be men, they will bless and appreciate the kindly care bestowed upon them, and be not only a credit to themselves, but to the human family at large.

WEIGHTS AND MEASURES, &c.

APOTHECARY'S WEIGHT.

20 grains	=	1 scruple	=	20 grains.
60 grains	=	1 drachm	=	3 scruples.
480 grains	=	1 ounce	=	8 drachms.
5,760 grains	=	1 pound	=	12 ounces.

The scruple and drachm are discarded in the new weights. Measures of capacity are used for liquids in mixing medicines. Formerly wine measure was employed, but now it is the imperial. The weight of the imperial minim of water is 91 grains, and is multiplied as follows :

60 minims	=	1 fluid drachm	=	60 minims.
480 minims	=	1 fluid ounce	=	8 fluid drachms.
9,600 minims	=	1 pint	=	20 fluid ounces.
76,800 minims	=	1 gallon	=	8 pints.

The fluid ounce is the measure of one ounce of water ; the pint, 1¼lbs. ; and the gallon, 10lbs. In prescriptions the weights and measures are generally expressed by signs or symbols, with Latin numerals affixed. These signs, with the Latin and English words which stand for them, are given below :

    m   Minim, 1-60th part of a fluid drachm.
    ℈ j   Scrupulus, a scruple.
    ℥ j   Drachma, a drachm.
    f ℥ j   Fluid drachma, a measured drachm.
    ℥ j   Uncia, an ounce (437.5 grains.)
    f ℥ j   Fluid uncia, a measured ounce.
    lb j   Libra, a pound (7,000 grains).
    Oj   Octarius, a pint.
    gr   Gramum, a grain.
    ss   Semis, half, affixed to any of the above signs.

The numerals j., i.j., iij., iv. v., etc., show the number of grains, ounces pounds, etc., to be taken ; thus, *m*lx, denotes 60 minims, ℥ vii. 7 drachms, and ℥ i. 1 ounce.

ON ADMINISTERING MEDICINES.

In prescribing or administering medicines, the following circumstances are always kept in view by medical men, and are of the utmost importance, viz., Age, Sex, Temperament, Habit, Climate, and Condition of the Stomach.

AGE.— See the table of proportionate doses given on the following page ; remembering that all medicines containing opium affect children more powerfully than adults.

SEX.—Women require smaller doses than men ; and are always more quickly affected by purgatives.

TEMPERAMENT.—Stimulants and purgatives more readily affect the florid or sanguine than the pale or phlegmatic ; consequently the former require smaller doses.

HABITS.--Persons in the habitual use of stimulants and narcotics require larger doses of such remedies to affect them, when laboring under disease, than others not so accustomed ; or those who have habituated themselves to saline purgatives, such as Epsom Salts, are more easily affected by such remedies.

CLIMATE.—Medicines act differently on the same person in summer and winter, and in different climates. Narcotics act more powerfully in hot than in cold climates, hence smaller doses must be given in the former ; but the reverse is the case with respect to calomel, consequently larger doses are required in hot climates.

CONDITION OF THE STOMACH.—The least active remedies operate violently on some persons, owing to a peculiarity of stomach or disposition of body connected with temperament. In giving medicine, the medical man always so regulates the intervals between doses that the following dose may be taken before the effect produced by the former is altogether effaced. By not attending to this rule, the cure is always commencing, but never rapidly proceeding—it may, indeed, have no effect at all. It is to be borne in mind at the same time, that some medicines, such as mercury, etc., are apt to accumulate in the system, and danger may thence arise if the doses be repeated too frequently. Aloes and castor oil acquire greater activity by use, so that the dose requires to be diminished. With due caution, and a proper attention to the doses ordered, no untoward circumstances need arise.

EMOTIONS AND PASSIONS OF THE MIND have a most powerful influence upon the disorders of the body. *Hope* is a mildly stimulating or tonic feeling, which is most beneficial in all cases. The influence of the *imagination* on disease has long been known, and the extraordinary cures we constantly hear of as effected by such absurd means as homœopathy, mesmerism, etc., are, in fact, all referable to the influence of the imagination over a diseased body or disordered mind.

DOSES PROPER FOR DIFFERENT AGES.

AGES.	PROPORTIONAL DOSES.	DOSE.
For an Adult	. . Suppose the dose ONE .	. as 1 drachm or 60 grains.
Under 1 year	. . Will require only 1-12th .	. " 5 grains.
" 2 years	. . ———————— ⅛ .	. " 8 "
" 3 "	. . ———————— ⅙ .	. " 10 "
" 4 "	. . ———————— ¼ .	. " 15 "
" 7 "	. . ———————— ⅓ .	. " 1 scruple or 20 grains.
" 14 "	. . ———————— ½ .	. " ½ drachm or 30 "
" 20 "	. . ———————— ⅔ .	. " 2 scruples or 40 "
Above 21 "	. . The full dose ONE .	. " 1 drachm.
" 65 "	. . Will require only ¾ .	. " 45 grains.
" 80 "	. . ———————— ⅔ .	. " 2 scruples or 40 grains

In the same manner for fluids divide the quantity suited for an adult by the above fractional parts. If a child under one year, the dose will be one-twelfth ; under two years, one eighth ; under three years, one-sixth, and so on.

APPROXIMATE MEASURE.

For the convenience of those who have not accurate measures at hand, we give the approximate quantities :

A teacup contains four fluid ounces, or one gill.
A wineglass contains two fluid ounces.
A tablespoon contains one-half fluid ounce.
A teaspoon contains one-eighth fluid ounce, or one drachm.
Sixteen large tablespoonsful make half a pint.
Eight " " one gill.
Four " " half gill.
Twenty-five drops are equal to one teaspoonful.

INDEX.

339

www.ingramcontent.com/pod-product-compliance
Lightning Source LLC
Chambersburg PA
CBHW021402210326

41599CB00011B/983